Fundamental Problems in Quantum Physics

Fundamental Theories of Physics

An International Book Series on The Fundamental Theories of Physics: Their Clarification, Development and Application

Volume 73

Fundamental Problems in Quantum Physics

edited by

Miguel Ferrero

University of Oviedo,
Oviedo, Spain

and

Alwyn van der Merwe

University of Denver,
Denver, Colorado, U.S.A.

SPRINGER-SCIENCE+BUSINESS MEDIA, B.V.

A C.I.P. Catalogue record for this book is available from the Library of Congress.

ISBN 978-90-481-4608-6 ISBN 978-94-015-8529-3 (eBook)
DOI 10.1007/978-94-015-8529-3

Printed on acid-free paper

CONTENTS

PREFACE

On the occasion of a visit to Egypt, Thales of Miletus, one of the Seven Wise Men of antiquity, was asked by the Pharaoh to estimate the height of the pyramid he had just constructed as his own tomb. Thales put a stick in the ground and measured the length of the shadow cast by the sun. Then, he measured the pyramid's shadow length, did some calculations, and finally, to the Pharaoh's astonishment, he told him the exact height of his pyramid.

What is surprising in this story, to the point that even today it can produce great astonishment if we meditate about it, is that Thales was able to work out, precisely, the value of something he could not measure directly: the pyramid's height. Is the human mind capable of knowing aspects and properties of the real world? Does it make sense to talk about material objects beyond our experimental operations, whose properties can be determined by the methods used in the sciences? Or, on the contrary, must we limit ourselves to speak about real objects only when they are measured?

Thales' achievement clearly exceeded that of the Pharaoh. Does that mean that Thales' thought process is more real than the pyramid? Can we go even further and become so amazed at our own intelligence as to attribute to it a deeper reality than the material object which provided the original stimulus?

Our astonishment at Thales' intellect is paralleled today by our astonishment at the power of quantum theory. The sheer accumulation of its exact predictions must surely make it the most successful theory in the history of humankind. In addition, it has stimulated an amazing series of technological developments.

And yet ... the intellect remains unsatisfied. For many physicists the theory contains, at least, strong conceptual difficulties. For others the apparent conclusions of the theory, about the reality of our physical world and the ways in which we discover that reality, remain philosophically unacceptable. A number of questions arise. Should we, perhaps, look for an interpretation of the formalism of quantum theory that adapts better to our usual mode of understanding? Is there an alternate theory, of which quantum theory is an approximation, which restores the principles of locality and realism, questioned in quantum theory? Is there any essential element missing in it? Or should we, on the contrary, abandon our outdated epistemology and try to explain instead, on the basis of a new quantum epistemology, how the classical world in which we live can emerge from a more fundamental quantum law? Or should we perhaps, leave out philosophical arguments altogether, and confine ourselves to the safe formal domain where just the calculus in itself is everything?

All of these possibilities were discussed, both from the theoretical and experimental points of view, at the International Symposium on Fundamental Problems in Quantum Physics, held in Oviedo (Spain) in the summer of 1993 in the pleasant surroundings of the Hotel de la Reconquista. The papers presented at this Symposium are collected in the present book.

We would like to express our gratitude to the contributors and to all the Institutions and Authorities that aided the Organizing Committee, in particular:

- Commission of the European Communities, DG XII (Brussels)
- Ministerio de Educación y Ciencia, DGICYT (Madrid)
- Presidente del Principado de Asturias
- Consejería de Educación, Cultura, Deportes y Juventud del Principado de Asturias
- Fundación Príncipe de Asturias
- Vicerrectorado de Investigación de la Universidad de Oviedo
- Vicerrectorado de Investigación de la Universidad Autónoma de Madrid
- Vicerrectorado de Investigación de la Universidad de Cantabria (Santander)
- Vicerrectorado de Estudiantes de la Universidad de Oviedo
- Dirección General de Turismo del Principado de Asturias

Last but not least, we would like to express our thanks to Professors A. F. Rañada, J. L. Sánchez-Gómez, and E. Santos as well as to Dr. Susana F. Huelga and Ms. Ana María Espino for their organizational activity, without which the symposium would not have been such a success.

The Editors

THE SUBJECT OF OUR DISCUSSIONS

Emilio Santos

Universidad de Cantabria
Santander, Spain

The subject of our discussions will be fundamental problems in quantum physics. It is a fact that all theories have problems, but maybe quantum theory is special in that its problems are of a fundamental nature, or they are more fundamental than in other cases. I am not concerned with mathematical problems, a subject which will be covered by some speakers in this Symposium,but with physical - or maybe philosophical - problems.

Sometimes it seems that we are still discussing the same things already discussed in the late twenties by the founding fathers. In fact, in our discussions about foundations we are still quoting Bohr versus Einstein, much as in the Middle Age philosophers quoted Plato versus Aristotle. (Incidentally, the essential subject seems to be the same, namely idealism versus realism.) However, the truth is that refined theoretical arguments and new experimental techniques have produced a substantial advance in the last two decades. But I think that, unfortunately, it is still true the famous sentence of Feynman, who said "nobody really understand quantum mechanics."

Soon after the discovery of quantum mechanics around 1925, people realized that the theory had two strange features, namely indeterminism and contextuality. Indeterminism, that is the fact that the maximum information available about the present is not sufficient to fully predict the future, is not strange by itself. Indeed, probability theory was invented just to deal with this situation. What is strange is the way that indeterminism appears in the new theory, because the evolution, that is Schrödinger´s equation, is perfectly deterministic whilst indeterminism manifests only in the process of measurement. Contextuality, that is the fact that the result of a measurement depends on the whole experimental arrangement, is also not strange by itself. Indeed, a standard rule of the scientific method is that experiments should be carefully controlled in order to be reproducible, which means that the whole context is important. Again, what is strange is the way in which contextuality appears, namely preventing any possible knowledge of some properties, for instance wave attributes, when one is measuring some other properties such as particle

M. Ferrero and A. van der Merwe (eds.), Fundamental Problems in Quantum Physics, 1–5.

attributes.

After a short period of intense activity, during which the apparently most obvious solutions were shown to fail, Bohr was able to find a kind of solution which seemed to solve all the problems. It became the basis of the Copenhagen Interpretation of quantum mechanics. The central idea was that quantum mechanics is actually not a theory of the microscopic systems themselves, but a theory about the possible relations between microsystems and observers (human beings?). The Copenhagen interpretation proved to be an extremely useful tool to predict, or at least reproduce, the results of the experiments by explicit calculations. However, the Copenhagen Interpretation gives rise to serious philosophical problems. In the first place it puts an artificial division on the material world: the observed system on one hand and the observer on the other (except if we assume that the observer is necessarily non-material, which raises even greater philosophical difficulties.) In the second place, the Copenhagen view prevents a realistic interpretation of the microworld: one is no longer allowed to speak about what "is" but only about what "can be found in a specific measurement." The positivistic, or pragmatic, philosophy inherent to the Copenhagen interpretation has been strongly criticized by the most conspicuous philosophers of science, for instance Sir Karl Popper to quote just one, and by great scientists including Einstein, Schrödinger and de Broglie.

In my opinion, the Copenhagen interpretation is very good provided it is just taken as a provisional solution, useful until the true, final, solution is found. But to know that we live in a provisional home is painful. As time elapsed and nobody found a good alternative, people preferred to believe that quantum mechanics is the final theory rather than wait for another one. For these people the good news came in the form of "theorems of impossibility of hidden-variables theories," the most celebrated proved by John von Neumann in 1932. After that, all activity on foundations was put to a rest for two decades, except for a few "old fashioned people" like Einstein and Schrödinger. At the same time these two decades saw a remarkable progress in the applications of quantum theory to new problems and even an impressive generalization of the theory, which culminated in the late forties with the development of relativistic quantum field theory. In particular, quantum electrodynamics became the theory with greatest predictive power in the whole history of science. All this seemed to be the final support for quantum theory supplemented with the Copenhagen interpretation which, by that time already was the "orthodox interpretation of quantum mechanics."

Renaissance of research on foundations came during the fifties and sixties, mainly due to the work of two men both of whom have recentely died , David Bohm and John Bell. In contrast, during this same period the orthodox view was reinforced with the development of quantum logics and more powerfull no-hidden-variables theorems. In any case the work done during these years, plus the recent development of new experimental techniques, has been the basis for the renewed interest in foundations that we have seen in the last thirty years.

Resting upon the work done by Bohm in the early fifties, two articles by Bell, published in 1965 and 1966, have been the starting point for the renaisance of the work in foundations during recent times. In one of the papers Bell clarified the subject of hidden-variables theories: contextual hidden-variables are compatible with quantum mechanics but noncontextual ones are not. In the other paper, the most celebrated one, he showed that local theories, which are partially contextual, are still incompatible with quantum mechanics. From that time, hundreds of papers have been devoted to the subject of quantum nonlocality, while a sharp debate has continued. Bell's theorem has been considered from irrelevant up to the greatest discovery of science. Some people have claimed that nonlocality is a generic property of quantum systems whilst others do not find any nonlocality in quantum theory. Still, some authors consider that Bell's theorem shows a dramatic conflict between quantum mechanics and relativity theory whilst others believe that even many classical systems may violate the Bell inequalities. The least that I can say is that a considerable confusion exists around the subject.

No less confusion exists with respect to the empirical tests of local hidden-variables theories against quantum mechanics, via Bell's theorem. When the possibility of real tests was studied, it was soon realized that the reliable experiments were very difficult. Then, a variety of constraints were proposed on the local hidden-variables theories that allowed, at least, to test restricted families of such theories. This gave rise to a first generation of experiments, performed during the seventies and early eighties, which refuted those restricted families of local theories and gave new confirmations of quantum mechanics in a domain where some doubts about its validity might arise. In the last decade, new optical experiments mainly based upon parametric down converters have accumulated new evidence for quantum mechanics and against these restricted local theories.

Unfortunately, the incorrect claim has been made many times that all local hidden-variables theories have been empirically refuted. In popular articles, and even in serious ones, this has been frequently combined with a fashionable emphasizing of the magical, rather than the rational, aspects of quantum mechanics. Only very recently the relevance of the loopholes involved in the performed experiments is becoming fully appreciated and a second generation of experiments are currently being proposed, which may really test quantum mechanics against local realism. These experiments are extremely difficult and it may even be that quantum effects not taken into account in "gedanken experiments", like the zeropoint fluctuations of the electromagnetic field, may prevent a truly reliable test. It has been even conjectured that local realism and quantum mechanics might be compatible, after all, for real experiments. In any case, the existence of local realist theories of the microworld is still an open possibility.

Another subject which has received considerable attention is the relation between quantum mechanics and gravitation or cosmology. The stupendous progress in observational astronomy, in particular from outer space, combined with the interest in very high energies, not achievable in Earth-based experiments, has directed the attention of particle theorists towards the processes taking place in the very early universe. Here, some combination of general relativity and quantum mechanics is unavoidable. Indeed the combination has already produced the inflationary scenario, which has dominated theoretical cosmology during the last decade. From our point of view, the most relevant consequence has been the increasing realization that the Copenhagen interpretation of quantum mechanics is not sufficient. In fact, when we deal with the universe as a whole, where is the observer external to the system?. Or can the observer be internal?. These difficult questions have displaced the Copenhagen interpretation from the central role it played during half a century and other interpretations are taking its place, in particular the Many-Worlds interpretation, which essentially means assuming the unrestricted validity of the quantum formalism. However, a satisfactory quantum gravity is not yet available. We are here confronted not only with the mathematical difficulties for quantizing a nonlinear theory, but with the problem of knowing which is more fundamental, quantization or space-time structure.

Maybe due to these problems with cosmology, or as a result of the strong criticisms by John Bell and other people, or simply by the maturity of our knowledge, it is a fact that measurement theory has become the center of attention of the workers in foundations. People no longer accept the dichotomy between observer and observed system, or between a normal evolution governed by Schrödinger´s equation and a sudden wave-packet collapse induced by the measurement. There is some general agreement that measurement is just a physical process, which should be understood in terms of the usual quantum formalism, without *ad hoc* measurement postulates. It is increasingly clear that our difficulties with the theory of quantum measurements came from the attempt to understand the "ideal measurement", which is in itself a self-contradictory concept. In fact, any real measurement involves an irreversible process of amplification. Now, irreversibility presuposes dissipation, and dissipation cannot be separated from fluctuation. Therefore, any real measurement involves noise and it cannot be ideal.

The central difficulty for a clear understanding of the measurement process is the problem of the lack of "objectification", that is the impossibility to explain how a coherent superposition of quantum states leads to sharp values for all the physical quantities in the macroscopic domain. This problem was clearly illuminated by Schrödinger with his celebrated example of the cat in a superposition state of alive and dead. A practical solution, which has been investigated recurrently at least during thirty years, is decoherence induced by the environment. In quantum mechanics we may say, paraphrasing the ecologists, that intelligent life will be impossible if we do not take care of the environment. However, we think that the "vacuum flucutations", or zeropoint noise, is an essential part of the environment, a fact not fully appreciated. In any case, although explaining the practical reasons for decoherence is certainly progress, the fundamental problem remains. Indeed, assuming the unrestricted validity of the superposition principle is not compatible with the cherished philosophical principle that things do exist in well defined, although maybe unknown, states. In this sense, several attempts have been made to modify, or restrict, the linear quantum evolution. Some of them will be presented at this Symposium.

Let me finish by summarizing my personal view. I think that quantum theory is just a stochastic theory not essentially different from classical ones like, say, the theory of Brownian motion. (Of course I do not mean that the basic stochastic processes are of Brownian type in the quantum case.) In any stochastic theory, noise is the central ingredient but it happens that noise seems to be absent from the simplified version of quantum theory which we call quantum mechanics of particles. Indeed, Schrodinger´s equation is fully deterministic. The problem is that people have tried to understand quantum mechanics by using highly idealized "gedanken" experiments, all of which share the neglect of the noise. Noise certainly appears in relativistic quantum field theory, with its formidable complexity, in the form of fluctuations of all fundamental fields. In particular, it is the essential constituent of what we naively call "the vacuum". Even the partial removal of that noise requires the heavy machinery of renormalization theory, but finally it is possible to get a mathematically tractable approximate theory which is the quantum mechanics of particles. In my opinion, it is not possible to understand elementary quantum mechanics just because it is an oversimplified idealization. But I hope that, in the future, we will be able to understand quantum fields as stochastic fields of some special nature. My conjecture is that the Hilbert space formalism is just an extremely efficient mathematical technique for dealing with these stochastic fields which, however, hides the physics. Our aim is therefore to decode what is encoded in the Hilbert space representation of quantum fields.

MEASUREMENT OF THE SCHRÖDINGER WAVE OF A SINGLE PARTICLE

Yakir Aharonov[1,2] **and Lev Vaidman**[1]

[1] *School of Physics and Astronomy*
Raymond and Beverly Sackler Faculty of Exact Sciences
Tel-Aviv University
Tel-Aviv, 69978 Israel

[2] *Physics Department, University of South Carolina*
Columbia, South Carolina 29208

We show that it is possible to measure the Schrödinger wave of a single quantum system. This provides a strong argument for associating physical reality with the quantum state of a single system, and challenges the usual assumption that the quantum state has physical meaning only for an ensemble of identical systems.

Key Words: quantum measurements, interpretation of quantum theory.

But you say Ψ is not real, I will ask: what *is* real in your theory? ... If your wave function is not real, you must tell me what is. – John Bell.

Today, the commonly accepted interpretation of the Schrödinger wave is due to Born. He proposed to interpret the wave intensity, not as the density of distribution of actual matter, as Schrödinger imagined, but as a probability density for the presence of a particle. According to Born, the Schrödinger wave can be tested only for an ensemble of particles. Schrödinger, however, wanted to believe that his wave represents a single particle: the wave is an extended object really moving in space. Several arguments were against this point of view:

(i) We have never seen the quantum state of a single particle in a laboratory. Indeed, while a wave is typically spread over a region of space, we never see a particle simultaneously in several distinct locations.

(ii) If we could see a quantum state, we could presumably distinguish it from any other quantum state, but the unitary time evolution of states

M. Ferrero and A. van der Merwe (eds.), Fundamental Problems in Quantum Physics, 7–13.

in quantum mechanics implies that it is impossible to distinguish between two different non-orthogonal states. Different outcomes of a measurement distinguishing these two states correspond to orthogonal quantum states of the composite system (measuring device plus particle). But, the scalar product between the initial states was not zero and remains nonzero under unitarity time evolution.

(iii) If we associate physical reality with a spread-out wave then the instantaneous "collapse" of the wave to a point during a position measurement seems to conflict with relativity.

In spite of these arguments we will show that Schrödinger was not so wrong after all. It is possible to associate a Schrödinger wave to a single particle. To prove this we will show that it is possible to measure the Schrödinger wave of a single particle [1]. We will demonstrate how the density $\rho = \Psi^*\Psi$ and the current $j = \frac{\hbar}{2i}(\Psi^*\nabla\Psi - \Psi\nabla\Psi^*)$ of a Schrödinger wave of a single particle can be measured. Then, density and the current will yield the wave itself.

The usual measurements assumed in argument (i) alter the Schrödinger wave and are not adequate; rather, here we will describe special *protective* measurements allowing us to measure ρ and j without changing the Schrödinger wave. In some cases energy conservation provides protection for the state, while in other cases we need a special protection procedure.

Let us consider a particle in a discrete nondegenerate energy eigenstate $\Psi(x)$. The standard von Neumann procedure for measuring the value of an observable A in this state involves an interaction Hamiltonian

$$H_{int} = g(t)pA, \tag{1}$$

coupling our system to a measuring device, or pointer, with coordinate and momentum denoted respectively by q and p. The time-dependent coupling $g(t)$ is normalized to $\int g(t)dt = 1$. The initial state of the pointer is taken to be a Gaussian centered around zero.

In standard impulsive measurements, $g(t) \neq 0$ only for a very short time interval. Thus, the interaction term dominates the rest of the Hamiltonian, and the time evolution $e^{-\frac{i}{\hbar}pA}$ leads to a correlated state: eigenstates of A with eigenvalues a_n are correlated to measuring device states in which the pointer is shifted by these values a_n. By contrast, the protective measurements of interest here utilize the opposite limit of extremely slow measurement. We take $g(t) = 1/T$ for most of the time T and assume that $g(t)$ goes to zero gradually before and after the period T. We choose the initial state of the measuring device such that the canonical conjugate p (of the pointer variable q) is bounded. For $g(t)$ smooth enough we obtain an adiabatic process in which the particle cannot make transition from one energy eigenstate to another, and, in the limit $T \rightarrow \infty$, the interaction Hamiltonian does not change the energy eigenstate. For any value of p, the energy of the eigenstate shifts by an infinitesimal amount given by first order perturbation theory:

$$\delta E = \langle H_{int} \rangle = \frac{\langle A \rangle p}{T}. \tag{2}$$

The corresponding time evolution $e^{-ip\langle A\rangle}$ shifts the pointer by the average value $\langle A\rangle$. (Here and below we will take $\hbar = 1$.) This result contrasts with the usual (strong) measurement in which the pointer shifts by one of the eigenvalues of A. By measuring the averages of a sufficiently large number of variables A_n, the full Schrödinger wave $\Psi(x)$ can be reconstructed to any desired precision.

As a specific example we take the A_n to be (normalized) projection operators on small regions V_n having volume v_n:

$$A_n = \begin{cases} \frac{1}{v_n}, & \text{if } x \in V_n, \\ 0, & \text{if } x \notin V_n. \end{cases} \tag{3}$$

The measurement of A_n yields

$$\langle A_n\rangle = \frac{1}{v_n}\int_{V_n} |\Psi|^2 dv = |\Psi_n|^2, \tag{4}$$

where $|\Psi_n|^2$ is the average of the density $\rho(x) = |\Psi(x)|^2$ over the small region V_n. Performing measurements in sufficiently many regions V_n we can reconstruct $\rho(x)$ everywhere in space. (Simultaneous measurement of all the variables A_n requires slower and weaker interactions, and thus takes more time.) For a real state the density $\rho(x)$ is itself enough to reconstruct the Schrödinger wave; we can fix the sign by flipping it across nodal surfaces.

In general case, however, in addition to measurements of the density $\rho(x)$, we have to measure current density. This time we also adiabatically measure the averages of

$$B_n = \frac{1}{2i}(A_n\nabla + \nabla A_n) \quad . \tag{5}$$

Indeed, $\langle B_n\rangle$ are the average values of the current $j = \frac{1}{2i}(\Psi^*\nabla\Psi - \Psi\nabla\Psi^*)$ in the region V_n. Writing $\Psi(x) = r(x)e^{i\theta(x)}$ with $r(x) = \sqrt{\rho(x)}$, we find that

$$\frac{j(x)}{\rho(x)} = \nabla\theta \quad ; \tag{6}$$

and the phase $\theta(x)$ can be found by integrating j/ρ.

For a charged particle the density $\rho(x)$ times the charge yields the effective charge density. In particular, it means that an appropriate adiabatic measurement of the Gauss flux out of a certain region must yield the expectation value of the charge inside this region (the integral of the charge density over this region). Likewise, adiabatic measurement of the Ampere contour integral yields the expectation value of the total current flowing through this contour in the stationary case.

Our discussion of the current of the particle is valid only for a Hamiltonian without vector potential. However, the eigenstates of such a Hamiltonian with a nonvanishing current are necessarily degenerate due to time reversal invariance. The method described above is appropriate only for nondegenerate eigenstates and, therefore, we have to consider problems with a vector potential **A**, for which we do have nondegenerate stationary states with non-zero current (e.g. the Aharonov-Bohm effect). Then, the definition of the (electric) current must be modified by replacement $\nabla \to \nabla - ie\mathbf{A}$. This replacement has to be done also for the definition of the observables B_n (Eq. (5)), and it leads to the obvious modification of the Eq.(6).

We have shown that stationary quantum states can be observed. This is our main argument for associating physical reality with the quantum state of a single particle. Since our measurement lasts a long period of time we do not have a method for measuring the Schrödinger wave at a given time. Thus, we have a direct argument for associating physical reality with stationary Schrödinger waves only over a *period* of time. The reader may therefore suspect that our measurements represent time-averaged physical properties of the system. Let us now present a few arguments explaining why, nevertheless, these measurements reflect properties of the Schrödinger wave at any given moment of time during the measurement.

An essential feature of our adiabatic measurement is that the state $|\Psi\rangle$ does not change throughout the experiment. Since the Schrödinger wave yields the complete description of a system and the interaction with the measuring device is constant throughout the measurement, we conclude that the action of the system on the measuring device is the same at any moment during the measurement.

The mathematical description of our measurement tells us the same: for any, even very short, period of time, the measuring device shifts by an amount proportional to $\langle A\rangle$, the expectation value of the measured variable, rather than to one of its eigenvalues a_n. Thus, expectation values, which mathematically characterize Schrödinger waves, can be associated with very short periods of time. In the instantaneous limit, expectation values and, therefore, the quantum state manifest themselves as properties of a quantum system defined at a given time. (Note, however, that pointer shifts during short time intervals are unobservable since they are much smaller than the uncertainty; only the total shift accumulated during the whole period of measurement is much larger than the width of the initial distribution, and therefore observable on a single particle.)

Moreover, suppose that (contrary to standard quantum theory), a system has a complete description that *does* change during the measurement process, and the (constant) Schrödinger wave we measure does not describe the system at a given time but represents only a time average of some hidden variables over the period of the measurement. Consider a model of a hydrogen atom in which the electron performs very fast ergodic motion in the region corresponding to the quantum cloud. The charge density might be either zero (if the electron is not there) or singular (if the electron is inside the infinitesimally small region including the space point in question). In spite of this fact, the measurement we have described will yield outcomes corresponding to a nonsingular charge density cloud. What it measures is the time average of the density, or how long a time the electron spent in a given place.

In order to see that this picture is inappropriate for the quantum case let us consider another example: a particle in a one-dimensional box of length L in the first excited state. The spatial part of the state is $\sqrt{2/L}\sin(2\pi x/L)$. The adiabatic measuring procedure described above will yield the Schrödinger wave density $(2/L)\sin^2(2\pi x/L)$. In particular, it equals zero at the center of the box. If there is some hidden position of the electron which changes in time such that the measured density is proportional to the amount of time the electron spends there, then half of the time it must be in the left half of the box and half of the time in the right half of the box. But it can spend no time at the center of the box; i.e., it must move at infinite velocity at the center. It is absolutely unclear what such an electron "position" would be. There *is* a theory [2] which introduces a "position" for a particle in addition to its Schrödinger wave; but according to this theory, the "velocity" of the particle in the given energy eigenstate vanishes: the position does not change at all. In the quantum picture the eigenstate of the particle in the box can be represented as a superposition of two running waves moving in opposite directions. The zero density at the center of the box is due to destructive interference – the phenomenon which cannot be reproduced in a classical ergodic model of a particle.

Our procedure is not applicable to degenerate energy eigenstates. The simplest way to deal with this case is by adding a potential (as part of the measuring procedure) to lift the degeneracy. This protection does not change the state. However, one can argue that it changes the physical situation. We can bring this change to a minimum by adding strong protection potential for a dense set of very short time intervals. Thus, most of the time the system has not only the same state, but also the original potential.

The procedure described above cannot measure properties of a state obtained by superposing several nondegenerate energy eigenstates. Applied to such a state, a measurement of A will yield shifts of the pointer corresponding to the expectation values of the variable A in the various energy eigenstates. In general, these values are distinct with differences greater than the initial uncertainty of the pointer position. Thus, after the interaction, the system and the measuring device are entangled. By "looking" at the measuring device we cause the Schrödinger wave to choose one of the energy eigenstates. Measurement of the Schrödinger wave – namely, measurement of the expectation values of the projection operators – causes collapse. A superposition of nondegenerate energy eigenstates is not protected by energy conservation: unitary evolution during the measurement leads to correlations between energy states and the states of the measuring device without changing the total energy, while collapse changes the energy itself. Nevertheless, we can measure even a superposition of energy eigenstates by a procedure similar to the one described above. We add a dense set of time-dependent potentials acting for very short periods of time such that the state at all these times is the eigenstate of the Hamiltonian together with the additional potential. Still, most of the time the system evolves under the free Hamiltonian. The proof of the efficiency of the above strong impulsive potentials is similar to the proof of the Zeno "paradox" in which a quantum system under a dense set of observation evolves in accordance with the evolution tested, and not according to the free Hamiltonian. In our case, the two evolutions are identical.

When measurements involve the above kind of protection, we have to know the state in order to prescribe the proper protection, i.e., the potential which makes the state a nondegenerate eigenstate. Indeed, it is impossible to measure an unknown unprotected wave of a single particle. One might object, therefore, that our measurement yields no new information, since the state is already known. However, we can separate the protection and measurement procedures: one experimentalist provides protection and the other measures the Schrödinger wave itself. Then the second experimentalist does obtain new information. The most important point, however, is that we directly measure properties of the Schrödinger wave of a single system using a standard measuring procedure. Our direct measurements of the density and the current of the Schrödinger wave challenge the commonly accepted notion that quantum states can be observed fully only when the measurement is performed on an ensemble of identical systems.

Consider now an apparent paradox arising from the measurement of Schrödinger wave. It is well known that even assuming instantaneous "collapse" of a quantum state, one cannot use the collapse for sending signals faster than light. At first, however, the possibility of measuring the value of the Schrödinger wave at a given location seems to allow such superluminal communication. Consider a particle in a superposition $\frac{1}{\sqrt{2}}(|1\rangle + |2\rangle)$ of being in two boxes separated by a very large distance. For this particle the expectation value of the projection onto the first box is $\langle P_1 \rangle = 1/2$. This value must be the outcome of a measurement performed on the first box. If, however, just prior to a measurement of the Schrödinger wave in the first box, someone opens and looks into the second box, causing collapse to a localized state $|1\rangle$ or $|2\rangle$, then the outcome of the measurement of the projection operator in the first box will drastically change: we no longer find $\langle P_1 \rangle = 1/2$ but rather 0 or 1 (depending on what is found in the second box). It seems, therefore, that measurements on one box can influence measurements on another box located arbitrarily far away.

However, this argument contains a flaw: The state $\frac{1}{\sqrt{2}}(|1\rangle + 2\rangle)$ is not a discrete nondegenerate eigenstate. Since there is no overlap between the states $|1\rangle$ and $|2\rangle$, the orthogonal state $\frac{1}{\sqrt{2}}(|1\rangle - |2\rangle)$ has the same energy. Thus, there is no natural protection due to the energy conservation, and an additional protection is needed. This protection, however, involves explicitly nonlocal interactions. These nonlocal interactions are the source of the alleged superluminal signal propagation. (A more subtle paradox of this sort is considered in another work [3].)

Let us come back to the three arguments against the realistic view of the Schrödinger wave presented in the beginning of the paper. First, we have shown that we *can* observe a quantum state. Although our discussion relied on Gedanken experiments, recent experimental work with so-called "weak links" in quantum circuits shows that slow adiabatic measurements of the Schrödinger wave can be performed in the laboratory [4]. Another proposals are discussed in Ref. (5).

The second argument is a correct statement, but it only implies that there is no single *universal* procedure for observing states. It still allows the possibility of an appropriate measuring procedure for any given state.

The last argument (iii) is the most serious one. Assume that the Schrödinger wave of a particle is nonvanishing only inside two separate boxes, and we find it in one of them. How did part of the wave move instantaneously from one box to another? We believe that a full answer to this argument requires a new approach to quantum theory [6]. The basic idea is that "reality" is defined not just by one Schrödinger wave, but by two waves. In addition to the usual wave evolving from a complete measurement in the past toward the future, we consider the wave evolving backwards in time from the complete measurement in the future. In other words, at present, say time t, the reality is not completely defined. The measurement at a time after time t is necessary to complete the description of a quantum system at time t. We have shown that expectation values of quantum variables and the quantum state itself have physical meaning, i.e., they are measurable for individual quantum systems. This result stands in sharp contrast to the standard approach in which the Schrödinger wave and expectation values are statistical properties of ensembles of identical systems.

ACKNOWLEDGMENTS

It is a pleasure to thank Daniel Rohrlich for helpful discussions. The research was supported in part by grant 425/92-1 of the the Basic Research Foundation (administered by the Israel Academy of Sciences and Humanities) and by grant PHY 8807812 of the National Science Foundation.

REFERENCES

1. Y. Aharonov and L. Vaidman, *Phys. Lett. A* **178**, 38 (1993).
2. D. Bohm, *Phys. Rev.* **85** (1952) 166.
3. Y. Aharonov and L. Vaidman, in *Quantum Control and Measurement*, H. Ezawa and Y. Murayama, eds. (North-Holland, Tokyo, 1992), p. 99.
4. T.P. Spiller, T.D. Clark, R.J. Prance, and A. Widom, *Prog. Low Temp. Phys.* **XIII** (1992) 219.
5. Y. Aharonov, J. Anandan, and L. Vaidman, *Phys. Rev. A* **47**, 4616 (1993).
6. Y. Aharonov and L. Vaidman, *Phys. Rev. A* **41** (1990) 11.

THE EMERGENCE OF CLASSICAL PROPERTIES FROM QUANTUM MECHANICS: NEW PROBLEMS FROM OLD

Leslie E. Ballentine
Physics Department
Simon Fraser University
Burnaby, B.C., V5A 1S6, Canada

The ongoing debates about the interpretation of quantum mechanics have often taken place in forums such as the measurement problem and Schrödinger's cat paradox. I urge that we move beyond those old forums, and consider instead the more general problem of how classical properties emerge from quantum mechanics. Quantum chaos appears as a special case of this general problem, being the study of the emergence of classical chaos from QM. Such a study quickly leads to the following conclusion: *the classical limit of a quantum state is typically an ensemble of classical orbits, and not a single classical orbit.* This fact enables us to understand the relation between several quantum and classical dynamical models. This study also reveals interesting connections between old and new problems.

Keywords: quantum mechanics, classical properties, chaos.

Central to the debates about the interpretation of quantum mechanics has been the question of the significance of the quantum state vector. The two principal views are:

(a) The state vector $|\psi\rangle$ provides a complete description of an *individual* system. A dynamical variable represented by the operator Q has a value (q, say) if and only if $Q|\psi\rangle = q|\psi\rangle$.
(b) The state vector describes the statistical properties of an *ensemble of similarly prepared systems.*

This debate has traditionally taken place within the context of Schrödinger's cat paradox and the analysis of the measurement process. Some useful conclusions have emerged from that

M. Ferrero and A. van der Merwe (eds.), Fundamental Problems in Quantum Physics, 15–28.

analysis, but if further progress is to be made, I believe we should move beyond the so-called "problem of measurement". In this paper I wish to discuss some newer problems, which turn out to have connections to old questions about the interpretation of quantum mechanics. The central theme of these problems is to understand *the emergence of classical properties from quantum mechanics.* I shall argue that an understanding of the old problems is helpful in solving the new, and that the study of the new problems can improve our understanding of the old.

1. ANALYSIS OF MEASUREMENT

The important result from the analysis of the measurement process is simple to derive. Let initial state of the object be $|q_i\rangle$, and the initial state of the measurement apparatus be $|A_0\rangle$. The coupling between them must be designed so that this initial state of the system (object + apparatus) evolves thus:

$$|q_i\rangle \otimes |A_0\rangle \rightarrow |\phi_i\rangle \otimes |A_i\rangle \,. \tag{1}$$

The final states of the apparatus that correspond to different values of q_i must be sufficiently different so that they are clearly distinguishable to the naked eye; such states are called *macroscopically distinguishable.* We may think of A_i as the eigenvalue of the "pointer position" of the apparatus. (It is of no present interest whether the final state of the object, $|\phi_i\rangle$, is different from the initial state.)

Consider next an initial state of the object that is a *superposition*, $|\psi\rangle = (|q_1\rangle + |q_2\rangle)/\sqrt{2}$. From (1) and the linearity of time evolution, it follows that this must lead to a final state of the system,

$$|\psi\rangle \otimes |A_0\rangle \rightarrow (|\phi_1\rangle \otimes |A_1\rangle + |\phi_2\rangle \otimes |A_2\rangle)/\sqrt{2}\,, \tag{2}$$

that is a *coherent superposition of macroscopically distinct apparatus "pointer position" states.* Such a state is also referred to as a *correlated* or *entangled* state of the object and apparatus. It should be emphasized that the inclusion of the *environment*, which is very fashionable today, makes no essential difference to the result. Instead of (2), we would then have

$$|\Psi\rangle\otimes|A_0\rangle\otimes|e_0\rangle \rightarrow$$
$$(|\phi_1\rangle\otimes|A_1\rangle\otimes|e_1\rangle + |\phi_2\rangle\otimes|A_2\rangle\otimes|e_2\rangle)/\sqrt{2}\,, \qquad (3)$$

where $|e_0\rangle$ is the initial environmental state, and $|e_1\rangle$ and $|e_2\rangle$ are final environmental states. This is also a coherent superposition of macroscopically distinct states, only the degree of entanglement is more complex that in (2). This simple result is worth stating as a theorem.

> *Measurement Theorem*: The *final state* of the entire system (object + apparatus + relevant environment) is, in general, a *coherent superposition of macroscopically distinct "pointer position" eigenvectors.*

The "measurement problem" arises because interpretation (a) denies that the pointer position in a state like (2) or (3) has any value, whereas we know from experience that at the end of a measurement the pointer position is at least *macroscopically* well defined. This has led to interminable discussions about "collapse" of the state vector in measurement, about which too much has been written.

The status of the problem can be summed up in a paradoxical couplet:

> *The measurement problem is insoluble;*
> *the measurement problem is solved.*

By this I mean, the measurement problem, *viewed as the problem of explaining how the "collapse of the state vector" comes about*, is insoluble, since such a "collapse" is incompatible with the known equation of motion. But the measurement problem, *viewed as seeking an interpretation of the formalism that is compatible with the existence of entangled macroscopic superpositions (Measurement Theorem)*, is solved by abandoning interpretation (a) in favor of the ensemble interpretation (b), which does not require any "collapse" process.

I have written about this resolution of the measurement problem elsewhere [1,2,3], and I do not wish to extend a discussion that has already gone on for decades. The essential problem is really not "measurement", but rather to understand *how classical properties emerge from quantum mechanics*, the "pointer position" of the apparatus being merely one example. In this paper, I wish to point out some other situations in which this problem can be studied, and to show how the insight derived from this resolution of the old measurement problem is helpful in understanding some new problems.

2. THE CLASSICAL LIMIT

The classical limit of quantum mechanics is often loosely described as $\hbar \to 0$. Since the quantum state function depends on $\hbar$, this limit involves a sequence of states that must be precisely defined. To do this, one should consider a well-defined physical situation in which the initial state is uniquely determined by the boundary conditions. This may be done by fixing the values of sufficiently many observables. In many cases, the appropriate limit can be expressed as $\hbar \to 0$, $n \to \infty$, with $\hbar n$ constant (n is a typical quantum number). That is to say, the quantum $\hbar$ must become small compared to the macroscopic action $\hbar n$. It is not satisfactory to restrict attention to a contrived sequence of wave functions, such as minimum uncertainty packets whose width in both position and momentum varies as $\sqrt{\hbar}$. The classical limit, described above, seldom leads to such wave functions. We shall see that, in most cases, *the classical limit of a quantum state is an ensemble of classical orbits,* not a single classical orbit. This fact can be viewed as an application of the insight gained from the measurement problem, the need for an *ensemble* interpretation of quantum states. Or it can be viewed as further support for the *ensemble* interpretation, from a source different from the measurement problem.

Many discussions of the classical limit are based on Ehrenfest's theorem, which states that, under certain conditions, the centroid of a wave packet state will follow a classical trajectory. But, as will be shown, those conditions are often *not* satisfied in the classical limit, and that limit of the wave function can not be associated with a single classical trajectory [4].

Consider a particle moving in one dimension in a scalar potential $V(x)$, which generates the force $F(x) = -\nabla V(x)$. Its Hamiltonian operator (operators are distinguished by a circumflex) is

$$\hat{H} = \hat{p}^2/2m + V(\hat{q}) , \qquad (4)$$

and the Heisenberg equations of motion are

$$d\hat{q}/dt = \hat{p}/m \,, \quad d\hat{p}/dt = F(\hat{q}) \,. \tag{5)(6}$$

We are interested in the equations satisfied by the mean values of position and momentum some state, $P = \langle\hat{p}\rangle$ and $Q = \langle\hat{q}\rangle$. The operators for the deviations from the mean values are

$$\delta\hat{q} = \hat{q} - \langle\hat{q}\rangle \,, \quad \delta\hat{p} = \hat{p} - \langle\hat{p}\rangle \,. \tag{7)(8}$$

Expanding (5) and (6) in powers of these operators and taking the average in some state then yields

$$dQ/dt = P/m \,, \tag{9}$$

$$\frac{dP}{dt} = F(Q) + \tfrac{1}{2}\langle(\delta\hat{q})^2\rangle \frac{\partial^2}{\partial Q^2} F(Q) + \cdots \,. \tag{10}$$

If $\langle(\delta\hat{q})^2\rangle$ and the higher-order terms are negligible, it is apparent that Q and P will obey the classical equations. It is therefore tempting to interpret the terms involving $(\delta\hat{q})$ as quantum corrections, whose smallness provides the criterion for the classical regime, but we shall see that that would be a mistake.

That the terms beyond $F(Q)$ in (10) are not purely quantum mechanical in origin is indicated by the fact that $\langle(\delta\hat{q})^2\rangle$ does not vanish in the limit of $\hbar \to 0$, $n \to \infty$, $\hbar n$=constant. (This is easily verified for harmonic oscillator or hydrogenic eigenstates.) In fact, $\langle(\delta\hat{q})^2\rangle$ is just a measure of the width of the probability distribution in configuration space, which need not vanish in the classical limit. We should, therefore, compare (10) with the statistical form of classical dynamics.

Let $\rho(q,p,t)$ be the probability distribution in phase space for a classical ensemble. It satisfies the Liouville equation,

$$\frac{\partial}{\partial t}\rho(q,p,t) = -\frac{p}{m}\frac{\partial}{\partial q}\rho(q,p,t) - F(q)\frac{\partial}{\partial p}\rho(q,p,t) \,. \tag{11}$$

From it, we can calculate the classical averages,

$$\langle q\rangle_c = \iint q\,\rho(q,p,t)\,dq\,dp \,, \tag{12}$$

$$\langle p \rangle_c = \iint p\, \rho(q,p,t)\, dq\, dp\ . \tag{13}$$

Differentiating these expressions with respect to t, using (11), and integrating by parts as needed, we obtain

$$d\langle q \rangle_c/dt = \langle p \rangle_c/m\ , \tag{14}$$

$$d\langle p \rangle_c/dt = \iint F(q)\, \rho(q,p,t)\, dq\, dp\ . \tag{15}$$

Expanding (15) in powers of $\delta q = q - \langle q \rangle_c$ then yields

$$\frac{d}{dt}\langle p \rangle_c = F(\langle q \rangle_c) + \tfrac{1}{2}\langle(\delta q)^2\rangle_c \frac{\partial^2}{\partial \langle q \rangle_c{}^2} F(\langle q \rangle_c) + \cdots\ , \tag{16}$$

where $\langle(\delta q)^2\rangle_c = \iint (\delta q)^2 \rho(q,p,t)\, dq\, dp$ is a measure of the width of the classical probability distribution. The significance of the terms involving δq is now clear: the centroid of a classical ensemble need not follow a classical trajectory if the width of the probability distribution is not negligible. Since the quantal equation (10) has exactly the same form as (16), it is apparent that the terms in (10) involving $\delta \hat{q}$ express the fact that the centroid of the quantum probability distribution does not follow a classical trajectory.

Thus violations of Ehrenfest's theorem are not always of quantum mechanical origin; a classical ensemble behaves similarly. We see here that the classical limit of a quantum state is typically an ensemble of classical orbits, and not a single classical orbit.

3. PARTICLE BETWEEN REFLECTING WALLS

The conclusion above is illustrated by the example of a particle bouncing between two walls, at $x = 0$ and $x = L$. A general time-dependent state function can be expanded in terms of the energy eigenfunctions,

$$\psi(x,t) = \sum_{n=1}^{\infty} c_n \sin(k_n x) \exp(-iE_n t/\hbar) , \tag{17}$$

where $k_n = n\pi/L$, and $E_n = (\hbar^2\pi^2/2mL^2)n^2$. Because all the frequencies in (17) are integer multiples of the lowest frequency, it follows that $\psi(x,t)$ is periodic, but its period,

$$T_{qm} = 4mL^2/\pi\hbar , \tag{18}$$

bears no relation to the classical period of a particle with speed v, $T = 2L/v$. The failure of $\psi(x,t)$ to oscillate with the classical period would be a problem if it were supposed to describe a single classical orbit. However ψ does make sense when compared to an ensemble of classical orbits, since the motion of the ensemble need not be periodic.

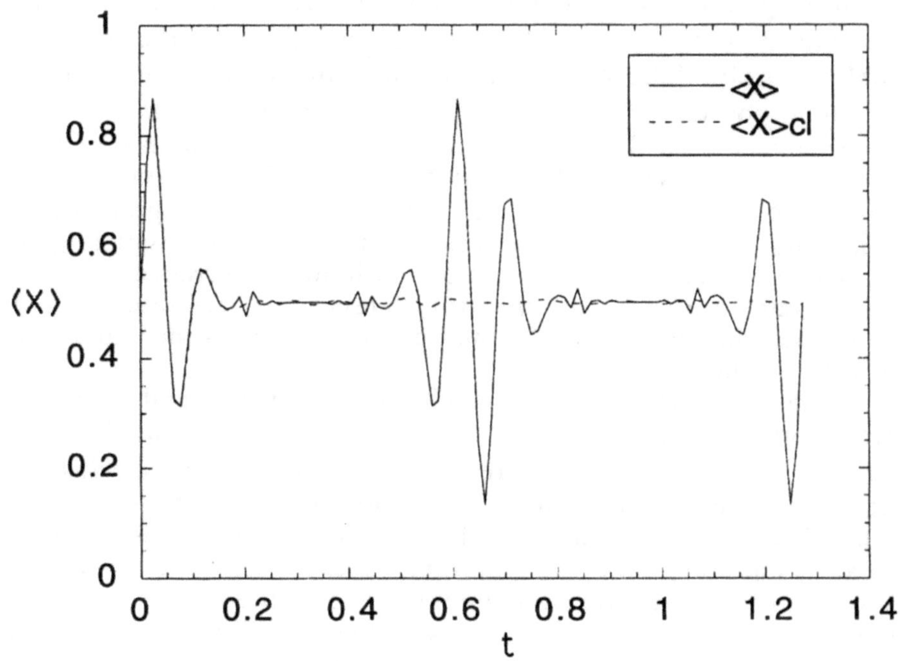

Fig.1. Average position of a particle confined to the unit interval, according to quantum theory (solid line) and classical ensemble theory (dotted line).

Let us consider an initial wave function of the form

$$\psi(x,0) = A(x)\exp(ikx)\,, \tag{19}$$

where $A(x)$ is a real amplitude function. The mean velocity of this state is $v = \hbar k/m$. The motion of this quantum state will be compared to that of a classical ensemble whose initial position and momentum distributions are equal to those of the quantum state (19), the initial phase-space distribution being the product of the position and momentum distributions.

We take the initial amplitude to be Gaussian,

$$A(x) = C\exp\{-[(x-x_0)/2a]^2\}\,. \tag{20}$$

This initial state has rms half-width $\Delta x = a$, and its mean position is taken to be $x_0 = L/2$. Results for $a = 0.1$, $v = 20$ (units: $\hbar = m = L = 1$) are shown in Fig. 1. The average position of the quantum state, $\langle x\rangle = \langle\psi(x,t)|x|\psi(x,t)\rangle$, exhibits a complex pattern of decaying and recurring oscillations that repeat with period T_{qm}. The average position of the classical ensemble closely follows the first quantum oscillations, but it decays to a constant value, $\langle x\rangle = L/2$, where it remains. The decay of the classical oscillation is due to the distribution of velocities in the ensemble, which causes it to spread and eventually cover the range $(0, L)$ uniformly. The spreading of the quantum wave function is very similar to the spreading of the classical ensemble. The later periodic recurrences of the quantum state are due to the interference of reflected waves and to the discreteness of the quantum spectrum, and are essentially nonclassical.

The time interval during which the classical and quantum theories agree well is shown in more detail in Fig. 2. Ehrenfest's theorem, which predicts $\langle x\rangle$ to follow a classical trajectory, is very inaccurate, even before the first reflection. But the failure of Ehrenfest's theorem does not indicate non-classical behavior; the quantum state and the classical ensemble are in close agreement, even though Ehrenfest's theorem is not applicable. The lower half of Fig. 2 shows that $\Delta x = (\langle x^2\rangle - \langle x\rangle^2)^{1/2}$ is also correctly given by the classical theory for $t \lesssim 0.14$. The non-monotonic behavior of Δx is caused by the folding of the ensemble upon itself when it is reflected from a wall.

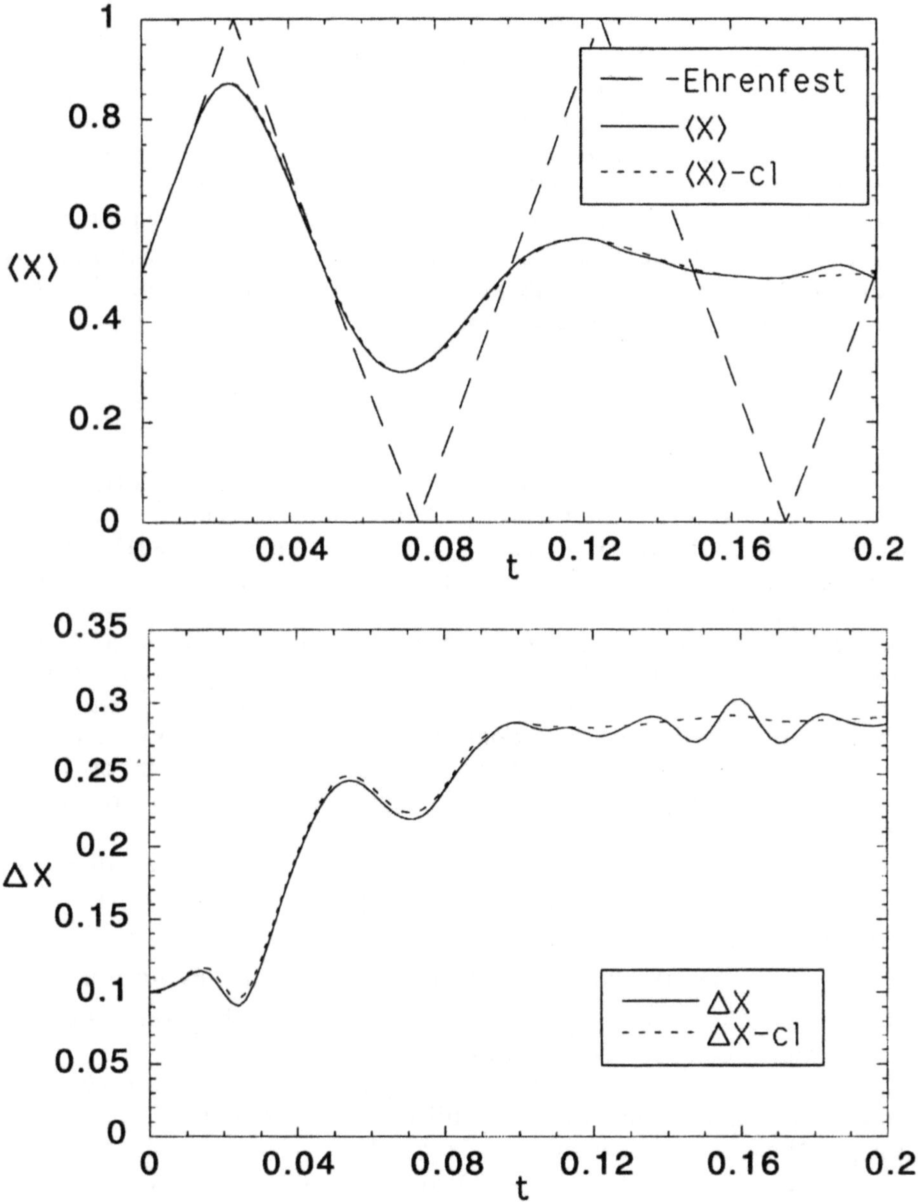

Fig. 2. Upper graph: average position according to quantum and classical theories compared with Ehrenfest's theorem (dashed line).
Lower graph: rms half-width of the position probability distribution, according to quantum theory (solid line) and classical theory (dotted line).

If we had expected the centroid of the wave packet to follow a single classical orbit (Ehrenfest's theorem), we would have underestimated the regime of quantum-classical correspondence by an order of magnitude. In this regime, the quantum and classical position probability densities are not necessarily the same in detail. In fact, the quantum probability density is approximately equal to the classical density *modulated by an interference pattern,* but the interference pattern is averaged out in calculating $\langle x \rangle$ and $\langle x^2 \rangle$. The classical limit is characterized, not by the absence of interference, but by the interference pattern being too fine to be resolved.

4. QUANTUM CHAOS

Chaos in classical mechanics [5] is usually characterized by extreme sensitivity to initial conditions. The separation between two initially close chaotic orbits will grow exponentially, and after a moderate time the two orbits will show no apparent relation to each other. Nonchaotic orbits, on the other hand, are organized into smooth families, and the separation between initially close orbits will grow slowly (often linearly with time). The name "quantum chaos" is loosely applied to the study of the quantum mechanics of systems that exhibit classical chaos. The central problem in the subject is to understand the emergence of classical chaos from quantum mechanics.

There is a simple argument that seems to prove that quantum mechanics can not exhibit chaos. Let $|\Psi_1(0)\rangle$ and $|\Psi_2(0)\rangle$ be two initially close states, such that $\langle\Psi_1(0)|\Psi_2(0)\rangle = 1 - \epsilon$, where ϵ is a small number. Unitarity of time evolution then implies that $\langle\Psi_1(t)|\Psi_2(t)\rangle = 1 - \epsilon$. Since the states do not separate, there can be no quantum chaos!

Some people have taken this argument as conclusive, but that view is shaken by the possibility of constructing a parallel argument that would seem to deny the existence of classical chaos. Let us describe classical states by probability distributions in phase space, which satisfy the Liouville equation. It is well known that the Liouville equation for a classical ensemble contains exactly the same information as does the Hamilton equations for the set of individual trajectories. Let $f_1(q,p,0)$ and $f_2(q,p,0)$ be two initially close phase space distributions. The overlap integral of these functions,

$$\{f_1(t)|f_2(t)\} = \iint f_1(q,p,t)\, f_2(q,p,t)\, dq\, dp\,, \tag{21}$$

has the properties of an inner product on phase space. Liouville's equation implies that this inner product is invariant,

$$\{f_1(t)|f_2(t)\} = \{f_1(0)|f_2(0)\}\,. \tag{22}$$

Hence, by this measure, nearby classical states do not separate from each other, and there should be no classical chaos!

The mathematics of these two arguments are correct but the conclusions are spurious. In the classical case, the error derives from confusing two meanings of the term "state": the individual state which is a single orbit, and the statistical state which is a probability distribution. The characterization of chaos as an exponentially rapid separation of nearby states applies only to the individual states. Chaos in the statistical states must be identified by another criterion, such as the growth of complex structures. In quantum theory, all states are subject to a statistical interpretation, and we see here that their time evolution is much more analogous to that of a classical statistical ensemble than to that of an individual orbit.

It is not the purpose of this paper to identify the signatures of chaos in quantum mechanics (see various conference proceedings, such as [6-8]), but only to point out that a new problem, to understand the relation of chaos to quantum theory, involves the old problem of interpretation of quantum states.

5. DYNAMICALLY-DRIVEN SPIN MODEL

Some old questions on the foundations of quantum mechanics came to light once again in the study of a model that was devised for an entirely different purpose. The model [9] consists of a localized spin j_1, a beam of spins j_2 polarized in the y direction, and a magnetic field B in the z direction. The beam particles interact, one at a time, with the localized spin. While a beam particle is in the interaction region, for a duration τ, the Hamiltonian of the two interacting spins is

$$H_{12} = a(\vec{S}_1 \cdot \vec{S}_2) + \vec{\mathbf{B}} \cdot (\vec{S}_1 + \vec{S}_2) \tag{23}$$

The beam affects the localized spin by transferring angular momentum to it, so the interaction tends to polarize the localized spin in the y direction. The magnetic field rotates any such polarization about the z axis. The competition between these two effects yields a very rich variety of behaviors [9-11].

We are interested here in the final state of the localized spin, after all transients have died out. In the classical model, the localized spin may exhibit a variety of time-dependences in its final state (depending on the values of a and B in (23)): stable fixed points, limit cycles, or chaos. In the quantum model, the density matrix of the localized spin (after tracing over the variables of the beam particles) always tends to a steady state that is independent of time. This poses a paradox. How can the rich variety of classical final states arise from the classical limit $(j \rightarrow \infty)$ of a quantum model that has no time dependence in its final state?

This paradox is resolved by recognizing that the classical limit of a quantum state corresponds to an *ensemble of classical orbits*, not to a single orbit. Dynamical information about this ensemble can be recovered by studying autocorrelation functions, such as $\langle S_y(0)\, S_y(t)\rangle$, even though the density matrix has no time dependence. By this method, its has been shown [11] that the quantum model has qualitatively different autocorrelations for values of the parameters (a, B) that correspond to different motions (regular or chaotic) in the classical model. The numerical results also show that the quantum correlations approach the classical correlations in the limit $j \rightarrow \infty$.

The success of these calculations brings us face to face with an old question in quantum theory. *What is the meaning of a correlation function like* $\langle Q(t_1)\, R(t_2)\rangle$ *if the two operators do not commute*? If the operators commute, then they possess a complete set of common eigenvectors, from which a non-negative joint-probability distribution can be calculated, and the correlation function is just the average of the product of the simultaneous eigenvalues of the two dynamical variables with that joint-probability distribution. But if the operators do not commute, then no such set of eigenvectors exists. Moreover, the correlation function is generally complex,

$$\langle Q(t_1)\, R(t_2)\rangle = \tfrac{1}{2}\langle Q(t_1)R(t_2) + R(t_2)Q(t_1)\rangle + \tfrac{1}{2}\langle [Q(t_1), R(t_2)]\rangle\,, \qquad (24)$$

the symmetrized term giving the real part, and the commutator giving the imaginary part.

The "orthodox" response to this problem has been to assert that two dynamical variables can not be simultaneously measured if their operators do not commute, therefore one should not discuss their correlations. Regardless of the truth of the premise of this argument, the conclusion is not appropriate. In fact, correlation functions like (24) arise naturally as the fundamental constituents of a dynamical theory, regardless of their measurability [12]. Moreover, one must be able to define a quantum correlation function in order to make contact with classical correlations in the appropriate limit. In my calculations, I assumed that the real part of (24) can be regarded as the quantum correspondent of the classical correlation function. This assumption seems reasonable, but it stands in need of a more fundamental justification.

6. CONCLUSION

I have argued that *the emergence of classical properties from quantum mechanics* is a useful focus for future research. The study of it has revealed connections between old problems in the interpretation of quantum theory and new problems associated with quantum chaos. In these investigations, it has proved essential to recognize that *the classical limit of a quantum state is typically an ensemble of classical orbits, and not a single classical orbit.* Perhaps these new results will be more effective than further discussions of the "measurement problem" in settling the question of an *individual* versus an *ensemble* interpretation of quantum states. Finally, these studies have brought to the fore an old problem to which no definitive answer has been given. What is the operational significance of a correlation function of noncommuting dynamical variables? Far from being "mere philosophy", the problems of the interpretation of quantum mechanics have an important bearing on some active areas of current research.

REFERENCES

[1] L. E. Ballentine, *Int. J. Theor. Phys.* **27**, 211-218 (1988).

[2] L. E. Ballentine, *Found. Phys.* **20**, 1329-1343 (1990).

[3] L. E. Ballentine, *Quantum Mechanics* (Prentice-Hall, Englewood Cliffs, New Jersey, 1990).
This book is now available only from the author.
[Fax: (604) 291-3592; e-mail: ballenti@sfu.ca].

[4] L. E. Ballentine, Yumin Yang, and J. P. Zibin, *Phys. Rev. A* (to be published).

[5] A. J. Lichtenberg and M. A. Lieberman, *Regular and Chaotic Dynamics* (Springer, New York, 1991).

[6] G. Casati, editor, *Chaotic Behavior in Quantum Systems* (Plenum, New York, 1985).
[7] H. A. Cerdeira, R. Ramaswamy, M. C. Gutzwiller, and G. Casati, editors, *Quantum Chaos* (World Scientific, Singapore, 1991).
[8] P. Cvitanovic, I. Percival, and A. Wirtzba, editors, *Quantum Chaos - Quantum Measurement* (1992, Kluwer, Dordrecht).
[9] L. E. Ballentine, *Phys. Rev. A* **44**, 4126-4132 (1991).
[10] L. E. Ballentine, *Phys. Rev. A* **44**, 4133-4141 (1991).
[11] L. E. Ballentine, *Phys. Rev. A* **47**, 2592-2600 (1993).
[12] J. E. G. Farina, *Am. J. Phys.* **45**, 1200 (1977). This paper gives a simple example in which the symmetrized correlation between the position and momentum operators arises, and is the natural analog of the corresponding classical correlation.

DEFORMATIONS OF SPACE-TIME SYMMETRIES AND FUNDAMENTAL SCALES

A. Ballesteros, F.J. Herranz,
M.A. del Olmo, and M. Santander

Departamento de Física Teórica
Universidad de Valladolid
E-47011, Valladolid, Spain

Quantum deformations of the (2+1)-dimensional Poincaré algebra are presented as symmetries of systems with a discrete direction. Dimensional properties of the deformation parameter are studied in connection with the general Cayley-Klein scheme from which these quantum algebras have been obtained.

Key words: quantum algebras, space-time symmetries, quantum mechanics.

1. INTRODUCTION

The concept of "deformation" appears quite naturally in physics; models are often improved by means of parameter-dependent corrections that preserve the theoretical framework of the "undeformed" theory. In a stronger sense, quantum theory itself can be mathematically described as a deformed version of classical mechanics on phase space [1], i.e., commutative classical observables are deformed into non commutative quantum ones with the Planck constant being the deformation parameter.

A kind of deformations coming from Lie algebras have recently appeared in the context of the quantum inverse scattering method [2]. These mathematical structures are called "quantum" (q-) groups and algebras, since some "quantization" procedure –in the standard sense– can be related with to [3]. Among their many interesting mathematical properties [4] it is worth remarking that the existence of a "deformed" principle for the composition of systems (tensor product of representations) constitutes

M. Ferrero and A. van der Merwe (eds.), Fundamental Problems in Quantum Physics, 29–34.

the cornerstone of the q–symmetry. The knowledge of the Lie algebra describing the symmetry of a certain physical system gives us the class of equations of motion compatible with it as well as many of their solutions. Likewise, these q–symmetries can be put into the same frame of mind, trying to characterize the physical systems linked with them.

When the deformation parameter is a real number some quantum algebras have been proven to describe symmetries of systems that exhibit a discretized direction [5–7]. In these cases, the deformation parameter plays the role of the fundamental scale of the system (lattice spacing for one-dimensional phonons and magnons [5,6], characteristic time for interactions in nuclei [7]). The quantum algebras associated to these systems are deformations of Poincaré and Galilei symmetries. To put forward this approach towards the characterization of higher dimensional lattices, a comprehensive mathematical description of this kind of algebras would be helpful. This program has been recently developed for the quantum analogues of (1+1), (2+1), and affine (3+1)–dimensional space-time symmetries (quantum kinematical algebras) within a Cayley-Klein (CK) approach [8–10]. We shall outline here some aspects of the (2+1)–d construction.

2. QUANTUM (2+1) CAYLEY-KLEIN ALGEBRAS

Let us consider a three-dimensional space-time as a symmetric homogeneous space of a group of inertial transformations (kinematical group). Three-dimensional CK geometries are defined as symmetric pseudo-orthogonal homogeneous spaces either of a $SO(p,q)$ group ($p+q = 4$ in this case) or of a non-semisimple one obtained from $SO(p,q)$ by contraction [11]. Since kinematical groups are of CK type, any (2+1) space-time with constant curvature turns out to be a particular case of three-dimensional CK geometries.

All the CK geometries of a given dimension and the algebras associated with their "groups of motion" can be simultaneously described and endowed with a set of involutions, contractions, and dualities that summarizes their algebraic and geometrical properties. Let J_{ij}, ($i < j$; $i,j = 0,1,2,3$) be a basis of a six-dimensional Lie algebra $\mathfrak{g}$ generating the groups of motions of our three-dimensional CK geometries. The Lie algebra $\mathfrak{g}$ can be proved to be determined by three real parameters κ_1, κ_2 and κ_3; its Lie brackets can be written in terms of $\kappa_{ij} = \kappa_{i+1}\kappa_{i+2}\cdots\kappa_j$

$(i < j;\ i,j = 0,1,2,3)$, as follows:

$$[J_{ij}, J_{lm}] = \delta_{im} J_{lj} - \delta_{jl} J_{im} + \delta_{jm} \kappa_{lm} J_{il} + \delta_{il} \kappa_{ij} J_{jm}, \quad (i \leq l,\ j \leq m). \tag{2.1}$$

The parameters κ_i $(i = 1,2,3)$ have a clear geometrical meaning as the "kind" (elliptical, parabolical or hyperbolical) of measure of separation between points, lines, and planes, respectively. Each of the coefficients κ_i can be separately scaled to $+1$, 0, or -1. So, there are $3^3 = 27$ essentially different CK geometries labelled by $(\kappa_1, \kappa_2, \kappa_3)$. Space-time geometries appear among them when an explicit physical meaning is given to the CK generators J_{ij}. This assignation is not unique and each possible election imposes different restrictions on the κ_i coefficients. Note also that, in the kinematical language, some involutions of the CK algebra are translated into parity and time-reversal automorphisms. A q–analogue of a certain Lie algebra is obtained by deforming its Lie brackets in such a way that the classical commutation relations are recovered in the limit $q \to 1$ and, at the same time, by finding a coproduct $\Delta : A \to A \otimes A$ consistent with these "quantum brackets" (Δ must be an algebra homomorphism in order to use it to build the deformed tensor product of representations). Such a deformation can be defined for the CK algebra (2.1) [9], i.e., all the 27 CK algebras can be simultaneously "quantized" as well as their CK underlying structure. In particular, the three new q–involutions transform the deformation parameter z $(q = e^z)$, and provide the key for its dimensional interpretation, as we shall see later. They are

$$\begin{aligned}
\Theta^q_{(0)} &: (J_{01}, J_{02}, J_{03}, J_{12}, J_{13}, J_{23}; z) \\
&\longrightarrow (-J_{01}, -J_{02}, -J_{03}, J_{12}, J_{13}, J_{23}; -z), \\
\Theta^q_{(1)} &: (J_{01}, J_{02}, J_{03}, J_{12}, J_{13}, J_{23}; z) \\
&\longrightarrow (J_{01}, -J_{02}, -J_{03}, -J_{12}, -J_{13}, J_{23}; -z), \\
\Theta^q_{(2)} &: (J_{01}, J_{02}, J_{03}, J_{12}, J_{13}, J_{23}; z) \\
&\longrightarrow (J_{01}, J_{02}, -J_{03}, J_{12}, -J_{13}, -J_{23}; -z).
\end{aligned} \tag{2.2}$$

3. Q–DEFORMED (2+1) POINCARÉ ALGEBRAS

If we take $H \equiv J_{01}$ as the time translation, $P_1 \equiv J_{02}, P_2 \equiv J_{03}$ as the space translations, $K_1 \equiv J_{12}, K_2 \equiv J_{13}$ are the boosts and $J \equiv J_{23}$

is the spatial rotation, the algebra with coefficients $(\kappa_1, \kappa_2, \kappa_3) = (0, -, +)$ corresponds to the (2+1) Poincaré one (under the same assumptions, the $(0, 0, +)$ case is the (2+1) Galilei algebra). Its quantum commutation relations read

$$\begin{aligned} &[J, K_1] = K_2, && [J, K_2] = -K_1 \cosh(zP_2), \\ &[J, P_1] = \tfrac{1}{z}\sinh(zP_2), && [J, P_2] = -P_1, \\ &[K_1, H] = P_1, && [K_2, H] = \tfrac{1}{z}\sinh(zP_2), \\ &[K_1, P_1] = H, && [K_2, P_2] = H, \qquad [K_1, K_2] = -J. \end{aligned} \tag{3.1}$$

The deformed tensor product of representations is given by

$$\begin{aligned} \Delta(P_2) &= 1 \otimes P_2 + P_2 \otimes 1, \\ \Delta(K_1) &= 1 \otimes K_1 + K_1 \otimes 1, \\ \Delta(H) &= e^{-\frac{z}{2}P_2} \otimes H + H \otimes e^{\frac{z}{2}P_2}, \\ \Delta(P_1) &= e^{-\frac{z}{2}P_2} \otimes P_1 + P_1 \otimes e^{\frac{z}{2}P_2}, \\ \Delta(K_2) &= e^{-\frac{z}{2}P_2} \otimes K_2 + K_2 \otimes e^{\frac{z}{2}P_2} \\ &\quad + e^{-\frac{z}{2}P_2}(\tfrac{z}{2}K_1) \otimes P_1 - P_1 \otimes e^{\frac{z}{2}P_2}(\tfrac{z}{2}K_1), \\ \Delta(J) &= e^{-\frac{z}{2}P_2} \otimes J + J \otimes e^{\frac{z}{2}P_2} \\ &\quad - e^{-\frac{z}{2}P_2}(\tfrac{z}{2}K_1) \otimes H + H \otimes e^{\frac{z}{2}P_2}(\tfrac{z}{2}K_1). \end{aligned} \tag{3.2}$$

Two central elements of this algebra can be easily derived:

$$\begin{aligned} C_1^q &= \tfrac{4}{z^2}\sinh^2(\tfrac{z}{2}P_2) + P_1^2 - H^2, \\ C_2^q &= \tfrac{1}{z}\sinh(zP_2)K_1 - HJ - P_1K_2. \end{aligned} \tag{3.3}$$

Finally, q–parity $\mathcal{P}^q$ and q–time reversal $\mathcal{T}^q$ are obtained from (2.2):

$$\begin{aligned} \Theta^q_{(1)} &\equiv \mathcal{P}^q : \{H \to H,\ \mathbf{P} \to -\mathbf{P},\ J \to J,\ \mathbf{K} \to -\mathbf{K};\ z \to -z\}, \\ \Theta^q_{(0)} \cdot \Theta^q_{(1)} &\equiv \mathcal{T}^q : \{H \to -H,\ \mathbf{P} \to \mathbf{P},\ J \to J,\ \mathbf{K} \to -\mathbf{K};\ z \to z\}. \end{aligned} \tag{3.4}$$

Note that zP_2 has to be dimensionless in order to obtain a homogeneous coproduct (3.2). This means that the deformation parameter has the dimension of length, in a way consistent with (3.4). A standard differential realization for the translation generators, if substituted in the first

q–Casimir, leads to a Klein-Gordon equation on a lattice in x_2 direction preserving t and x_1 as continuous variables:

$$\begin{aligned}\mathcal{C}_1^q\Psi(x_1,x_2,t) = \tfrac{1}{z^2}\big[\Psi(x_1,x_2+z,t) + \Psi(x_1,x_2-z,t)\\ - 2\Psi(x_1,x_2,t)\big] + \partial^2_{x_1}\Psi(x_1,x_2,t) - \partial^2_t\Psi(x_1,x_2,t).\end{aligned} \tag{3.5}$$

On the other hand, the CK algebra $(0,+,-)$ corresponds also to Minkowski space-time, provided that $P_1 \equiv J_{01}$, $P_2 \equiv J_{02}$, $H \equiv J_{03}$, $K_1 \equiv J_{13}$, $K_2 \equiv J_{23}$, $J \equiv J_{12}$. This algebra gives rise to a "time-like" q–deformed Poincaré algebra which is just the (2+1)–dimensional version [12] of the κ-Poincaré algebra of Lukierski *et al.* [13]. Now, the deformation parameter z has the character of time ($\mathcal{P}^q(z) = z$ and $\mathcal{T}^q(z) = -z$). A "second order" Casimir of this quantum algebra is

$$\mathcal{C}_1^q = \tfrac{4}{z^2}\sinh^2(\tfrac{z}{2}H) - P_1^2 - P_2^2. \tag{3.6}$$

Some physical implications of its associated "time lattice" have been explored so far (a deformed field theory [13], the quantum Dirac equation and its associated Coulomb problem [14,15], the links between this Poincaré deformation and quantum gravity [16] as well as an astrophysical limit of the deformation of special relativity given by (3.6) [17]). However, the most intriguing fact is the strong mathematical reluctance of the quantum kinematical algebras already known to allow a discretization in more than one space-time direction. A hint to overcome this difficulty could come from considering a quantization of the whole conformal symmetry of a given space-time with constant curvature. For the (1+1)–dimensional cases, this possibility seems to be promising: their (classical and quantum) conformal algebras are included as (2+1) CK algebras. In particular, the $so(2,2)_q$ Anti-DeSitter quantum algebra obtained from $(\kappa_1,\kappa_2,\kappa_3) = (+,-,+)$ is a deformation of the conformal algebra of (1+1) Minkowski space that exhibits a kind of global lattice induced from a discrete hyperbolical rotation together with a discrete dilation.

ACKNOWLEDGMENT

This work has been partially supported by a DGICYT project (PB91–0196) from the Ministerio de Educación y Ciencia de España.

REFERENCES

[1] F. Bayen, M. Flato, C. Fronsdal, A. Lichnerowicz, and D. Sternheimer, *Ann. Phys.* **111** (1978) 61.

[2] N. Yu. Reshetikhin, L.A. Takhtadzhyan, and L.D. Faddeev, *Leningrad Math. J.* **1** (1990) 193.

[3] V.G. Drinfeld, *Proceedings of the International Congress of Mathematics* (MRSI, Berkeley, 1986) 798.

[4] P.P. Kulish, "Quantum Groups and Quantum Homogeneous Spaces," preprint 1993.

[5] F. Bonechi, E. Celeghini, R. Giachetti, E. Sorace, and M. Tarlini, *Phys. Rev. Lett.* **68** (1992) 3178.

[6] F. Bonechi, E. Celeghini, R. Giachetti, E. Sorace, and M. Tarlini, *J. Phys. A: Math. Gen.* **25** (1992) L939.

[7] F. Bonechi, E. Celeghini, R. Giachetti, E. Sorace, and M. Tarlini, *Phys. Lett.* **280B** (1992) 180.

[8] A. Ballesteros, F.J. Herranz, M.A. del Olmo, and M. Santander, *J. Phys. A: Math. Gen.* **26** (1993) 5801.

[9] A. Ballesteros, F.J. Herranz, M.A. del Olmo, and M. Santander, *J. Phys. A: Math. Gen.* **27** (1994) 1283.

[10] A. Ballesteros, F.J. Herranz, M.A. del Olmo, and M. Santander, *J. Math. Phys.* **35** (1994) 4928.

[11] M. Santander, F.J. Herranz, and M.A. del Olmo, *Proceedings of the XIX ICGTMP, Anales de Física, Monografías.* Vol. 1.I, p. 455 (CIEMAT/RSEF Madrid 1993).

[12] S. Giller, P. Koshinski, M. Majewski, P. Maslanka, and J. Kunz, *Phys. Lett.* **286B** (1992) 57.

[13] J. Lukierski, H. Ruegg, and A. Nowicky, *Phys. Lett.* **293B** (1992) 344.

[14] A. Nowicky, E. Sorace, and M. Tarlini, *Phys. Lett.* **302B** (1993) 419.

[15] L.C. Biedenharn, B. Mueller, and M. Tarlini, preprint, 1993.

[16] M. Maggiore, *Phys. Rev.* **49D** (1994) 5182.

[17] G. Domokos and S. Kovesi-Domokos, Johns Hopkins University preprint, 1993.

ASPECTS OF QUANTUM REALITY

S. Bergia

Department of Physics, University of Bologna
I.N.F.N., Sezione di Bologna
Bologna, Italy

The world of quantum phenomena has always presented great difficulties for those who feel that a real world should lie behind the appearances. Nonlocality makes the difficulty even greater. A new formulation of quantum mechanics, due to Roncadelli, suggests the possibility of interpreting in a somewhat physical terms the particle quantum paths, thus adding a further realistic feature to the de Broglie-Bohm causal interpretation. However, neither Roncadelli's universal noise nor his basic stochastic equation allow an interpretation of quantum mechanics in terms of physical processes taking place in the space-time continuum. As I will argue, this makes it very unlikely that attempts at implementing a fully physical view of quantum behaviour as a classical stochastic process will ever succeed.

Key words: quantum ontology, Fresnel white noise, stochastic formulation of quantum mechanics.

1. INTRODUCTION

I am aware of various recent studies which cast serious doubts on the opinion, widespread in the scientific community, that experiments have conclusively refuted local realism.[1] It is of course perfectly possible that their authors are right, and that our mourning over local realism is premature. I will however adopt here the majority's viewpoint, that is, I will assume the conclusiveness of the experiments.

The world of quantum phenomena has always presented great difficulties for those who feel that a real world should lie behind the appearances. Nonlocality[2] makes the difficulty even greater. Of course, one could always take the attitude of giving up the search for an underlying reality. However, although no epistemology can *force* up on us a realistic attitude, I feel very

M. Ferrero and A. van der Merwe (eds.), Fundamental Problems in Quantum Physics, 35–42.

much in the same way R. Penrose has recently expressed as follows:

> I have taken for granted that any "serious" philosophical viewpoint should contain at least a good measure of realism. It always surprises me when I learn of apparently serious-minded thinkers, often physicists concerned with implications of quantum mechanics, who take the strongly subjective view that there is, in actuality, *no* real world 'out there' at all! The fact that I take a realistic line wherever possible is not meant to imply that I am unaware that such subjective views are often seriously maintained—only that I am unable to make sense of them. (Penrose 1989, p. 299)

I think that those who cannot help feeling this way should make a strenuous effort to implement a view of the quantum world "as if it really was there," even more so after the recent developments. Those who would like such a "view of quantum reality" to be non self-contradictory and to be describable in terms of physical processes taking place in the space-time continuum (I will call this, for lack of a more suitable term, a new mechanistic (NM) view) should not hide to themselves that the task has become a formidable one and will require a great deal of ingenuity: Simple-minded ideas will just not do.

An attempt along this direction consists in trying to read quantum mechanics (QM) as some kind of Brownian motion. This view is encouraged by Nelson's stochastic formulation of quantum mechanics (Nelson 1966). Stochastic mechanics (SM) makes use of classical probabilistic concepts, and in facts interprets QM in terms of a classical stochastic process. However, the formalism of SM does not allow an immediate NM interpretation, since its classical limit does not correspond to any of the possible formulations of classical dynamics.

A recent reformulation of QM due to Roncadelli (Roncadelli 1992; for a detailed account, see Roncadelli and Defendi 1992), also considers a universal noise, which determines the particles random trajectories, as the origin of the quantum behaviour. As we shall see, Roncadelli's formulation is not subject to the above criticism. After briefly reviewing it, pinpointing its differences with SM (Sec. 1), I will stress (Sec. 2) its close relationship with the de Broglie-Bohm formulation (or causal interpretation of QM—for recent accounts, see Vigier 1987, Dewdney 1987). Since Roncadelli's formulation suggests the possibility of interpreting in a somewhat physical terms the particles quantum paths, it seems to add a further realistic feature to the causal interpretation. I will in fact argue (Sec. 3) that, among the existing formulations, it is the one that gives expression to most aspects of quantum reality. However, neither Roncadelli's universal noise nor his basic stochastic behaviour as a classical stochastic process will ever succeed. Therefore, even though Roncadelli's formulation goes some way along the direction of providing *a* "quantum ontology,"[3] it tells us that any such ontology must necessarily be non mechanistic, not even NM.

2. RONCADELLI'S FORMULATION OF QUANTUM MECHANICS

Roncadelli's formulation starts from considering the formal close analogy between QM and the theory of classical Markovian stochastic processes (CMSP), an analogy which, however, holds between classical *probabilities* and quantum *amplitudes.*

This formal analogy invites to look for a formulation of QM analogous to the Fokker-Planck (FP) formulation holding for a CMSP; the FP equation, however, would not hold for the quantum *probability*, as in Nelson's formulation, but rather for the probability *amplitude*, i.e., for the wave function. And, in fact, it is easily shown (see, for instance, Roncadelli and Defendi 1992) that Schrödinger's equation can be writen as a FP equation, although with a purely imaginary diffusion coefficient given by

$$\epsilon = \frac{i\hbar}{2m}, \tag{1}$$

where the velocity field $\mathbf{v}$ is replaced by the potential of the velocity dependent forces, $\boldsymbol{\Omega}$ according to

$$\mathbf{v} \to -\frac{\boldsymbol{\Omega}}{m}. \tag{2}$$

On the other hand, a CMSP can be alternatively described in two other equivalent ways, that is, as a Wiener process (the transition probability is here expressed as a functional integral) or in terms of a "Langevin" equation (I will later justify the use of the inverted commas).

In the light of the above analogy, one expects two further formulations of QM to correspond to the Wiener and the "Langevin" formulations. The first one, as one easily guesses, is Feynman's formulation. Roncadelli has analysed in detail the latter, which is the one that interests us here.

I recall for completeness the "Langevin" formulation of CMSP. Let us consider the FP equation in the absence of dissipation and of emission and absorption of particles, in which case it reduces to the continuity equation

$$\frac{\partial P}{\partial t} = -\nabla[P\mathbf{v}] \tag{3}$$

for the probability density P. This equation is the fluid counterpart of the equation

$$\frac{d\mathbf{x}}{dt} = \mathbf{v}\left(\mathbf{x}(t), t\right), \tag{4}$$

describing the particle aspects.

The basic idea of the "Langevin" formulation is that the effect of dissipation may be simulated by adding a noise term to the right-hand side of Eq. (4), which then becomes the stochastic differential equation

$$d\xi(t) = \mathbf{v}\,(\xi(t), t)\, dt + \eta(t) dt. \tag{5}$$

I have used the inverted commas to denote this "Langevin" equation to stress the fact that it is customary to call Langevin equation a stochastic differential equation in which the unknown is not a configurational variable but a velocity component. If the noise term in Eq. (5) is a Wiener process, this formulation is proved to be equivalent to the FP formulation.

The quantum equivalent of Eq. (5) is obtained, by replacement of the velocity field $\mathbf{v}$ in terms of $\boldsymbol{\Omega}$ according to Eq. (2), as

$$d\xi(t) = -\frac{1}{m}\boldsymbol{\Omega}\,(\xi(t), t)\, dt + \eta(t) dt, \tag{6}$$

where the properties of the noise term $\eta(t)$ shall have to be specified.

One may observe that, even in the absence of the noise term, Eq. (6) has no immediate interpretation in classical mechanics. However, Roncadelli has shown that it admits such an interpretation if one performs a suitable canonical transformation on the system described by the action S. He has shown that the equation can then be cast in the form

$$d\xi(t) = \left[\frac{1}{m}\nabla S(\mathbf{x}, t) - \boldsymbol{\Omega}(\mathbf{x}, t)\right]_{\mathbf{x}=\xi(t)} dt + (\hbar/m)^{1/2}\eta(t) dt, \tag{7}$$

where the noise term has been redefined by extraction of a suitable factor.

Unlike Eq. (6), Eq. (7) allows an immediate physical interpretation in the classical limit, where it expresses the guidance or characteristic equation determining the trajectories departing from the surfaces of given action in a Hamiltonian-Jacobi dynamical scheme. It coincides in fact with the equation describing the particle counterpart (see Eq. (4)) of the continuity equation (3) holding in the case of a fluid satisfying the condition

$$\mathbf{v}\,(\xi(t), t) = \left[-\frac{1}{m}\nabla S(\mathbf{x}, t) - \boldsymbol{\Omega}(\mathbf{x}, t)\right]_{\mathbf{x}=\xi(t)}. \tag{8}$$

(Hereafter the term $\boldsymbol{\Omega}\,(\xi(t), t)$ will be neglected; Eq. (8) then expresses the irrotationality of the velocity field.) The completion of a dynamical scheme is achieved in terms of the Hamilton-Jacobi equation

$$\frac{\partial S}{\partial t} = -\frac{|\nabla S|^2}{2m} + V. \tag{9}$$

Equations (7) and (9) provide a formulation of QM equivalent to Schrödinger's formulation under suitable conditions for the noise term. Whereas, in the case of a CMSP, the noise must be a Gaussian white noise, with the *probability* distribution

$$P[\eta(t)] \sim \exp\left\{-(1/4D)\int_{-\infty}^{+\infty}\sum_i \eta_i(t)\eta_i(t)\right\}, \tag{10}$$

Roncadelli has shown that, in the quantum case, it must be the so-called Fresnel white noise, with the *amplitude* distribution

$$A[\eta(t)] \sim \exp\left\{(1/2)(im/\hbar)\int_{-\infty}^{+\infty}\sum_i \eta_i(t)\eta_i(t)\right\}. \tag{11}$$

The comparison with Nelson's formulation is easily carried out. The latter is based as well on a stochastic differential equation of the "Langevin" type; however, the noise is characterized by a *probability* distribution; on the other hand, the classical limit of the stochastic equation has no immediate dynamical meaning.

3. COMPARISON WITH THE DE BROGLIE-BOHM CAUSAL INTERPRETATION

Both the Roncadelli formulation and the de Broglie-Bohm causal interpretation are rooted in the Hamilton-Jacobi formulation of classical dynamics. The formalism on which the de Broglie-Bohm interpretation is based consists of a continuity equation for the probability density ρ, with $\mathbf{v} = \frac{1}{m}\nabla S$ (irrotationality condition), and of the Hamiltonian-Jacobi equation (9), in which the quantum potential terms, $\frac{\nabla^2\sqrt{\rho}}{\sqrt{\rho}}$, is added in the right-hand side.

In Roncadelli's formulation, the Hamilton-Jacobi is unaltered, and the quantum behaviour results from adding the noise term $(\hbar/m)^{(1/2)}\eta(t)dt$ on the right-hand side of the particle counterpart of the continuity equation.

One would tend to think that the latter approach "explains" the former, in the sense that it indicates in quantum fluctuations the origin of the quantum potential. However, *no explicit relation can be found between quantum potential and noise.* The reason is that the noise has to do with the imaginary part of the Schrödinger equation, the quantum potential with the real part (as is well known, the de Broglie-Bohm approach arises from the decomposition of the Schrödinger equation into a real and an imaginary part, giving rise to the two coupled equations for action and probability density). From a more physical point of view, note that *the quantum potential contributes to determine the surfaces of equal action*, and thus the trajectories are completely determined by the quantum potential and the initial conditions. On the contrary, *fluctuations have nothing to say about surfaces of equal action*, which in this approach are completely determined by classical mechanics.

The connection between the two formulations arises from the fact that *the two approaches lead eventually to the same quantum paths.* From this point of view, it is worth stressing that Roncadelli's formulation is equivalent to Feynman's. Moreover, "generalized Feynman paths" controlled by S, joining (x', t') with (x'', t''), are introduced, which may be interpreted as (quantum) random walks $\xi\,(t; x', t'; [\eta(\cdot), S(\cdot)])$ satisfying the condition $\xi\,(t''; x', t'; [\eta(\cdot), S(\cdot)]) = x''$.

4. ONTOLOGICAL ASPECTS

Roncadelli's formulation suggests the possibility of interpreting in a somewhat physical terms the particles quantum paths, and thus seems to add a further realistic feature to the causal interpretation.[4] From the foregoing discussion, it could be continued that, among the existing formulations, it is the one that gives expression to most aspects of quantum reality.

However, *neither Roncadelli's universal noise nor his basic stochastic equation allow a NM interpretation.* On the one hand, *the universal Fresnel noise has an amplitude, and not a probability, distribution*, with a significant factor of i in the exponent (Eq. (11)). It seems very unlikely that a mechanism whatsoever producing such an effect could ever be conceived. On the other hand, *the "Langevin" equation (5)*, in spite of having a classical counterpart, *is not dynamical*, in the sense that its classical limit has nothing to do with Newton's law of motion.

If we accept the view the Roncadelli's formulation goes as close as possible to a NM representation of quantum reality, it seems improbable, notwithstanding some promising general indications (Garbaczewski and Vigier 1992, Garbaczewski 1992, 1993), that attempts at implementing a fully physical view of quantum behaviour as a classical stochastic process will ever succeed. I am here specifically referring to approaches such as that of stochastic electrodynamics (SED)[5] and of stochastic geometrodynamics, as one may call it,[6] whose basic tool is a Langevin equation proper, whose classical limit *coincides with* Newton law of motion, and whose noise term is also classical, in the sense that it is characterized by a *probability* distribution.

Therefore, in conclusion, even though Roncadelli's formulation goes some way along the direction of providing a "quantum ontology," in the sense alluded to in the Introduction, it tells us that any such ontology must necessarily be non mechanistic, not even NM.

ACKNOWLEDGMENTS

I wish to acknowledge conversations with M. Roncadelli and F. Cannata. The responsibility of the epistemological viewpoints expressed in the article is, of course, only mine.

Bergia, S., Cannata, F., Pasini, A. (1989), *Phys. Lett.* **137A**, 21.

Cetto, A. M., de la Peña, L. (1991), *Found. Phys. Lett.* **4**, 73.

Cufaro Petroni, N., Dewdney, C., Holland, P., Kyprianidis, A., Vigier, J. P. (1984), *Phys. Lett.* **106A**, 368; (1985a) *Lett. Nuovo Cimento* **42** 285; (1985b) *Phys. Rev. D* **32**, 1375.

Cushing, J. T., McMullin, E. (1989), *Philosophical Consequences of Quantum Theory. Reflections on Bell's Theorem* (University of Notre Dame Press, Notre Dame). De la Peña, L., Cetto, A. M. (1993), "Planck's law as a consequence of the zeropoint radiation field," Preprint IFUNAM FT93-023, July 1993.

Dewdney, C. (1987), "Calculations in the causal interpretation of quantum mechanics," in Honig, Kraft, and Panarella 1987.

Ferrero, M., Santos, E. (1985), *Phys. Lett.* **108A**, 373.

Garbaczewski, P. (1992, 1993), *Phys. Lett. A* **162**, 129; **172**, 208.

Garbaczewski, P., Vigier, J. P. (1992), *Phys. Rev. A* **167**, 445.

Honig, W. M., Kraft, D. W., and Panarella, E., eds. (1987), *Quantum Uncertainties - Recent and Future Experiments and Interpretations* (Plenum, New York).

Jarrett, J. P. (1989), "Bell's theorem: A guide to the implications," in Cushing and McMullin 1989.

Marshall, T. W., Santos, E., Selleri, F. (1983), *Phys. Lett.* **98A**, 5.

Marshall, T. W. (1983), *Phys. Lett.* **99A**, 163; (1984) *Phys. Lett.* **100A**, 225.

Marshall, T. W., Santos, E. (1985), *Phys. Lett.* **107A**, 164.

Nelson, E. (1966), *Phys. Rev.* **150**, 1079.

Penrose, R. (1989), *The Emperor's New Mind* (Oxford University Press, Oxford).

Roncadelli, M. (1992), "Langevin approach to quantum mechanics," prepreint FN/T-92/43.

Roncadelli, M., Defendi, A. (1992), in *I cammini di Feynman* (Quaderni di Fisica Teorica, Università degli Studi di Pavia, Dipartimento di Fisica Nucleare e Teorica).

Rosales, J. L., Sánchez Gómez, J. L. (1989), *Phys. Lett. A* **142**, 207.

Santos, E. (1992), "Does quantum mechanics predict violations of Bell in-

equalities?", in van der Merwe, Selleri, and Tarozzi, (1992).
Stapp, H. P. (1989), "Quantum nonlocality and the description of nature," in Cushing and McMullin (1989). Van der Merwe, A., Selleri, F., and Tarozzi, G., eds. (1992), *Bell's Theorem and the Foundations of Modern Physics* (World-Scientific, Singapore).
Vigier, J.-P. (1987), "Theoretical implications of time-dependent double resonance neutron interferometry," in Honig, Kraft, and Panarella (1987).

NOTES

1. For example, Marshall, Santos, Selleri 1983, Marshall 1983, 1984, Marshall and Santos 1985, Ferrero and Santos 1985, Santos 1992, Garuccio, this conference. I am here freely using the current terminology; for a thorough discussion of what is actually implied by the results of the Bohm-EPR class of experiments, I refer the reader to some of the essays collected in Cushing and McMullin 1989, in particular to Jarrett.
2. Or whatever is implied by the experimental results—see footnote 1.
3. This means, roughly speaking, a specific view about the character of the physical world as exhibited by quantum phenomena; one cannot expect such a view to be uniquely determined by the phenomena. For a thorough discussion of the subject I refer the reader to Stapp 1989.
4. In the latter, particle paths are determined by the quantum potential, which does not admit an immediate physical interpretation.
5. For a historical account, see de la Peña and Cetto 1993. Recently Cetto and de la Peña have proposed and started developing a new version of SED (Cetto and de la Peña 1991).
6. Here also there have been promising indications; see, for instance, Cufaro Petroni *et al.* 1984, 1985a,b; Rosales and Sánchez Gómez (1989); in these approaches the subquantum background of SED, provided by the zero-point e.m. field, is replaced in terms of metric fluctuations; some of the difficulties encountered in these attempts have been discussed by Bergia, Cannata, Pasini 1989.

KOCHEN-SPECKER DIAGRAM OF THE PERES-MERMIN EXAMPLE

Adán Cabello

Departamento de Física Teórica
Universidad Complutense
28040 Madrid, Spain

To show that the Peres [1] and Mermin [2] "incompatible results of quantum measurements" for a two spin-1/2 particles system are just a particular Kochen-Specker [3] example in four dimensions, we build a "KS-diagram" of such a case.

Key words: contextuality, Kochen-Specker theorem, Peres-Mermin example.

1. THE KOCHEN-SPECKER THEOREM

The Kochen-Specker theorem [3] basically proves the incompatibility between quantum mechanics and noncontextuality.

Consider the following assumptions characteristic of quantum mechanics:

(QM1) *(Mathematical characterization of the observables).* In QM each physically observable magnitude has associated a self-adjoint operator in a Hilbert space.

(QM2) *(Possible results of a measurement).* The possible results of measuring that observable correspond with the eigenvalues of the operator.

(QM3) *(Compatibility ⇔ commutability).* Two observables are compatible (they can be measured simultaneously with infinite precision) if their corresponding operators are commutative, $[A, B] = AB - BA = 0$.

(QM4) *(Relationships between observables ⇒ Relationships between results).* Given A and B, two compatible observables, and being a and b the results of measuring A and B, if the observables satisfy certain functional

M. Ferrero and A. van der Merwe (eds.), Fundamental Problems in Quantum Physics, 43–46.

relation, $f(A, B) = 0$, the results must satisfy the same relation, $f(a, b) = 0$.

Kochen and Specker show that if the dimension of the space of states is higher than two, then the previous assumptions are incompatible with

(5) *(Noncontextuality)*. If one observable A is compatible with two others, B and C, but B is not compatible with $C([A, B] = 0, [A, C] = 0$, but $[B, C] \neq 0)$, the result of measuring A is independent of whether A is measured together with B or together with C.

KS construct an explicit counter-example for a spin-1 system. It consists of a set of 117 observables of the type S_n^2 (the square of the spin component in the direction n). Two observables of this kind are compatible if their respective directions are orthogonal. For a spin-1 system, one must satisfy

$$S_k^2 + S_l^2 + S_m^2 = 2I \tag{1}$$

($\hbar = 1$), where k, l, m are mutually orthogonal directions. Since the eigenvalues of S_n^2 are 1 and 0 and they have to satisfy (1), every set of three orthogonal directions must have one 0 and two 1's. KS prove that the construction showed in Fig. 1 is realizable on the sphere.

In Fig. 1 each vertex represents an observable S_n^2 (but only the directions i, p, j, k, q, r, are in their real positions on the sphere), a and i represent the same direction, as well as b and j, and c and k. Two points are joined if they correspond to orthogonal directions (if they represent compatible observables). Thus, triangles represent three mutually compatible observables. Because the eigenvalues of S_n^2 are 1 and 0 and due to the relation (1), in the vertex of each triangle there must be two 1's and one 0. But it is impossible to fill the whole diagram with this premise in a consistent way. This is basically the proof of the incompatibility between QM and non-contextuality of Kochen and Specker.

2. THE PERES-MERMIN EXAMPLE

More recently Peres [1] and Mermin [2] have proposed a new example of incompatible results that requires just 9 observables for a two spin 1/2 particles system. We build a "KS-diagram" of such a case to show that it is just a particular Kochen-Specker example in a four-dimensional space.

In Fig. 2 each circle represents an observable. We call $x1$ the observable $\sigma_x^{(1)} \otimes I^{(2)}$, xy the observable $\sigma_x^{(1)} \otimes \sigma_y^{(2)}$, and so on. Two circles are joined if their associated operators commute. Thus, a triangle with circles in its vertex represents a mutually commutative set of operators (a mutually compatible set of observables), as in KS's original diagram. The difference

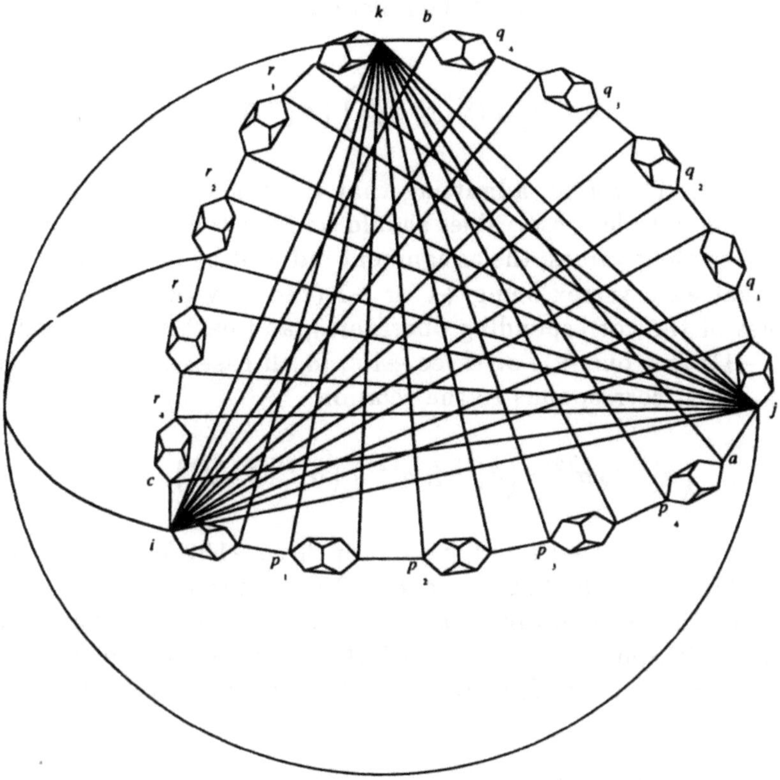

Fig. 1. Three-dimensional version of the original Kochen-Specker diagram.

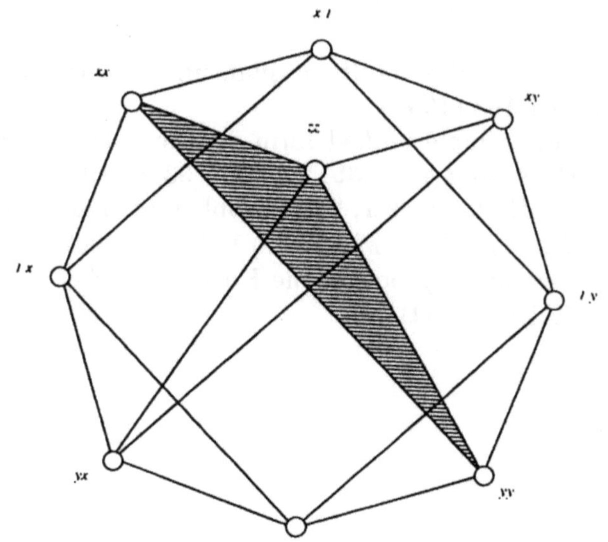

Fig. 2. Proposed diagram for the Peres-Mermin example.

with the KS proof is simply that the functional relations between observables are relations like

$$\left(\sigma_x^{(1)} \otimes I^{(2)}\right)\left(I^{(1)} \otimes \sigma_y^{(2)}\right)\left(\sigma_x^{(1)} \otimes \sigma_y^{(2)}\right) = I, \qquad (2)$$

instead of (1). Therefore, admitting that the functional relations between (mutually compatible) observables should be reflected in the functional relations between their values, the assignment induced by (2) in the KS diagram requires that each vertex being $+1$ or -1 (the only possible values for the eigenvalues of the corresponding operators), and each triangle can have either two -1's and one $+1$, or three $+1$'s, in all cases except one (the filled triangle), which corresponds to the relation

$$\left(\sigma_x^{(1)} \otimes \sigma_x^{(2)}\right)\left(\sigma_y^{(1)} \otimes \sigma_y^{(2)}\right)\left(\sigma_z^{(1)} \otimes \sigma_z^{(2)}\right) = -I, \qquad (3)$$

which needs either two $+1$'s and one -1, or three -1's. With this assignation we cannot fill the whole diagram in Fig. 2 in a consistent way.

In fact, by decomposing the 9 observables in the PM example, we can build a set of 24 one-dimensional projectors that is not "colourable" (with one 1 and three 0's for each tetrad of orthogonal projects) [4].

ACKNOWLEDGEMENTS

I wish to thank Susana G. Fdez. Huelga for letting me use her computer to write the first draft of this article, and the organizers for inviting me to the symposium.

1. A. Peres, "Incompatible results of quantum measurements," *Phys. Lett. A* **151**, 3-4, 107-108 (1990).
2. N. D. Mermin, "Simple unified form for the major no-hidden-variables theorems," *Phys. Rev. Lett.* **65**, 27, 3373-3376 (1990).
3. S. Kochen, and E. P. Specker, "The problem of hidden variables in quantum mechanics," *J. Math. Mech.* **17**, 1, 59-87 (1967).
4. A. Peres, "Two simple proofs of the Kochen-Specker theorem," *J. Phys. A* **24**, 4, L175-L178 (1991).

ZEROPOINT WAVES AND QUANTUM PARTICLES

A. M. Cetto and L. de la Peña

Instituto de Física, UNAM
Apartado postal 20-364, 01000 México, D.F.

It is proposed that the particles acquire their undulatory properties as a result of their interaction with the random zeropoint radiation field. The modes of very high frequency, up to a cutoff ω_c fixed by the particle's effective structure, induce a fine vibration which can be considered as the nonrelativistic version of *zitterbewegung*; internal consistency demands that $\omega_c = mc^2/\hbar$. The zeropoint field waves of this frequency that interact strongly with the moving particle form a modulated wave with de Broglie's wavelength; this wave is described by the stationary Schrödinger equation. Another effect of the assumed upper limit for the particle's response frequency is the appearance of an attractive Casimir force, which may counterbalance the Coulomb force of the charged particle and thus explain its stability. A simple hypothesis may therefore help to explain two entirely different phenomena.

Key words: stochastic electrodynamics, zeropoint field, de Broglie wavelength, Casimir forces.

1. INTRODUCTION

The idea that there exists a fluctuating zeropoint radiation field (zpf) and that it is closely connected to the phenomenon of quantization has gradually gained strength since the original proposals of Planck and Nernst. [1] In the last decades, in particular, within the framework of stochastic optics the possible connection of the zpf with the quantum properties of light has been studied, [2] while for the rest, stochastic electrodynamics (SED) has been exploring the behaviour of microsystems under the action of this field. [3]

Can SED offer a physical explanation for the peculiar wave properties of matter? It makes sense to expect a positive answer, considering that the theory includes in a fundamental way an undulatory element, namely

M. Ferrero and A. van der Merwe (eds.), Fundamental Problems in Quantum Physics, 47–56.

the electromagnetic waves that constitute the vacuum. The continuous action of this field on the particles can be expected to impress on the latter a wave behaviour under appropriate conditions, without this implying that the particles lose their essentially corpuscular character.

In the following pages we study this possibility. We start by recalling the dynamical effects produced by the zpf on a particle which responds to the modes of all frequencies up to a maximum value ω_c, assumed to be fixed by the particle's effective structure. The particle is seen to vibrate under the effect of this field, with wave number $k_c = \omega_c/c$. It is suggested that a relativistic version of this calculation would have produced the *zitterbewegung*, a fine oscillation with ω_c given (approximately) by Compton's frequency. The coupling of the vibrating particle to the field modes of frequency ω_c, is shown to generate in the laboratory frame a modulated carrier wave with a wavelength given by de Broglie's formula. The time-independent equation for this wave is the stationary Schrödinger equation. Heuristic arguments applied to some simple examples show that the quantization rules for stationary states emerge from the condition of stationarity of the modulated wave.

The Schrödinger equation embodies thus the dual nature of quantum systems, by describing both the ensemble of particles and the corresponding modulated wave.

A further interesting consequence of the particle having an effective structure and an associated cutoff frequency ω_c, refers to the Casimir effect on the particle. The results of the calculations are briefly mentioned at the end of this paper, and the possibility is discussed of finding in this effect an explanation for the stability of the electron, contrary to what is usually believed. Thc introduction of a cutoff with a physical meaning rather than a mathematical trick, seems therefore to have far reaching consequences.

2. THE VIBRATING PARTICLE

In this section we recall very briefly some results on the dynamics of a particle under the action of the zpf modified by boundary conditions. For a detailed exposition we refer the reader to Ref. [4]. In a nonrelativistic treatment one can start with the equation

$$m\ddot{\mathbf{x}} = e\mathbf{E}_r + e\mathbf{E}_s, \tag{1}$$

where $\mathbf{E}_r$, $\mathbf{E}_s$ are the electric components of the particle's radiation reaction and the stochastic background field, respectively. In empty space the former would have the familiar form $\sim \dddot{\mathbf{x}}$, but in the general case it assumes a more complicated form, that can be expressed as:

$$\mathbf{E}_r = -2\pi e \sum_{n,\sigma} \mathbf{G}_{n\sigma}(\mathbf{x}) \int_0^\infty \mathbf{G}^*_{n\sigma}(\mathbf{x}(t-s)) \cdot \dot{\mathbf{x}}(t-s) \exp(i\omega_n s) ds,$$

where $\mathbf{G}_{n\sigma}$ are orthonormal functions, solutions of $\nabla^2 \mathbf{G}_{n\sigma} + k_n^2 \mathbf{G}_{n\sigma} = 0$ under the appropriate boundary conditions. The indices n, σ refer as usual to the wave vector and polarization, respectively. In terms of the same functions, the vacuum field can be written

$$\mathbf{E}_s = i \sum_{n,\sigma} \sqrt{\pi \hbar \omega_n} \mathbf{G}_{n\sigma}(\mathbf{x}) a_{n\sigma}(t) + \text{c.c.},$$

where $a_{n\sigma}(t) = a^0_{n\sigma} \exp(i\omega_n t)$, with $a^0_{n\sigma}$ statistically independent random amplitudes that average to zero and whose square average is $\left\langle |a^0_{n\sigma}|^2 \right\rangle = 1$, so that the field has an average energy $\frac{1}{2}\hbar\omega_n$ per normal mode, as is usual in SED (and QED). For simplicity it is assumed that the $a^0_{n\sigma}$ have fixed amplitude and random phase, *i.e.*, $a^0_{n\sigma} = \exp(i\theta_n)$. In the nonrelativistic approximation already implicit in writing Eq. (1), the stationary solution is

$$\dot{\mathbf{x}} = \mathbf{v} + \sum_{n,\sigma} (f_n \mathbf{G}_{n\sigma} a_{n\sigma} + \text{c.c.}), \tag{2}$$

with $\mathbf{v}$ constant and $f_n = (e/m)\sqrt{\pi\hbar/\omega_n}$ to lowest order in $\alpha = e^2/\hbar c$. The average kinetic energy associated to the stochastic motion in Eq. (2) is then:

$$T_s = \frac{1}{2} m \left\langle (\dot{\mathbf{x}} - \mathbf{v})^2 \right\rangle = \frac{\pi e^2 \hbar}{m} \sum_{n,\sigma} \frac{|\mathbf{G}_{n\sigma}|^2}{\omega_n}. \tag{3}$$

For the calculation of T_s one needs the explicit form of $\mathbf{G}_{n\sigma}$, which depends on the geometry of the objects that impose the boundary conditions. For simplicity we consider a perfectly conducting plate; then, at a distance x of the plate, the calculation gives

$$T_s = \frac{\alpha(\hbar\omega_c)^2}{2\pi mc^2} \left[1 + \frac{\cos^2 k_c x}{(k_c x)^2} \right], \tag{4}$$

where an upper bound for the frequency has been introduced, to take into account the decoupling of the particle from the high-frequency field by virtue of its assumed effective structure. The first term in Eq. (4) is constant and represents therefore a (radiative) contribution to the proper energy; it can actually be identified with the contribution of the $\mathbf{A}^2$-term in the hamiltonian of the particle immersed in the free vacuum field (see, e.g.,

Ref. [5]). The second term represents a vibration in space with the high wave number $k_c = \omega_c/c$ and with an amplitude that decreases with the distance to the plate. Note that if a relativistic motion had been assumed, with velocity c, the result would have been an oscillation with frequency ω_c. Now it happens that in the end, for reasons of internal consistency of SED (see Sec. 3), the cutoff frequency *has* to be set at Compton's value

$$\omega_c = mc^2/\hbar, \tag{5}$$

which corresponds in order of magnitude to the *zitterbewegung* frequency. This suggests therefore to view the *zitterbewegung* as a relativistic vibration sustained by the high-frequency modes of the vacuum field. We thus verify that as a result of the vibrations, the particle acquires an effective structure with a radius of the order of Compton's wavelength,

$$\lambda_c = h/mc, \tag{6}$$

in agreement with the assumption introduced in the calculation.

Although the changes in the structure of the field have been assumed to be produced by a plane conducting plate, it is easy to see that the particle itself could have been considered responsible for these changes (although the calculations would have become much more complicated). This point of view will be reinforced in Sec. 5. Also, it must be mentioned that similar results are derived for a neutral particle, modelled as a small dipole of natural oscillation frequency much smaller than ω_c. The conclusion is, therefore, that the vacuum field modified by the particle induces a sustained high-frequency vibration on it, that in its turn endows the particle with an effective structure.

3. ORIGIN OF DE BROGLIE'S WAVELENGTH

From Eq. (4) it is clear that the high-frequency vibration obtained does not depend on the state of translational motion of the particle; the same result holds for a particle moving under the action of an external force, as long as the variations of the systematic velocity are smooth compared with the fine oscillations, so that the approximations involved in the solution (2) are still valid. We can therefore conclude that the particle has acquired an internal "clock" of constant frequency ω_c in its proper frame. As it vibrates with this frequency, it will interact selectively with the corresponding vacuum field modes that sustain this motion. The precise description of such interaction would require a more detailed analysis, which has not been done; a heuristic discussion, also within the nonrelativistic framework, may be found in Dodd's work. [6] Further, we refer the reader to the work by Kracklauer, [7] where a physical reasoning similar to the present one can be found.

We now sketch a simple kinematic derivation, parallel to that of de Broglie's theory: [8] consider the one-dimensional problem of a particle that moves with respect to the laboratory, with a velocity v along the $x-$direction, and pay attention to the two (plane) waves of frequency ω_c in the proper frame that propagate in both directions along the $x-$axis; such waves are of the form $\varphi_\pm = a_\pm^0 \exp i(\omega_c t \mp k_c x) + \text{c.c.}$, with $a_\pm^0 = \exp(\pm i\theta_c)$, and they superpose to form a standing wave of wavelength $2\pi/\omega_c$. In the laboratory frame, however, these waves have frequencies $\omega_\pm = \gamma\omega_c(1 \pm \beta)$, where $\gamma^{-1} = \sqrt{1-\beta^2}$ and $\beta = v/c$, as usual; their superposition is a travelling wave of the form

$$\varphi(v) = 4\cos(\gamma\omega_c t - k_B x + \theta)\cos(\omega_B t - \gamma k_c x), \tag{7}$$

with

$$\omega_B = \gamma\beta\omega_c = ck_B. \tag{8}$$

Equation (7) represents a modulated carrier wave, the wavelength of the modulation being given by

$$\lambda_B = \frac{2\pi}{k_B} = \frac{\lambda_c}{\gamma\beta} = \frac{h}{p}, \tag{9}$$

where $p = m_0\gamma v$ is the linear momentum of the particle and Eq. (6) has been used for λ_c. This modulation with de Broglie's wavelength is the spatial expression of the Doppler shift on the two original waves, when viewed from the laboratory frame; it is clearly not a result of a Lorentz transformation. This explains of course the simple dependence of de Broglie's wavelength on the *relative* velocity of the particle with respect to the laboratory. Further, note that the phase of the modulation travels with the particle's velocity: $\omega_B/\gamma k_c = v$.

Let us now pay attention to the carrier: its group velocity is given according to Eqs. (7) and (8) by $v_g = \partial(\gamma\omega_c)/\partial k_B = v$, it oscillates with a frequency $\gamma\omega_c$, and its phase velocity is $u = \gamma\omega_c/k_B = c^2/v$. Hence, from the equation for the amplitude ϕ of this wave,

$$\frac{\partial^2\phi}{\partial x^2} - \frac{1}{u^2}\frac{\partial^2\phi}{\partial t^2} = 0,$$

we obtain, with $\frac{\partial^2\phi}{\partial t^2} = -(\gamma\omega_c)^2\phi$, ω_c given by Eq. (5), and $p^2(x) = 2m(E-V)$, in the nonrelativistic case:

$$\frac{\hbar^2}{2m}\frac{\partial^2\phi}{\partial x^2} + (E-V)\phi = 0. \tag{10}$$

Thus the stationary Schrödinger equation provides a description in configuration space of the modulated carrier wave. Note that the exact form of the coefficient of $\partial^2\phi/\partial x^2$ was determined by the value chosen for ω_c, Eq. (5). If we had kept ω_c as a free parameter, the coefficient in Eq. (10) would have been instead: $mc^4/2\omega_c^2$. By using this equation to calculate *e.g.* the energy of the ground state of the harmonic oscillator and comparing with the well-known solution of SED (which coincides with the quantum solution), we would have obtained Eq. (5) for ω_c; this is what we mean by saying that ω_c is fixed by internal consistency.

According to this simple analysis there are two scales of interest: [7] one is associated to the very fine (sub-quantum) oscillations, and the other is the much larger scale of de Broglie's wavelength and all the associated wave phenomena (*i.e.*, the quantum scale).

The present undulatory approach suggests establishing contact with the quantization rules. A couple of simple examples serve to illustrate how such rules can emerge from the condition of stationarity of the modulated wave. Let us first consider a particle in an infinite one-dimensional potential well of width L. We assume that it has reached a stationary state of motion, characterized by an energy E. For this state of motion to be maintained, the accompanying modulation must form a stationary pattern and the particle must encounter this wave in the same conditions after each complete orbit (the details of the carrier wave are unimportant, due to its extremely small wavelength). From Eq. (7), this means that the phase of the stationary modulation must change by an integral multiple of 2π, i.e., $\triangle\Phi = 2Lk_B = 2\pi n$, whence according to Eq. (9): $p = mv = nh/2L$, or $E = n^2h^2/8mL^2$, which is the well-known quantum-mechanical result.

The same argument can be formally applied to other instances in which the particle describes a closed orbit with velocity v. Take for instance a circular motion with radius r around a fixed axis: the periodicity condition on the wave reads $2\pi r k_B = 2\pi n$, whence the angular momentum becomes quantized: $l = mvr = n\hbar$. This is just the old heuristic example of de Broglie.

Whenever the particle performs a bounded periodic motion along a closed orbit, one can express the condition of stationarity as $\triangle\Phi = 2\pi n$. In the case of variable velocity one has $k_B = k_B(x)$ (where x is the displacement along the orbit), whence $\triangle\Phi = \oint k_B(x)dx$; the condition of quantization becomes therefore

$$\oint p(x)dx = nh, \tag{11}$$

which is the Wilson-Sommerfeld rule in its simplest form (restricted to cases in which the wave is reflected an even number of times along the orbit). Equation (11) is expected to give exact results only when the classical path of the particle (given by $p(x)$) coincides with the path available for the accompanying wave; this happens, for instance, when particle and wave are reflected by a perfectly rigid wall. In shallow potentials, instead, the partial penetration of the wave can contribute to its phase increment whereas it does not contribute to the classical phase integral.

We have greatly simplified the discussion by considering one-dimensional motion only. It is clear that for a more general analysis of quantization it would be necessary to consider the complete set of zeropoint waves interacting with the particle in all directions, and to derive the corresponding modulated wave, which is in general a difficult task.

4. WAVES AND CORPUSCLES

The wave described by Eq. (10) can have a considerable spatial extension; in fact, for a free particle or a stationary coherent beam of well-defined velocity the wave train is in principle infinitely long, in the absence of barriers. Even in a more realistic picture, both the length and the width of a wave packet can be made to extend over a good number of wavelengths. The particle, on the other hand, maintains its essentially pointlike character –or its finite dimension, in the case of a complex structure such as an atom– and moves in space following a trajectory, however complicated it may be. As seen above, the continuous coupling of the particle to its modulated wave can have important effects on the particle dynamics; the appearance of quantization of the bounded states of motion is a most dramatic example. The modulated wave will of course be also subject to the usual phenomena characteristic of waves, such as diffraction, interference, refraction, etc., and these effects, in their turn, will become evident through the behaviour of the particle, as it interacts significantly with the components of its associated wave.

This strong association of particle and wave explains why the Schrödinger equation, which has been derived here in de Broglie's spirit as a wave equation, serves simultaneously for a statistical description of the particle. The normal concept of a Schrödinger equation referring to the particle's probability amplitude in configuration space, is entirely consistent with this scheme. In fact, as is shown in Ref. [4], under certain dynamical conditions that are specific of the SED system (but do not hold for other stochastic systems), it is possible to derive the Schrödinger equation for the probability amplitude ψ, whose stationary form coincides with the wave equation (10)

$$E\psi = -\frac{\hbar^2}{2m}\frac{\partial^2\psi}{\partial x^2} + V\psi. \tag{12}$$

The fact that the solutions of Eqs. (10) and (12) must satisfy the same mathematical requisites of continuity and square integrability, implies that in a stationary situation the probability amplitude ψ must be proportional to the wave amplitude ϕ at each point. Where exactly in the ϕ-wave will the particle be found in each instance is a matter of chance, at this level of analysis; it is clear that only a statistical description for the particle can match the wave description.

5. CASIMIR ENERGY OF THE PARTICLE

This paper started by considering the modification of the structure of the vacuum field by the presence of matter, and calculating the average dynamical effect of this modified field; the high-frequency vibrations of the particle where thus obtained. Many papers have been devoted instead to the calculation of the change in the energy content of the field that arises from the modification of its structure, normally called the Casimir energy. In such calculations, a cutoff parameter is in general introduced in order to avoid divergencies, but an effort is made to obtain final results that are independent of the cutoff. For macroscopic bodies this technique has given in general correct results. [9] For instance, it predicted successfully an attractive Casimir force between two parallel conducting plates, that decreases with the distance between plates.

Now according to the above discussion, the cutoff frequency ω_c introduced in the calculation of Eq. (4) is a physical parameter, as it represents the largest frequency at which the particle is still able to respond to the electromagnetic field. Further, we have seen that this cutoff is related to the finite effective structure of the particle, and that a self-consistent calculation within SED gives $\omega_c = mc^2/\hbar$. It seems therefore interesting to find out if the introduction of such a physical cutoff throws some new light on the old problem of the Casimir energy of a small particle; especially since a central motivation for the numerous and cumbersome Casimir-effect calculations of recent years has been precisely the possibility of applying the results to systems of very small dimensions, such as electrons, or confined quarks and gluons. Inspired by the result obtained for the parallel plates, Casimir himself had suggested the possibility that if such an attractive force is obtained also in the case of a spherical configuration, it could account for the Poincaré stress that is needed to counterbalance the repulsive Coulomb force of the charge. [10] The usual calculations, however, have all yielded an outward force in this case. [9,11]

When the frequency cutoff is duly introduced in the calculations, one can see from the study of typical configurations [12] (i.e., the infinite parallel plates, or the hollow sphere) that the expressions derived for the Casimir energy are regular functions $E_c(y)$ of the product $y = ak_c$, where

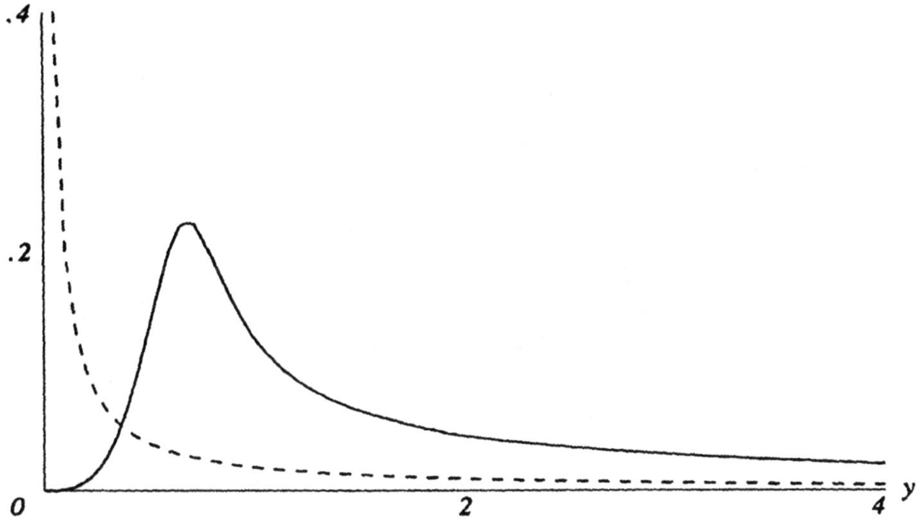

Fig. 1. Approximate shape of the Casimir energy curve for the spherical conducting shell, in units of $(\frac{1}{2}\hbar\omega_c)$, as a function of $y = ak_c$. Broken line: Coulomb energy. (Taken from Ref. [12].)

a represents a typical dimension of the system (the separation between plates, or the radius of the sphere). For very small objects ($a \to 0$), a large value for k_c does not imply $y \gg 1$. In the usual calculations, however, the approximation $y \gg 1$ is always taken at some point; the results of such calculations are therefore not generally valid for small objects.

In Fig. 1 the approximate shape of the energy curve for the spherical conducting shell (with cutoff) is drawn as a function of y (see Ref. [12]). Note that for $y < 1$, E_c is an increasing function of y; hence, for instance, if $k_c = mc/\hbar$, the Casimir force becomes attractive for a shell of radius smaller than Compton's wavelength. A more detailed calculation shows that this inward pressure can be sufficient to compensate for the electrostatic repulsion of the charge, whose energy decreases as a^{-1}; this opens up the possibility of explaining the stability of the electron as a result of the balance between Casimir and Coulomb forces on the particle, as originally suggested by Casimir.

This work was supported in part by Conacyt through grant No. 0682-E9208.

REFERENCES

1. M. Planck, *Verh. Deut. Phys. Ges.* **13**, 138 (1911); W. Nernst, *Verh. Deut. Phys. Ges.* **18**, 83 (1916).

2. T. W. Marshall and E. Santos, *Found. Phys.* **18**, 185 (1988).
3. L. de la Peña, in *Stochastic Processes Applied to Physics and Other Related Fields*, B. Gómez et al., eds. (World Scientific, Singapore, 1983).
4. L. de la Peña and A. M. Cetto, *Found. Phys.* **25** (1995).
5. A. M. Cetto and L. de la Peña, *Phys. Rev. A* **37**, 1952 (1988).
6. J. D. Dodd, J. Phys. **B 8**, 157 (1975), *Eur. J. Phys.* **4**, 205 (1983).
7. A. F. Kracklauer, *Phys. Essays* **5**, 226 (1992).
8. L. de Broglie, *Une tentative d'interprétation causale et non-linéaire de la mécanique ondulatoire* (Gauthier-Villars, Paris, 1956).
9. G. Plunien, B. Müller, and W. Greiner, *Phys. Rep.* **134**, 87 (1986).
10. H. B. G. Casimir, *Physica* **19**, 946 (1956).
11. T. H. Boyer, *Phys. Rev.* **174**, 1764 (1968); B. Davies, *J. Math. Phys.* **13**, 1324 (1972); K. A. Milton, L. de Raad and J. Schwinger, *Ann. Phys.* (N.Y.) **115**, 388 (1978).
12. A. M. Cetto and L. de la Peña, *Nuovo Cimento* **108B**, 447 (1993).

RESULTS OF ATOM INTERFEROMETRY EXPERIMENTS WITH POTASSIUM

John F. Clauser

Department of Physics
University of California, Berkeley
Berkeley, California 94720

Results of high flux atom interferometry experiments with potassium in generalized Talbot-Lou configurations are presented. The interferometer consists of a sequence of three planar vacuum-slit diffraction gratings, microfabricated from silicon nitride membranes. Interference fringes are sensed by measuring the transmission of atoms on a hot-wire as a function of grating relative position. Different spatial Fourier components in the diffraction pattern are resonant in the interferometer at different atomic velocities. When a laser cooled slow beam is incident, various different diffraction patterns are observed as a function of atomic velocity, selected via the tuning of cooling lasers. In an alternative "Heisenberg Microscope" configuration an incident thermal beam produces a velocity average over different fringe patterns that averages over and washes out the high frequency Fourier components. In this configuration AC modulated laser light passes through the interferometer. Via the Doppler shift, it is scattered only by atoms in a narrow velocity band. Since imaging of the fluorescent light *could* determine which slit an atom passes, the laser destroys, and thereby reveals via the AC modulation, the associated high-frequency fringe contribution.

Key words: atom, interferometer, matter wave, de Broglie wave.

1. INTRODUCTION

Following suggestions by Altshuler and Frantz [1], Dubetskii *et al* [2], and Clauser [3], there is recent interest in experimentally demonstrating de Broglie wave interference exhibited by the propagation of whole neutral atoms [4–8]. Atom interferometry experiments, in essence, are striking de Broglie-wave variants of Young's two-slit experiment. The experiments

M. Ferrero and A. van der Merwe (eds.), Fundamental Problems in Quantum Physics, 57–71.

described here are a N-slit extension of this. Given my own earlier experimental and theoretical work on Bell's theorem [9], I do not agree with R. Feynmann's assertion that the two-slit experiment contains the *whole* mystery of quantum mechanics, but I do agree that it embodies a big chunk of it!

An atom such as potassium is a very complicated thing. It contains electrons, virtual photons and a nucleus, further comprised of nucleons and mesons, in turn full of quarks, gluons, and lots of virtual stuff, much of which we probably have not yet conceived. How do big whole atoms exhibit spatial de Broglie wave interference? The Hamiltonian for such a beast is a sum over the kinetic and internal potential energies of the atom's constituent parts. By the *magic* of the center-of-mass transformation, this sum can be split into a sum of two parts [10]. The first part is the Hamiltonian studied by spectroscopists. The second part represents the kinetic energy of a moving point particle, whose mass is that of the whole atom. Atom interferometry experiments exploit the latter part, and/or both parts. Yet despite an atom's hopelessly complicated composite structure, it remains undisturbed as a whole atom, even after it passes simultaneously through more than one slit at a time. How does it do so? Beats me!

2. INTERFEROMETRY USING A COLD SLOW ATOM SOURCE

This paper summarizes the results of two different experiments demonstrating atom spatial interference. See Refs. 4 for more details. The apparatus for the first experiment is diagramed in Fig. 1. Two copropagating potassium beams pass simultaneously through an atom interferometer and are detected by surface ionization on a hotwire [11]. The first is temporally continuous (DC), and has a thermal velocity distribution broadly peaked at about 540 m/sec. It acts as a "parent" for the second beam. The second beam is temporarily chopped (AC) and synchronously detected to allow it to be distinguished it from the first. The AC beam is slow ($v \approx 182$ m/sec) and cold. Figure 2a shows two possible computer simulated velocity profiles for the slow beam. Due to peculiarities of the cold beam's production and detection methodology, its velocity profile has phase-reversed wings. The DC thermal beam is produced by scattering near the oven's exit slit, which is displaced from the interferometer's axis. It is about 130 times more intense than the slow beam. The latter is velocity selected by using AC modulated (chopped ON and OFF at 6 Hz) laser light, incident on the thermal beam at 20°. Scattering of about 7 photons by each slow atom deflects it out of the low velocity portion of the parent beam's thermal distribution, onto the interferometer's axis. The parallel component of the laser's incidence provides

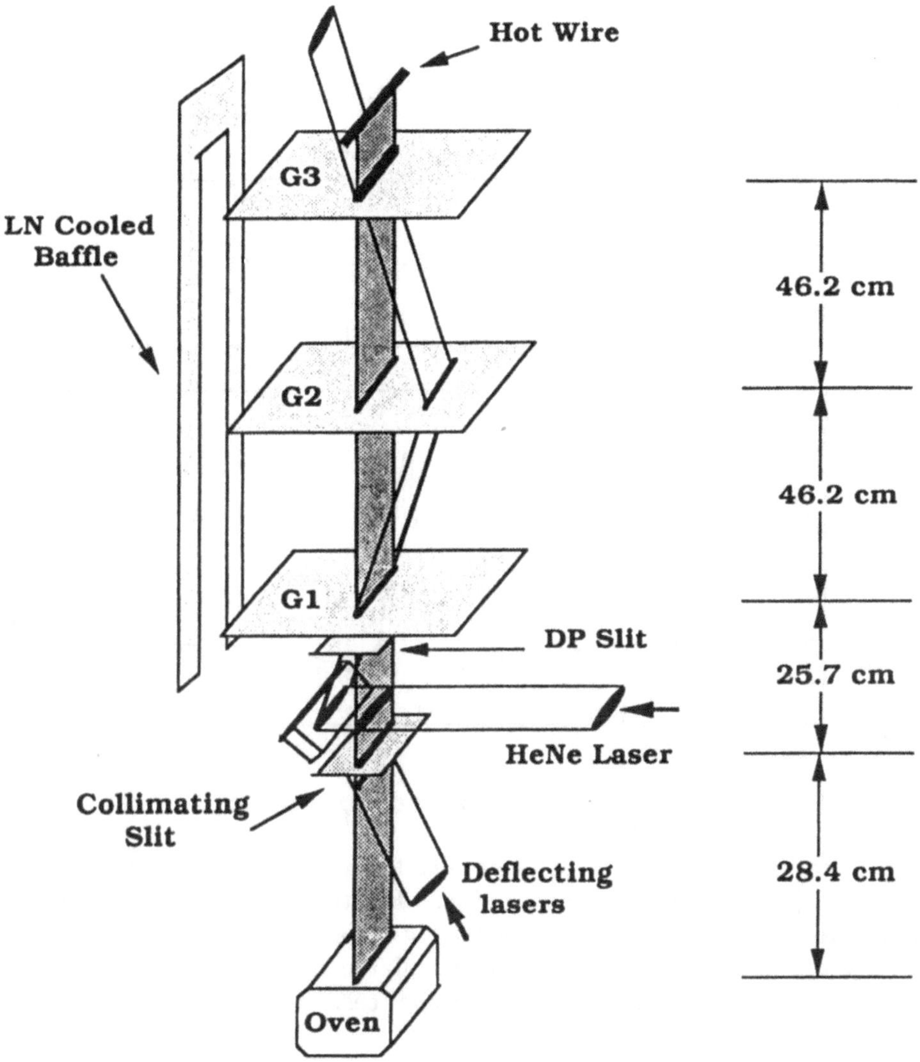

Fig. 1. Geometry of the experimental apparatus used for experiments with a cold slow incident atomic beam. A HeNe laser beam passing through the gratings is used for rotational alignment of the interferometer.

Doppler velocity selection, while the perpendicular component provides momentum transfer for deflection.

Typical lateral spatial profiles of the slow (AC) and thermal (DC) beams are shown in Fig. 3. Perpendicular heating of the slow beam by random recoil of fluorescent photons, predicted by Einstein [12] and sought experimentally by Frisch (unsuccessfully, without the benefit of laser tech-

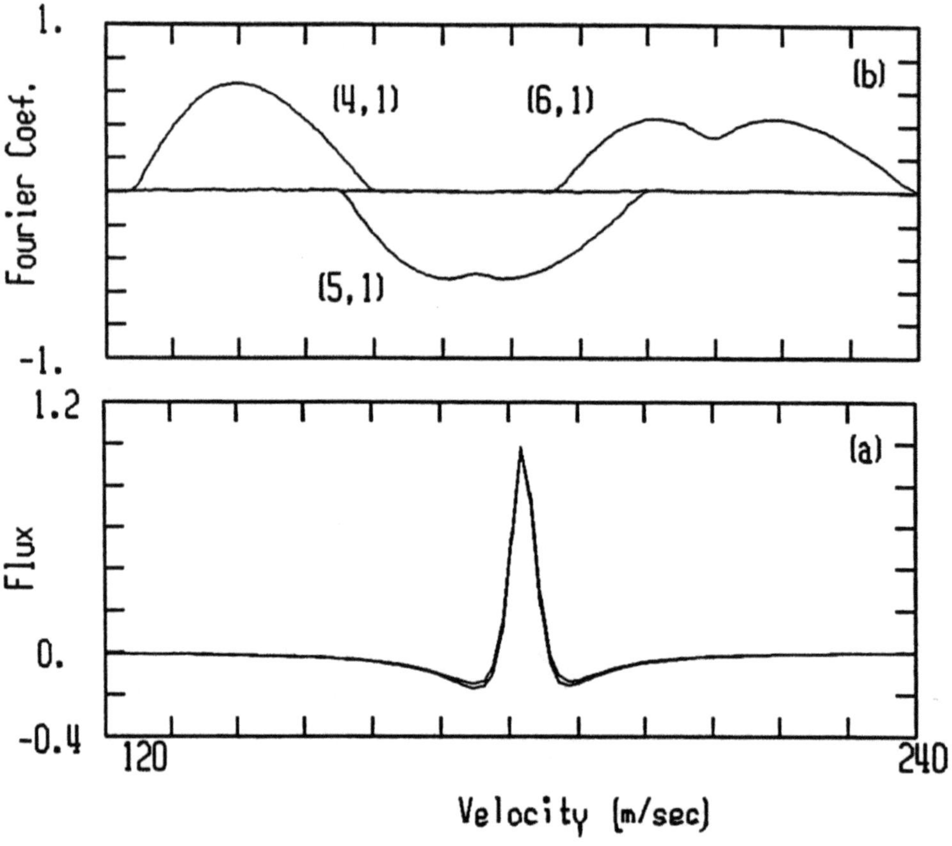

Fig. 2. (a) Two possible velocity profiles for our source, assuming a parallel velocity spread of about 10–12 m/sec. Positive and negative signs correspond to in-phase and phase-reveresed AC signal levels. (b) 4th, 5th, and 6th spatial harmonic content of the calculated fringe pattern at the respective (m, n) = (4,1), (5,1) and (6,1) resonances.

nology) [13], is evident in the data. The effective source brightness for cold slow (182 m/sec) atoms is about 4×10^{15} atoms $\mathrm{cm}^{-2}\mathrm{sr}^{-1}\mathrm{sec}^{-1}$, yielding a maximum transmitted and detected current of roughly 4×10^{5} atoms per sec.

The interferometer consists of a sequence of three parallel planes, each containing a rectangular vacuum-slit transmission grating, G_1, G_2, and G_3, respectively. The spacings between the grating planes are $R_1 = R_2 =$ 46.2cm. The gratings are micro-fabricated from 1μ thick silicon-nitride membranes supported by silicon frames, with parallel slits etched through the membranes. Gratings G_1 and G_3 have the same period, $a_1 = a_3 = 16.2\mu$, and have 22 and 75 slits, respectively. Grating G_2 has 111 slits, with a period, $a_2 = 8.1\mu$. All three are 8.5 mm long and have an average open fraction

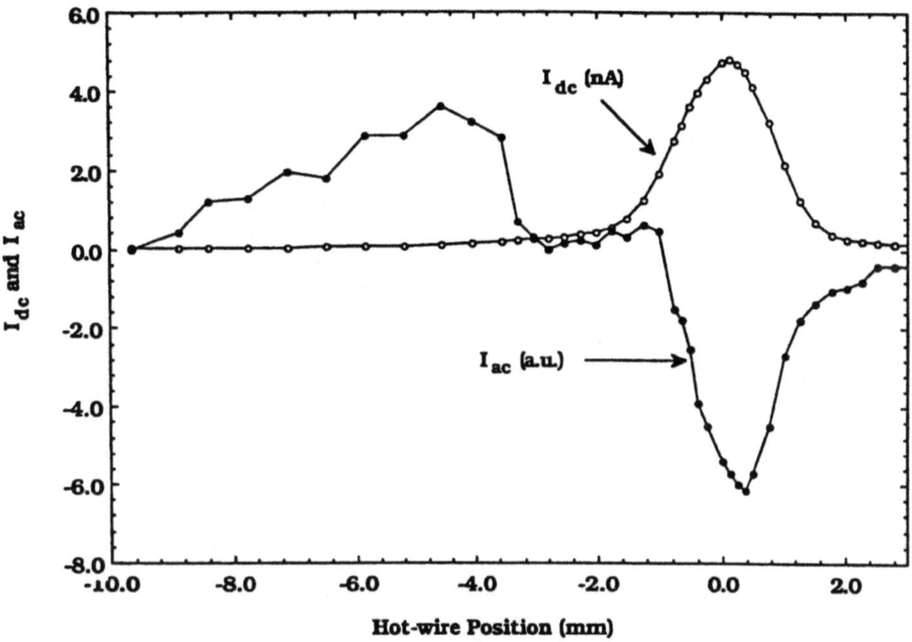

Fig. 3. Spatial profiles of the DC parent thermal beam and AC chopped cold slow beam by scanning the position of the hot-wire detector. These profiles were taken at a laser tuning to select atoms at 295 m/sec, with high laser power, and with no intervening interferometer. For the first interferometry experiment the oven's position is offset so that the deflected AC beam is along the interferometer axis.

of about $s/a = 1/8$, where s is the slit width.

For an interferometer with the above geometry Clauser and Reinsch pointed out that de Brolige-wave fringes will be formed on and masked by G_3 via the generalized Talbot-Lau effect [14]. Fringes are sensed by monitoring the variation of the transmission in response to slowly scanning G_2's position. An interfering path set within the beam envelope consists of nested diamonds, starting at one source slit of G_1, and forming an interference pattern on G_3. Solid-angle acceptance is enhanced (by a factor of almost 10^7 over the atom interferometer geometry used by Keith *et al.* [7]) by incoherent addition of the current from many source slits on G_1, each providing many such nested diamonds at all possible skew angles between the slits of G_1 and G_2.

With low velocity monochromatic illumination an interference pattern is thus formed. The fringe pattern (and the transmitted current) contains various spatial harmonics of the geometric shadow Moiré period. Each harmonic will be resonant in the interferometer at a different atomic ve-

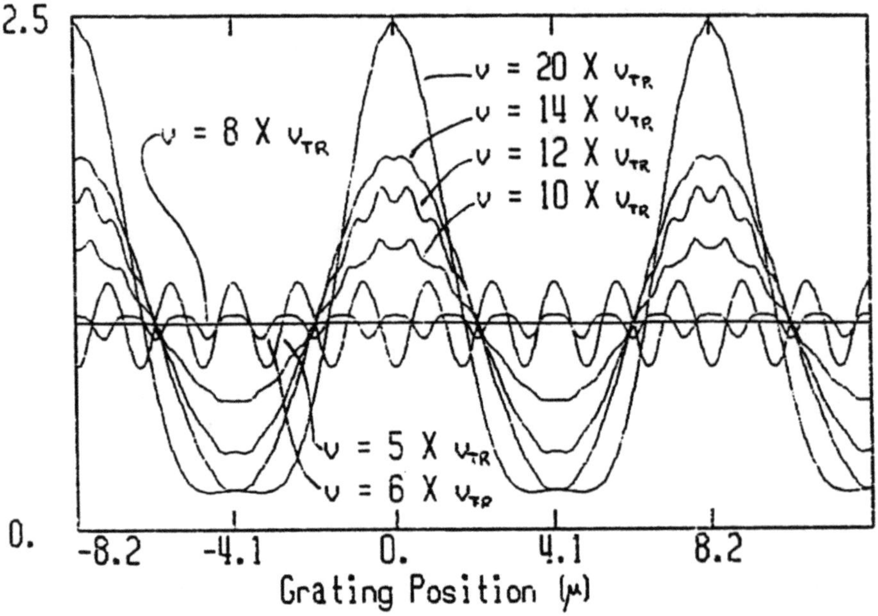

Fig. 4. Calculated normalized interferometer transmission as a function of and grating displacement, for various atomic velocities, 175, 210, 280, 350, 419, 489, and 699 m/sec, corresponding to 5, 6, 8, 10, 12, 14, and 20 times v_{TR}, respectively.

locity. Figure 2b shows the calculated amplitudes for the 4th, 5th and 6th harmonic components of a Fourier decomposition of the pattern as a function of atomic velocity. A resonance for the m'th harmonic occurs at $\lambda_{dB}/\lambda_{TR} = n/m$ when m and n approximate small integers. Here, λ_{dB} is the atoms' de Broglie wavelength, $\lambda_{TR} \equiv a_2^2/\rho$ is the interferometer's Talbot-Rayleigh wavelength, and $\rho \equiv R_1 R_2/(R_1+R_2)$ is the interferometer's reduced length. For our geometry the associated resonant atomic velocity is $m/n \times V_{TR}$, where $V_{TR} \approx 35$ m/sec. Phase reversal of the $(m,n) = (5,1)$ resonance (5th harmonic) is a consequence of the fact that the product, $m \times n$, is odd.

Figure 4 shows for various atomic velocities the calculated interferometer transmission as a function of G_2 displacement, averaged over the finite slit widths G_1 and G_3. At velocities above $a_2/s_2 \times v_{TR} \approx 8\ v_{TR} =$ 278 m/sec, the oscillatory fringe structures give way to the $(n \to 0)$ geometric shadow Moiré. The calculated AC and DC signals at any grating position are given by a weighted integration of the transmission over atomic velocity. For the AC signal, a weighting by the velocity profile of Fig. 2a samples a narrow range of the velocities where high frequency oscillatory fringe structures occur. A thermally distributed weighting washes out the high-frequency structures and yields a diffraction limited geometric Moiré for the DC fringe pattern.

Figure 5a shows the measured (solid) and calculated (dashed) DC signal as a function of grating position, while Fig. 5b shows the simultaneously measured AC signal in response to a single scan of G_2's position. Each data point represents about a 4 sec integration. Due to a start-up hysteresis in our piezoelectric translator, the G_2 displacement axis suffers from a non-linear distortion for negative values, evident in both the AC and DC signals. Despite this distortion a fifth spatial-harmonic component is evident in the AC signal. Since the laser tuning selects atoms near the $(m, n) = (5, 1)$ resonance, this harmonic is expected to dominate. However, given the v^3 weighting of atoms in the parent beam, the phase-reversed high-velocity broad Lorentzian wing of the profile contributes significant sixth harmonic at opposite phase that enhances end fringes and washes out central fringes. Figure 5b also shows the associated calculated AC fringe patterns for the two assumed incident AC velocity profiles of Fig. 2a.

Since the interferometer is highly velocity selective, a small change in the excitation spectrum has a strong effect on the shape of the pattern. Had we instead assumed a broad non-reversed velocity profile in our simulation, the strong peaks would be shifted by half a DC fringe period. It also has a significant effect on the vertical offset of the pattern (suppressed in Fig. 5b), since the 5th and 6th (modest visibility) harmonics have opposite phase. Cancellation of the associated opposite vertical offsets enhances the apparent AC fringe visibility. Thus, although the observed AC fringe visibility is about 80% referenced to the blocked-laser AC signal level, that of each velocity component is probably lower. Calculation assuming various s/a values indicates that the asymmetry and finite visibility of the DC fringe pattern are due dominantly to quantum-mechanical diffraction by G_2 and more weakly to geometrical averaging by G_1 and G_3.

3. "HEISENBERG MICROSCOPE" DECOHERENCE ATOM INTERFEROMETRY

Recently, Walls *et al.* [15] analyzed a problem, analogous to that of the "Heisenberg microscope," for freely propagating atoms with well defined momenta that form de Broglie-wave fringes in a Young's two-slit interferometer. They consider a situation wherein both slits are simultaneously illuminated with light that is resonant with an atomic transition, and calculate the resulting atomic fringe visibility as a function of slit separation. They predict that when the slits are separated sufficiently, so that a Heisenberg microscope viewing the reemitted fluorescence can image this light to determine which slit an atom passes, then the atomic fringe visibility will vanish.

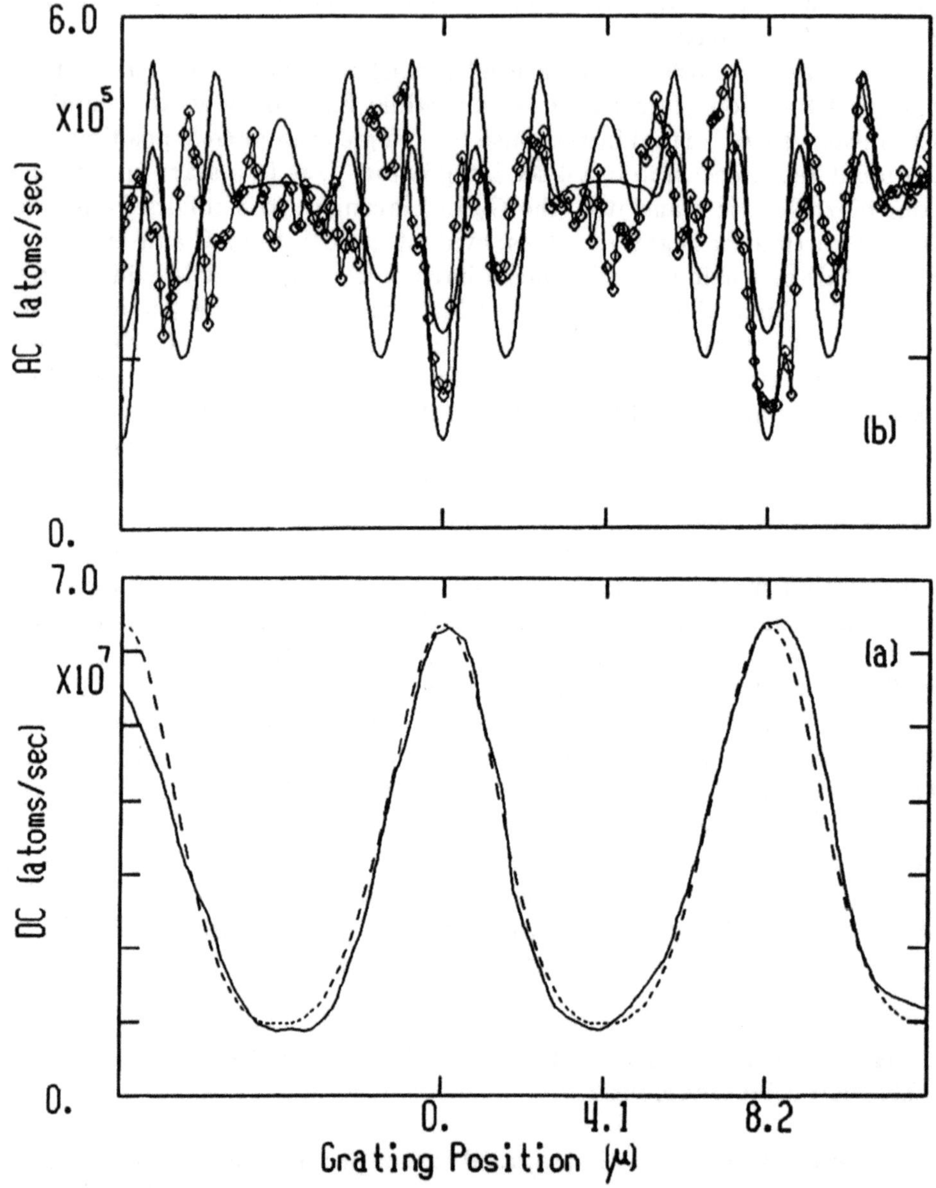

Fig. 5. (a) Measured (solid line) and calculated (dashed line) DC signal as a function of grating relative displacement. Zero level corresponds to the laser-blocked condition. (b) Simultaneously measured AC signal (diamond points) and calculated AC signals (with the vertical offsets suppressed), for the different assumed velocity distributions of Fig. 2a.

But when the slit spacing is comparable to the optical wavelength, such a determination by the microscope exceeds its resolving power, and then the interference pattern will persist. Further, the presence of the microscope is unnecessary for the predicted visibility dependence on slit spacing to obtain (none is present in our experiment). Only the microscope's illumination need be present.

The above predictions can be tested. Indeed, Sterr *et al.* [8] destroyed atom interference fringes by passing high intensity resonant laser light through their atom interferometer. In that experiment many photons were scattered off of each atom, and while their atomic paths have an amplitude for being physically separated in space, actually the paths are continuously distributed in space and are not clearly localized by a scattering event. In our experiment we destroy atom interference fringes by the scattering of a single low energy photon by an atom. Thus, with no microscope illumination (and no scattering) amplitudes for an atom's passage simultaneously through more than one physically different slit provide quantum interference and produce a multiple slit interference pattern. With illumination, however, a scattering can localize an atom's path to a region smaller than the slit spacing. Given the multiply connected geometry an atom, thus localized, can pass through only one slit. Hence the scattering can be used to determine which slit the atom passes, whereupon no fringe pattern will form.

The above process is used by our second atom interferometry experiment. A high spatial frequency interference fringe pattern is revealed by its destruction. Now only a thermal potassium beam is transmitted through the same velocity selective atom interferometer used above. Atoms at characteristic resonant atomic velocities (and/or de Broglie wavelengths) form fringe patterns that contain high spatial-frequency Fourier components. As indicated in Fig. 5a, the thermal velocity distribution produces an average over these components that washes out and hides the high frequency fringes. AC modulated laser light now passes through the interferometer near G_2. Since imaging of the fluorescent light *could* be used to determine which G_2 slit an atom passes, the contribution to the averaged pattern by atoms at the laser's Doppler shifted wavelength is removed. That component is thus AC modulated and detected. Potassium's hyperfine structure effectively limits the number of scatterings per atom to about one via the high probability that following a scattering the atom will optically pump and thereafter be transparent to the laser radiation [16]. To further assure only one scattering per atom in an atom's flight time through the laser beam, and to provide a narrow effective laser bandwidth, the laser is attenuated heavily.

The apparatus, shown in Fig. 6, is a modification of that of Fig. 1. The oven is now located on the axis and the deflecting laser light is not used. Instead, "microscope illumination" consists of a highly attenuated AC chopped laser beam passing through the interferometer at near anti-

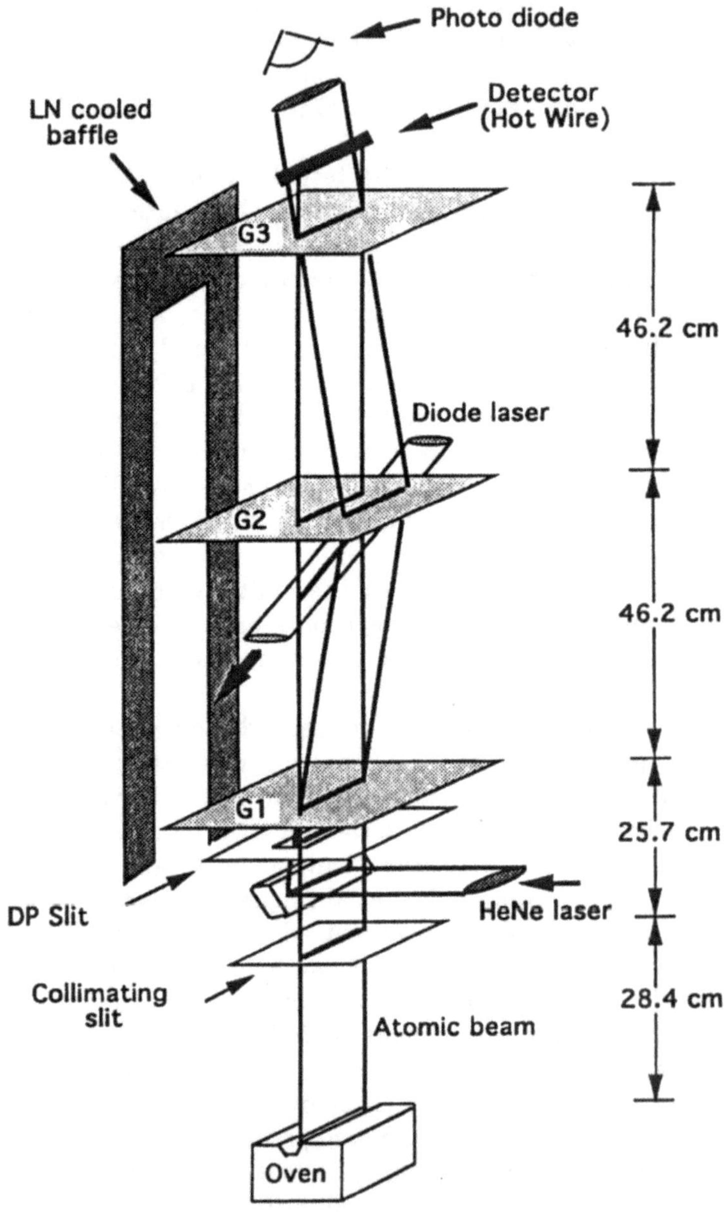

Fig. 6. Geometry of the experimental apparatus for the Heisenberg microscope experiments. The photomultiplier used to measure fluorescence intensity is not shown.

parallel incidence (20°) to the beam, immediately below G_2. When the laser is (OFF), then interference fringes are formed, but the beam's thermal velocity average prevents their direct observation. When the laser is ON, it resonates with two different velocity groupings of atoms, since K^{39} has two atomic ground-state hyperfine levels ($F = 1,2$). $F = 2$ atoms will be resonant only for tunings with $v_{F=1}$ greater than 351 m/sec, the velocity Doppler equivalent of the hyperfine resonance spacing.

We calculate the laser-ON transmission probability, assuming a kinematical scattering near G_2 of one photon by each atom, assume a classical atomic trajectory, and use the point-wise momentum-transfer photon scattering model by Einstein [12] in his discussion of the kinematics required for thermal equilibrium to obtain when a gas is irradiated by thermal light. We first consider a line source of atoms at G_1 and the scattering of one incident circularly polarized photon near G_2, and then calculate the transmission by a single G_2 slit. Summing the result over the various slits in G_1, G_2, G_3 and averaging over the associated G_1 and G_3 slit widths yields the laser-on transmission probability, given the scattering of one photon. The transmitted AC current is given by the difference between this transmission probability and that for no laser light, averaged over the thermal velocity distribution, and weighted by the probability that an atom will scatter one (and only one) photon in an atom's transit through the laser beam. This probability has two resonant near-Lorentzian components separated by 351 m/sec, corresponding to and weighted by the thermal hyperfine level populations, 40% $F = 1$ and 60% $F = 2$.

To observe interference fringes, the laser is tuned to resonate with $F = 1$ atoms at a velocity of 211 m/sec, i.e. 6 times the Talbot-Rayleigh velocity, $v_{TR} \approx 35$ m/sec, corresponding to AC fringes at the sixth spatial harmonic of the geometric shadow period. While holding the laser tuning constant, G_2 is scanned laterally, and the DC and AC signals are recorded simultaneously. These signals are shown as a function of G_2 displacement, Δx, in Figs. 7a and 7b, along with the calculated DC and AC currents. The sixth spatial harmonic associated with the Talbot-Lou resonance is evident in the AC signal's data.

To observe the velocity dependence of the AC signal, G_2's position is held fixed and the laser tuning is swept. The measured signals obtained when the gratings are positioned for minimum and maximum transmitted DC current are shown in Figs. 8a and 8b, along with the calculated signals. Agreement between the calculated and measured signals appears to be quite good.

For comparison, in Fig. 8c we show the laser excited fluorescence intensity measured using a photomultiplier and the same laser incidence angle (but with no gratings present) in response to a similar laser frequency scan. We note that the hyperfine structure is not resolved in the fluorescence spectrum. By contrast, the transmission spectrum of Fig. 8a displays two well-resolved peaks whose the spacing corresponds to the hyperfine struc-

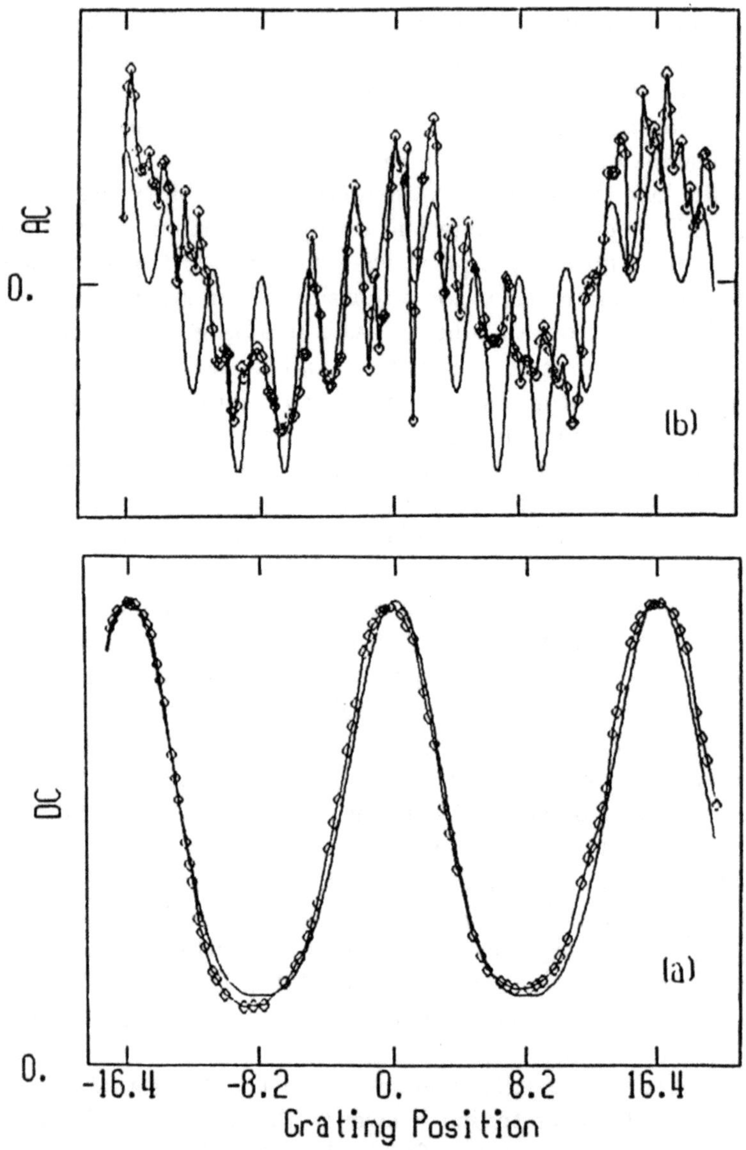

Fig. 7. Calculated and observed (diamond points) atomic currents as a function of grating relative displacement, Δx (microns). (a) DC signal due to thermal beam illumination, (b) AC signal showing the destroyed fringes for atoms at 211 m/sec. To give better agreement between the observed and calculated AC patterns, a small DC signal cross-talk component is added to the calculated AC signal.

ture. Evidently we have constructed what amounts to an atom interference

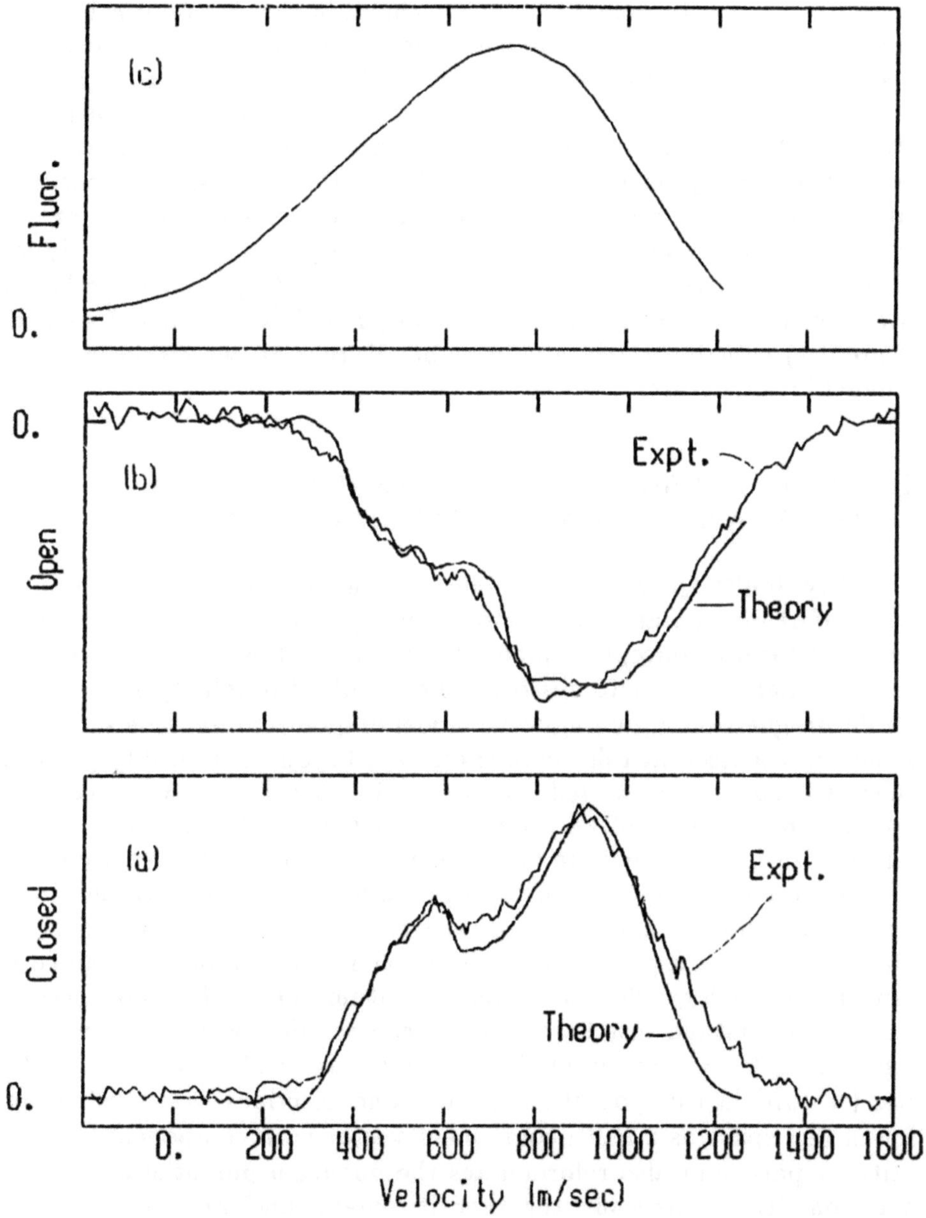

Fig. 8. (a,b) Calculated and observed atomic current as a function laser tuning, in terms of $v_{F=1}$ [via Eq. (1)], for fixed grating relative displacement. (a) AC current at minimum DC current, (b) at maximum DC current. (c) Fluorescence intensity as a function laser tuning, $v_{F=1}$.

filter, whose velocity selectivity allows us to narrow the effective transmitted velocity range, so as to provide an improvement in the atom-optical spectral resolution. When the gratings are positioned for minimum DC transmission, then neither the laser-ON nor laser-OFF conditions transmit atoms at high velocity. However, the reasons for these two high-velocity cut-offs differ. In the laser-ON condition only low velocity atoms have sufficient scattering angle to reach a G_3 open slit. A cut-off via this process then occurs at an atomic velocity of about 740 m/sec. In the laser-OFF condition only atoms with velocities below $s_2/a_2 \times v_{TR} = 278$ m/sec have sufficiently long de Broglie wavelength so that two adjacent G_2 slits produce overlapping constructive interference at an open G_3 slit. The different cut-off velocities (and profiles) effectively create a "pass-band" that allows resolution of the K^{39} hyperfine structure.

4. HOW DO THESE EXPERIMENTS HELP US TO UNDERSTAND QUANTUM MECHANICS?

They don't! Quantum mechanics is already tough enough to understand. Unfortunately, atom interferometry experiments further exacerbate attempts to form a conceptual model for quantum dynamics. For example, these experiments appear to destroy the conceptual simplicity and elegance of the de Broglie-Bohm guiding-wave model of quantum dynamics and variants thereof discussed at this conference, even beyond the locality problems these models face, *vis à vis* Bell's Theorem. In normal quantum mechanics, the relative magnitudes of an atom's de Broglie wavelength, λ_{dB}, and its so-called "size," $\approx a_0$, are quite unimportant. Indeed, in the experiments of Refs. 6–8, we have $\lambda_{dB} \ll a_0$, in our experiments [4], $\lambda_{dB} \approx a_0$, and in the experiments of Ref. 5, $\lambda_{dB} \gg a_0$. However, via Hamilton-Jacobi theory, in guiding-wave models a particle's translational quantum motion is guided via its "surfing" down a gradient of Bohm's quantum potential. It now becomes difficult to understand just how an atom can so effectively "surf" when its surfboard length ($\approx a_0$) is many times longer than the length of the guiding ripples ($\approx \lambda_{dB}$) in this potential. Even if one describes the wave's propagation in real space as a multi-component vector with components for each constituent part, and also reformulates the quantum potential similarly so that all constituent parts surf together, it remains unclear how one does so and maintains the wholeness of an atom.

Common with everyday experience, one finds no difficuly recognizing that an object may exist (*remain*) at the same position at many different times. These experiments now will force us to think more deeply about the symmetry (and dualism) between space and time. Here, by passing through more than one slit at a time, an atom appears to exist at the same time at many different positions! The English language seems to have no existing verb that is the space-time dual to *remain.*

ACKNOWLEDGEMENTS

This work was supported by ONR Grant N00014-90-J-1475 and the Firm J. F. Clauser and Associates, Walnut Creek, California. I also acknowledge my thank coworkers S. Li, M. Reinsch, and assistance from G. Garfein, and the staff and students at the UC Berkeley Microstructures Lab.

REFERENCES AND FOOTNOTES

1. S. Altshuler and L. M. Frantz, US Patent # 3,761,721.
2. B. Ya. Dubetskii *et al.*, *Pis'ma, Zh. Eksp. Teor. Fiz.* **39** (11), 531 (1984).
3. J. F. Clauser, *Physica B* **151**, 262 (1988); US Patents # 4,874,942 and # 4,992,656.
4. J. F. Clauser and S. Li, *Phys. Rev. A* **49**, R2213 (1994); *Phys. Rev. A*, Sept. 1994.
5. F. Shimizu *et al.*, *Phys. Rev. A* **46**, R17 (1992); M. Kasevich and S. Chu, *Phys. Rev. Lett.* **67**, 181 (1991).
6. O. Carnal and J. Mlynek, *Phys. Rev. Lett.* **66**, 2689 (1991); F. Riehle *et al.*, *Phys. Rev. Lett.* **67**, 177 (1991).
7. D. W. Keith *et al.* , *Phys. Rev. Lett.* **66**, 2693 (1991).
8. Sterr *et al.*, *Appl. Phys. Lett. B* **54**, 341 (1992).
9. See. J. F. Clauser and A. Shimony, *Rep. Prog. Phys.* **41**, 1881 (1978), and J. F. Clauser, in *Foundations of Quantum Mechanics*, T. D. Black *et al.*, eds. (World Scientific, Singapore, 1992), p. 168.
10. Most quantum mechanics texts don't even display this magic. It can be found, however, in A. Messiah, *Quantum Mechanics* (Wiley, New York, 1962), Vol. II, pp. 395 and 412.
11. S. Li and J. F. Clauser, *Phys. Rev. A* **49**, 2702 (1994).
12. A. Einstein, *Phys. Z.* **18**, 121 (1917), in B. L. van der Waerden, *Sources of Quantum Mechanics* (Dover, New York, 1968), p. 63.
13. O. R. Frisch, *Z. Phys.* **86**, 42 (1933).
14. J. F. Clauser and M. W. Reinsch, *Appl. Phys. B* **54**, 380 (1992), and references therein.
15. D. Walls, *et al.*, in *Foundations of Quantum Mechanics*, T. D. Black *et al.*, eds. (World Scientific, Singapore, 1992), p. 157.
16. Optical pumping is avoided in the cold incident beam experiment above by using a cycling transition, and by applying two laser frequencies for the deflecting light.

ON THE UNCERTAINTY RELATIONS

J.R. Croca

Faculdade de Ciências, Universidade de Lisboa
Departamento de Física
Campo Grande, Ed. C1, 1700 Lisboa, Portugal

An overall view of some possible interpretations and forms for the uncertainty relations and certain aspects of quantum nonlinear theories correlated with the local analysis by wavelets are presented. Also two possible experiments that may test the general validity of the usual Heisenberg uncertainty relations are discussed.

Key words: uncertainty relations, meaning of Fourier spectral analysis, local wavelet analysis, X-ray interferometry, foundations of quantum mechanics.

1. INTRODUCTION

Since Bohr[1] interpretation of the Heisenberg uncertainty relations in terms of the Fourier analysis, there was a general acceptation of them, at least on its mathematical form. It is true that there were and still are lots of arguing on the actual meaning[2] of those relations, nevertheless almost[3] everybody seems to agree with its actual mathematical formulation.

Some, like de Broglie and many others, have always felt uneasy with some physical implications[4] of those relations, like for instance the infinite spreading of the matter wave packets, given by the known relation

$$\Delta x(t) \approx \frac{\hbar}{2m} t, \tag{1}$$

If this relation holds true for all cases, then an electron ejected by the Sun, described by a wave packet with a longitudinal size $\Delta x_0 \approx 10^{-8} cm$, when arriving at the Earth would have a dimension greater than the Earth's diameter!

M. Ferrero and A. van der Merwe (eds.), Fundamental Problems in Quantum Physics, 73–82.

Those, and other reasons induced some to believe that the master equation for full describing the quantum phenomena must be a nonlinear equation. These kind of equations have nondispersive localized solutions in space and time, which are very well suited for describing the evolution of nondispersive causal particles. Recently some authors, like Mackinnon[5], Gueret[6], Vigier[7], and many others have proposed finite nondispersive, soliton-like , forms for the wave function solutions of a general master wave equation.

If quantum phenomena really need for full explanation a more general theory formulated in non linear terms, then the usual nonlocal Fourier transforms will most likely need to be replaced by a different set of transforms. The recent development of the theory of wavelets[8], or finite waves, seems to indicate that a local analysis, both in time and space, of the functions is possible. These facts open the way for a more general set of uncertainty relations, derived, not as Bohr has done from the Fourier nonlocal transforms, but from a local analysis. Of course, these new uncertainty relations must in the most cases coincide with the usual ones.

2. A NEW FORM FOR THE UNCERTAINTY RELATIONS

In order to derive mathematically a different set of uncertainty relations[9], it is convenient to assume:

- 1 Quantum phenomena need for full explanation a nonlinear theory.
- 2 The nonlocal Fourier analysis, in a more complete theory, need to be replaced by a local wavelet type analysis.
- 3 Free quantum particles, are in general small, and can be localized, at any time, in a finite nondispersive volume.

Let us consider an unidimensional finite source of approximate size σ_s, each point of which emits monochromatic pulses of width σ_p described by a finite wave of the form

$$g(x,t;\varepsilon,k) = A\exp\left[-(x-\varepsilon-vt)^2/2\sigma_p^2\right]\exp[i(kx-\omega t)]. \qquad (2)$$

It is also assumed that each point of the unidimensional source consists of many oscillators emitting pulses at wave numbers k that are Gaussian, distributed with average k_0,

$$f(x) = \alpha\exp\left[-(k-k_0)^2/2s_k^2\right], \qquad (3)$$

and that the spacial distribution of these emitters at each instant is also approximately Gaussian, with a width σ_s of the line source

$$s(\varepsilon) = \beta \exp\left[-\varepsilon^2 / 2\sigma_s^2\right]. \tag{3'}$$

The total pulse from this source will then be

$$\psi(x) = A\alpha\beta \int_{-\infty}^{+\infty}\int_{-\infty}^{+\infty} s(\varepsilon) f(k) g(x,t;\varepsilon,k) d\varepsilon dk, \tag{4}$$

which upon integration on ε gives

$$\psi(x,t) = A\alpha\beta\sqrt{2\pi}\int_{-\infty}^{+\infty} \exp\left[-\frac{(x-vt)^2}{2(\sigma_p^2+\sigma_s^2)} - \frac{(k-k_0)^2}{2\sigma_k^2} + i(kx-\omega t)\right] dk \tag{5}$$

Now assuming that for the chosen range of k the velocity of each pulse is practically constant, one is allowed to write

$$\omega = vk, \quad v \approx c^{te}, \tag{6}$$

and by substitution in (5) it yields

$$\psi(x,t) = \gamma \frac{2\pi\sigma_p\sigma_s\sigma_k}{(\sigma_p^2+\sigma_s^2)^{1/2}} \exp\left[-(x-vt)^2 / 2\sigma_x^2\right] \exp[i(kx-\omega t)], \tag{7}$$

where

$$\sigma_x^2 = \left[\sigma_k^2 + 1/(\sigma_p^2+\sigma_s^2)\right]^{-1}. \tag{8}$$

This expression (8) represents a generalized dispersion relation. Although σ_s has been used to describe the approximate size of the one.dimension source, it could as well be regarded as representing the coherence length of a point-like source with a coherence time τ, such that $\sigma_s = v\tau$.

Some remarks can be made on the new dispersive relation (8), considering the following cases:

(a) sources with a relative "large" coherence length σ_s,
(b) wide-band sources, i.e. sources with "large" σ_k,
(c) truly monochromatic sources, $\sigma_k \approx 0$,

(d) very sharp pulses, σ_p "small."

In cases (a) and (b), sources with large σ_s $(1/(\sigma_p^2 + \sigma_s^s) \approx 0)$ and/or relatively wide-band $(1/(\sigma_p^2 + \sigma_s^2) \ll \sigma_k^2)$, the general dispersion relation (8) becomes the usual one

$$\sigma_x \sigma_k = 1. \tag{9}$$

In cases (c) and (d), since in these two cases $(1/(\sigma_p^2 + \sigma_s^2) \gg \sigma_k^2)$, one gets

$$\sigma_x = \sqrt{\sigma_p^2 + \sigma_s^2}. \tag{10}$$

As can be seen from Fig.1, in most practical instances relation (8) and (9) do give the same results.

The essential difference between the new dispersion relation (8) and the usual (9) lies then in the fact that, with the former, one can have a source with practically no dispersion $(\sigma_k \approx 0)$ giving a finite pulse, whereas with (9) the size of the pulse grows indefinitely.

It was seen that starting from very general assumptions for quantum "particle" represented by finite, nondispersive waves, it was possible, by pure formal calculations, to obtain a more general set of uncertainty relations that avoid the shortcomings of the usual ones. The question of the physical meaning of such relations remains, of course, an open question. In situations of this kind only experiment can decide which is the better formula for the uncertainty relations

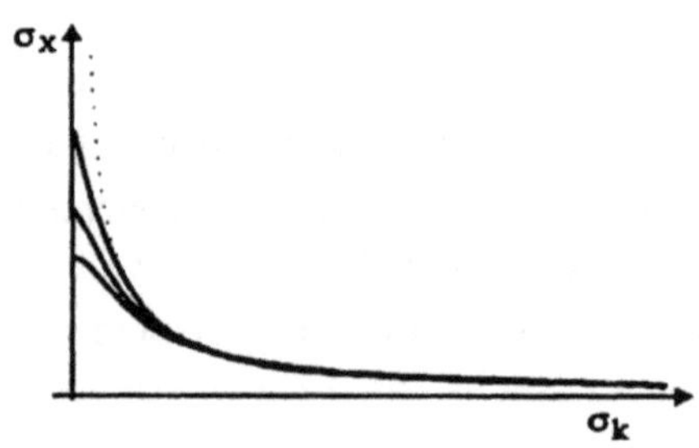

Fig.1. Usual Heisenberg dispersion relations (dotted line) and generalized dispersion relations (full) for different values of $(1/(\sigma_p^2 + \sigma_s^2))$.

3. EXPERIMENTS TO TEST THE GENERAL VALIDITY OF HEISENBERG UNCERTAINTY RELATIONS

There seems to be many ways to investigate the general validity of the usual uncertainty relations, however in the present work I shall discuss only two possible experiments[10,11].

In quantum mechanics, as it is well known, the mean life of an excited state τ_0 is identified with the coherence time of the emitted particle. When taken to its ultimate consequences, together with the uncertainty relations this leads to some strange conclusions. For instance, it is known that the 5+ state of the nucleus $^{126}_{51}Sb$ has a mean life of 1140 s and decays to the 8- state emitting gamma photons of 17.7 kev[12]. Consequently, this leads to a photon wave packet of the same coherence time $\tau_0 = \Delta t_0 \approx 1140s$. From this coherence time it is possible to calculate the approximate coherence length, which gives a longitudinal size for the wave packet very large, in fact it is greater than the diameter of Earth's orbit.

Since the above inference seems to defeat the physical feeling, which assumes the single gamma photon as a very small entity, it would be interesting to submit the conclusion to an experimental validation.

An experiment to test the general validity of the Heisenberg uncertainty relations, based on those ideas, seems possible with the modern methods of X-ray interferometry. Essentially the experiment consists in using an X-ray in which the source is a suitable gamma emitter of reasonable energy and mean life. X-ray sources, even if the same energy of the gamma emitters, are not suitable because the mean life of the excited states are too short. The interferometer, shown in Fig.2, has a shutter placed in one arm, that chops the passing wave packet, with an initial coherence time Δt_0, in minute pieces with a coherent time $\Delta t <<<< \Delta t_0$. Now the question is: What are the predicted results for the interference pattern at the detector?

The answer to the question depends on the assumptions made on the nature of the quantum particle.

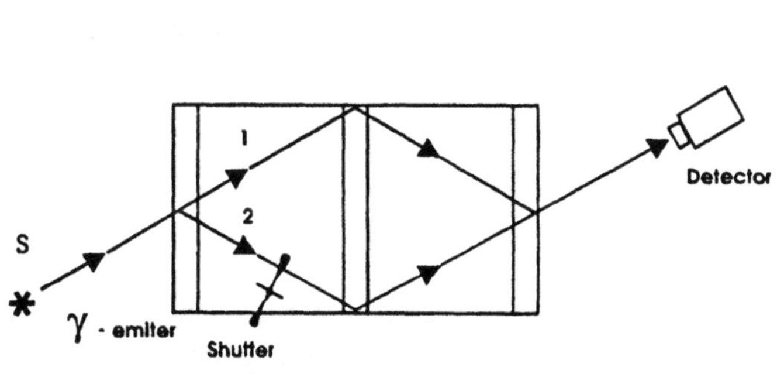

Fig.2. Proposed experiment. X-ray interferometer with a shutter in one arm.

3.1. Quantum Particles are Small Entities

In this case the action of the chopper on the beam two, when the time interval between two cuts is much greater than the "true" coherence time of the particle, is to absorb more or less wavelets. This situation is depicted in Fig.3.

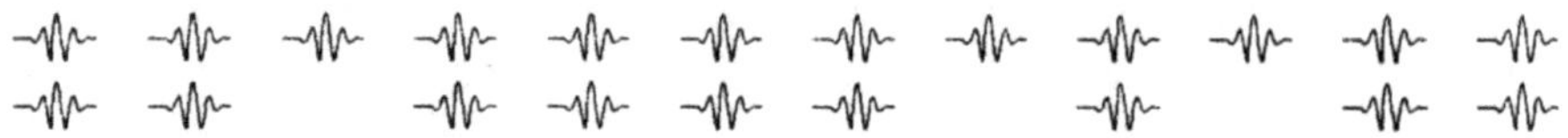

Fig.3. Overlapping of small waves.

It is possible to show[10] that the visibility for this case is given by

$$\gamma_s = \frac{2\alpha}{1+\alpha}, \tag{11}$$

where α is the absorption factor, such that $I_2 = \alpha I_1$, $(0 \leq \alpha \leq 1)$. When the rate of incident particles is small one may have $\alpha \approx 1$ which means a visibility practically one. That is for small particle emission rate the action of the chopper on the arm two of the interferometer does not greatly change the visibility of the interference pattern.

3.2. Heisenberg Uncertainty Relations $\Delta E \, \Delta t \geq h$ Holds True in All Cases

In such circumstances the action of the chopper on the beam two of the interferometer is to change the initial minimum energy dispersion of the packet from ΔE_0 to ΔE so that $\Delta E >>>> \Delta E_0$, since the coherence time of the chopped pieces is much smaller than the initial coherence time ($\Delta t <<<< \Delta t_0$). Therefore at the overlapping region one wave practically monochromatic superimposes to many small wave packets, see Fig. 4, producing a specific interference pattern.

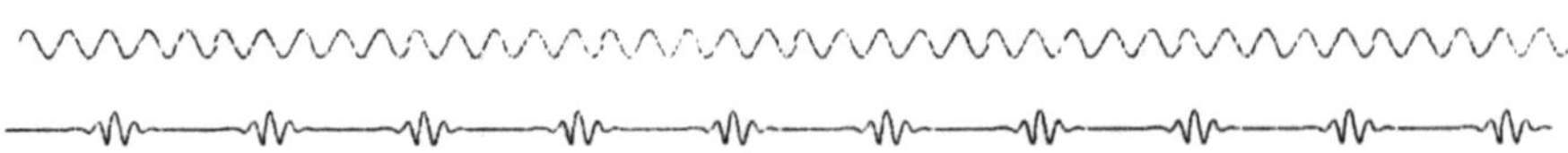

Fig.4. Overlapping of two waves: one practically monochromatic, the other composed of many small wave packets with large energy dispersion.

In this situation two cases may be considered:

(a) The action of the chopper on the beam produces small wave packets with a random relative phase, the most likely case. Then no interference pattern is to be expected at the detector.
(b) It is assumed, with Gähler and Golub[13] in what they call *diffraction in time*, that chopper is made in such a way that it maintain the relative phase among the small wave packets nearly constant. It is possible to show, under reasonable approximations, that, even in this very optimistic situation, the visibility γ_u is smaller than γ_s, obtained under the assumption that the quantum particles are small entities.

The other possible experiment to test the usual uncertainty relations is based on the spreading of the matter wave packets, given by formula (1), $\Delta x \approx [\hbar / m(\Delta x_0)]t$. According to this formula, for instance, for Auger electrons produced in a monolayer of vacuum-deposited gold atoms on a perfect crystal surface, as proposed by Scheer et all.[14], the uncertainty of the initial position Δx_0 can be of the order of the atom layer, say $\Delta x_0 = 10$Å, which by substitution in the spreading formula (1) gives after one second $\Delta x \approx 100 km$. Meaning that, after one second, the initial wave packet of about 10 Å increases to a size of 100 km. This conclusion is very striking. Even so the usual interpretation is perfectly able to deal with it. In this contest it means that after one second it is possible detect an electron somewhere within a distance of about 100 km, with a velocity distribution equal to the minimum initial one $\Delta v \geq \Delta v_0$. Since

$$\Delta x_0 \Delta v_0 \geq \frac{\hbar}{2m}, \tag{12}$$

which, after substituting Δx_0, gives $\Delta v \geq \Delta v_0 \approx v_M - v_m \approx 10^7$ cm s^{-1}.
Other explanation for the spreading of the matter wave packets is possible in context of a non linear theory where each single quantum free particle is described by a finite wave of constant shape.

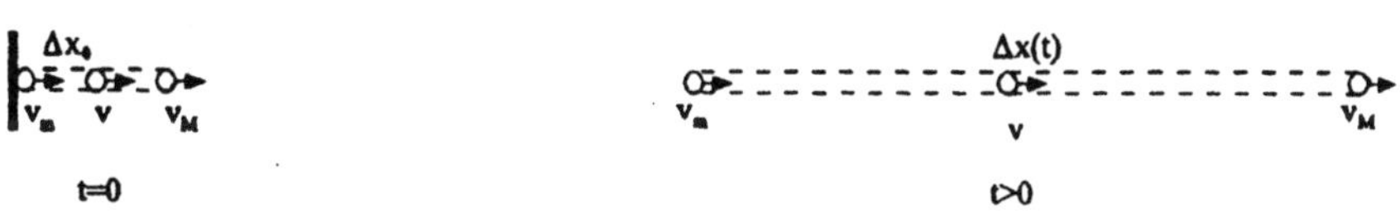

Fig.5. At time t=0, particles with different velocities occupy an initial length Δx_0; as the time increases, they will be spread over a much larger length $\Delta x(t)$.

In this natural model, the spreading of the "wave train" is only a way of saying that an initial burst of many particles, of constant small size, increase the distance between each other to a value that depends only of the elapsed time and of their respective range of velocities Δv_0.

Therefore, in this model, an electron "wave train" of 100 km means that the possible distance between the faster and slower electron after one second is of that magnitude as depicted in Fig.5.

In order to see which is the right model consider the following experiment: An electron source emits, at fixed known time intervals t_0, electrons with an uncertainty in the position $\Delta x_0 \approx 100\,\text{Å}$, which according to the Heisenberg relations correspond to a minimum energy dispersion $\Delta v_0 \approx 10^{-6}\,cm.s^{-1}$. A millisecond later the initial wave packet will have spread to a size of 10 *m*. Suppose now, that from each electron wave packet, and always for the same elapsed time $t=1$ *ms*, one "slices " a piece of, say, $\delta x = 100$ Å. This slice from the wave packet can in principle be made with an electromagnetic field, triggered by the emission of the electron from the source, as seen in Fig.6.

Now, if one measures the velocity dispersion δv of these small slices of length δx from the expanded electron packet, what is to be expected?

According to the Heisenberg uncertainty relations, one must always have, $\delta x\, \delta v \geq \hbar / 2m$; therefore as $\delta x \approx \Delta x_0$, and $\Delta x_0\, \Delta v_0 \cong \hbar / 2m$, this implies that $\delta v \geq \Delta v_0$. That is, the velocity dispersion of the arriving electron, from the small slice, will be equal or greater than the initial minimum dispersion Δv_0. One may conclude that when one "slices" the expanded electron wave train, one interacts with all the Fourier components of the packet, modifying them in such a way, that at the end the predicted velocity distribution must be at least equal to the initial one. If it could be otherwise, it would be contra the Heisenberg relations and therefore against the usual quantum mechanics.

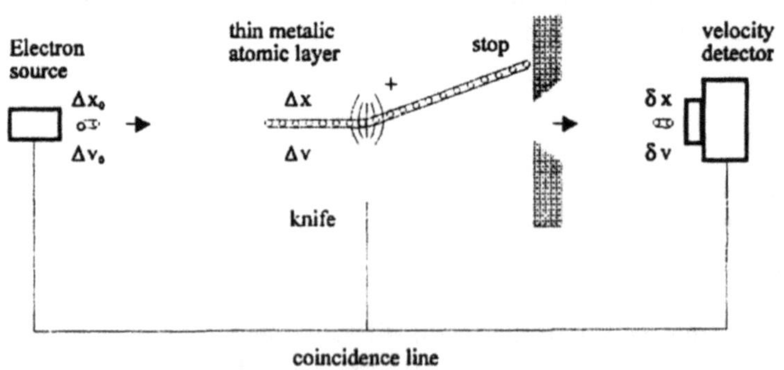

Fig.6. Schematic representation of the proposed experiment.

Assuming that the quantum phenomena need for full explanation a non linear theory and therefore that the quantum particle must be described by a finite local wave, the spreading of the "wave packet" has no real meaning for the individual electrons. It is only a mathematical description for the average separation between particles of different velocities. One is really slicing nothing. What happens is a selection from the expanded "packet" of particles that have a smaller range of velocities. Only those particles that fall in the selected smaller velocity range have the chance to be detected. Therefore if the nonlocal Fourier spectral decomposition is only a mathematical device for the average particle dispersion, a smaller velocity dispersion is to be predicted $\delta v \ll \Delta v_0$, as shown in Fig.7.

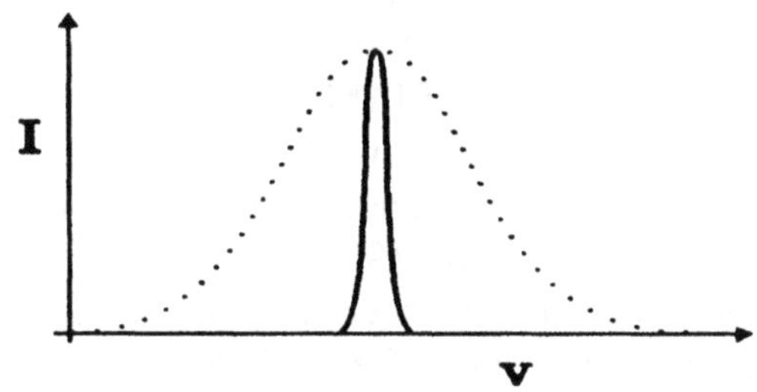

Fig.7. Predicted results for the velocity distribution of the electrons from the slice δx of the expanded wave packet. Usual theory (dotted line) vs nonlinear theory (solid line).

Because $\delta x \approx \Delta x_0$ and $\delta v \ll \Delta v_0$ ($\Delta x_0 \Delta v_0 \cong \hbar / 2m$), one concludes that $\delta x \delta v \ll \hbar / 2m$. the experiment violating the Heisenberg uncertainty relations.

ACKNOWLEDGMENTS

I want to thank my friend Prof. M. Ferrero for is invitation for the International Conference at Oviedo.

BIBLIOGRAPHY

1. N. Bohr, (1928) - *Como Lectures, Collected Works*, Vol. 6 (North-Holland, Amsterdam, 1985).
2. K. Popper, *Quantum Theory and the Schism in Physics* (Hutchinson, London, 1972).

3. J. Hilgevoord and J.B.M. Uffink, in *Michrophysical Reality and Quantum Formalism,* A van der Merwe, F. Selleri, and G. Tarozzi, eds. (Kluwer Academic, Dordrecht, 1988).
4. L. de Broglie, *Les Incertitudes d'Heisenberg et l'Interprétation Probabiliste de la Mécanique Ondulatoire* (Gauthier-Villars, Paris, 1988); English translation: *Heisenberg's Uncertainties and the Probabilistic Interpretation of Wave Mechanics,* A Van der Merwe (Kluwer Academic, Dordrechet, 1990).
5. L. Mackinnon, *Found. Phys.* **8**, 157(1978); *Problems in Quantum Physics,* L. Kostro, A. Posiewnik, J. Pykacz, and M. Zukowski, eds. (World-Scientific, Singapore, 1988).
6. Ph. Gueret and J.P. Vigier, *Lett. Nuovo Cimento* **38,** 125(1983).
7. J.P. Vigier, *Phys. Lett. A,* **135**, 99(1989).
8. A. Grossmann and J. Morlet, SIAM *J. Math. Anal.* **3,** 723(1989); *Decomposition of Function into Wavelets of Constant Shape and Related Transforms* (World-Scientific, Singapore, 1980). C.K Chui, *An Introduction to Wavelets,* (Academic, New York, 1992).
9. J.R. Croca and A. Rica da Silva, "Uncertainty relation for the superposition of finite monochromatic pulses", subm. for publication.
10. J.R. Croca and P. Amorim, "Coherence time of a particle and the mean life of a state, a possible test of the general validity of the Heisenberg uncertainty relations", subm. for publication.
11. J.R. Croca, *Physics Essays* 1, 71(1990).
12. C.M. Lederer and V.S. Slinley, eds., *Table of Isotops* (Wiley, New York, 1978).
13. R. Gäheler and R. Golub, *Z. Phys. B*, **56**, 5(1989); **Physica B, 151**, 135 (1988).
14. J. Scheer and M. Schmidt, *Phys. Lett. A,* **115,** 322(1986).

CONTINUOUSLY DIAGONALIZED DENSITY OPERATOR OF OPEN SYSTEMS

Lajos Diósi

KFKI Research Institute for Particle and Nuclear Physics
H-1525 Budapest 114, POB 49, Hungary

We showed several years ago that the density operator of Markovian open systems can be diagonalized continuously in time. The resulting pure state jump processes correspond to quantum trajectories proposed in recent quantum optics calculations or, at fundamental level, to exact consistent histories.

Key words: open systems, jump process, quantum trajectories, consistent histories.

The quantum state of an open quantum system — actually a subsystem of a closed one — cannot be described by deterministically evolving state vectors ψ. A given pure initial state

$$\rho = P \quad (= \psi\psi^\dagger) \tag{1}$$

turns into *mixed* state immediately after the system has interacted with its environment:

$$\rho \rightarrow \rho' \neq U\rho U^\dagger . \tag{2}$$

Consider the *diagonalization* of the mixed density operator ρ' :

$$\rho' = \sum_n p'_n P'_n . \tag{3}$$

It is well known that the eigenstates P'_n can be interpreted as possible pure states of the system:

$$\rho' = P'_n \ , \tag{4}$$

with the corresponding probability p'_n; $(n = 1, 2, \ldots)$. In such a way a pure state representation can be maintained even after the interaction. If, as usual, interactions occur repeatedly then the *stochastic jump* (4) must be introduced repeatedly after each interaction.

We restrict ourself to open systems of permanent idealized interaction with the environment, resulting in *Markovian* evolution equation for the density operator:

$$d\rho/dt = \mathcal{L}\rho \equiv -i[H, \rho] + \ldots \ , \tag{5}$$

M. Ferrero and A. van der Merwe (eds.), Fundamental Problems in Quantum Physics, 83–87.

where ... stands for terms representing non-unitary evolution. These terms make any given initial pure state

$$\rho(t_0) = P(t_0) \tag{6}$$

mixed during an arbitrary (short) period ϵ. For time $t_1 = t_0 + \epsilon$, the resulting mixed density operator

$$\rho(t_1) = e^{\epsilon\mathcal{L}} P(t_0) \tag{7}$$

may be diagonalized:

$$\rho(t_1) = \sum_n p_n(t_1) P_n(t_1). \tag{8}$$

So, at $t = t_1 + 0$, we can restore the system's pure state as

$$\rho(t_1 + 0) = P_{n_1}(t_1), \tag{9}$$

with probability $p_{n_1}(t_1)$; $(n_1 = 1, 2, 3, \ldots)$. Again, at $t_2 = t_1 + \epsilon$, we get mixed density operator

$$\rho(t_2) = e^{\epsilon\mathcal{L}} P_{n_1}(t_1), \tag{10}$$

and we diagonalize it:

$$\rho(t_2) = \sum_n p_n(t_2) P_n(t_2). \tag{11}$$

At $t = t_2 + 0$, we restore pure states

$$\rho(t_2 + 0) = P_{n_2}(t_2), \tag{12}$$

with probability $p_{n_2}(t_2)$; $(n_2 = 1, 2, 3, \ldots)$. And so on, for $t_3 = t_0 + 3\epsilon, \ldots, t_\nu = t_0 + \nu\epsilon$.

Accordingly, one has constructed a stochastic process for the *conditional* quantum state $\rho(t)$ of the open system. The conditional $\rho(t)$ jumps into a pure state at times $t_0, t_1, \ldots, t_\nu$, while it is getting slightly mixed between the jumps. In average, the process recovers the *unconditional* density operator of the system, satisfying the ensemble evolution equation (5).

If we let ϵ go to zero (at $\nu\epsilon = const.$) then the conditional state $\rho(t)$ will be pure all the time. We proved the existence of this limit in 1985 [1]. The limiting process is a generalized *Poisson* (jump) *process* for the pure state

$$\rho(t) \equiv \psi(t)\psi^\dagger(t). \tag{13}$$

The analytic expressions of the pure state jump process contain two nonlinear operators: the *frictional* Hamiltonian H_ψ and the *transition rate* operator W_ψ; see [2] and [3]. Then, the pure state satisfies the frictional Schrödinger equation

$$\frac{d}{dt}\psi(t) = -iH_{\psi(t)}\psi(t) \tag{14}$$

for most of the time, apart from the discrete orthogonal jumps

$$\psi(t+0) = \psi_n(t) \tag{15}$$

to the nth eigenstate of the current transition rate operator $W_{\psi(t)}$. The transition rate of the jump is equal to the nth eigenvalue $w_n(t)$ of $W_{\psi(t)}$.

The stochastic average of the pure state density operator (13) recovers the unconditional density operator and satisfies the ensemble evolution equation (5).

References 2 and 3 show jump process' equations for the general evolution equation (5). Here we consider the simplest Lindblad [4] structure:

$$d\rho/dt = -i[H,\rho] + F\rho F^\dagger - \frac{1}{2}\{F^\dagger F, \rho\} \tag{16}$$

where F is the only Lindblad generator. Let us relate it to concrete (open) physical systems and enlist typical cases:

—in spontaneous emission, $F = const. \times |0><1|$,
—in damped cavity oscillation, $F = const. \times a$,
—in pumped laser, $F = const. \times a^\dagger a$,
—in Brownian motion, $F = const. \times q + i\ const. \times p$,
—in Stern-Gerlach apparatus, $F = const. \times \sigma_z$.

The coupling constants set the strenghts of the environmental interactions.

As can be be shown [2,3], the frictional Schrödinger equation (14) takes the form

$$\begin{aligned}\frac{d\psi}{dt} &= -iH\psi + \frac{1}{2}\left(<F^\dagger> F - H.C.\right)\psi \\ &\quad - \frac{1}{2}\left(F^\dagger - <F^\dagger>\right)\left(F - <F>\right)\psi + \frac{1}{2}\Delta_F^2\psi,\end{aligned} \tag{17}$$

where $\Delta_F^2 = <F^\dagger F> - <F^\dagger><F>$. The above deterministic evolution of the state vector happens to be interrupted by the orthogonal jumps (15) which, as follows from [2] and [3], take the form

$$\psi \rightarrow \frac{1}{\sqrt{w}}\left(F - <F>\right)\psi. \tag{18}$$

The rate w of the above transition, appearing in the normalization factor, is just equal to the Lindblad generator's quantum spread Δ_F in the current state ψ. (If I had more than one Lindblad generator in the evolution equation (16) then more than one outcome would exist for the jump, each with its partial transition rate w_n; see in [2,3].)

In 1992 Dalibard *et al.* [5] considered the simple quantum optical Bloch equation which corresponds to our evolution equation (16) with the special choice $F = \sqrt{\Gamma}|0><1|$. The authors construct a pure state jump process, similar to ours. Their frictional Schrödinger equation reads

$$\frac{d\psi}{dt} = -iH\psi - \frac{1}{2}F^\dagger F\psi + \frac{1}{2} <F^\dagger F> \psi \tag{19}$$

while their jump is

$$\psi \rightarrow \frac{1}{\sqrt{w}} F\psi. \tag{20}$$

The jump rate w is equal to $< F^\dagger F >$.

No doubt, the jump process (19,20) of Dalibard *et al.* is a bit simpler to implement on a computer as compared to our jump process (17,18). From theoretical point of view, however, the orthogonal jump process which we have obtained by the continuous diagonalization of the density operator seems to be more justified. First of all, as shown in [1], the orthogonal jump process is ***observable***. Let us consider the Hermitian operator

$$\mathcal{O} = \frac{1}{w}(F - < F > \psi\psi^\dagger(F^\dagger - < F^\dagger >) \tag{21}$$

which is actually $1/w$ times the transition rate operator itself. Its eigenvalues are 0 or 1. It is shown in [1] that a ***continous observation*** of $\mathcal{O}$ is possible. No quantum Zeno paradox will enter because the Markovian dynamics represents stonger effects than the continuous observation. This latter leads just to the orthogonal jump process (17,18). For most of time, the observed value of $\mathcal{O}$ (21) is 0 while the state vector obeys to the nonlinear Schrödinger equation (17). For some random instants, however, the observed value may be 1: then the state vector has just performed the orthogonal jump (18).

That the above observability of the jump process is fundamental has got new support recently. There exists an alternative interpretation of quantum mechanics in terms of *consistent histories* [6] instead of von Neumann measurements. In 1993, Paz and Zurek [7] suggested that the successive diagonalizations (6-9) and(10-12) of the density operator led to an *exact* consistent set of quantum histories for a Markovian (open or sub-) system. From that, it is straightforward to see the continuous diagonalization, too, leads to exact consistent histories. Actually, the orthogonal jump process (14,15), defined uniquely for any Markovian (sub-) system, generates exact consistent histories [3]. These histories may be *the* classical content of the given quantum dynamics.

A possible lesson is that early results achieved in the frames of the standard von Neumann measurement theory turn to be crucial if transformed into the language of new interpretations. The old (von Neumann) and the new (consistent history) languages both tell us the same interpretational problems and thus wait for common solutions [8].

I am deeply indebted to the organizers of the Symposium for inviting me to participate and to give this talk. This work was supported by the Hungarian Scientific Research Fund under Grant OTKA 1822/1991.

REFERENCES

1. L. Diósi, *Phys. Lett.* **112A**, 288 (1985).
2. L. Diósi, *Phys. Lett.* **114A**, 451 (1986).

3. L. Diósi, *Phys. Lett.* **185A**, 5 (1994).
4. G. Lindblad, *Commun. Math. Phys.* **48**, 119 (1976).
5. J. Dalibard, Y. Castin, and K. Mølmer, *Phys. Rev. Lett.* **68**, 580 (1992).
6. R. B. Griffiths, *J. Stat. Phys.* **36**, 219 (1984).
7. J. P. Paz and W. Zurek, *Phys. Rev.* **D48**, 2728 (1993).
8. L. Diósi, *Phys. Lett.* **280B**, 71 (1992).

THE HAZY SPACETIME OF THE KÁROLYHÁZY MODEL OF QUANTUM MECHANICS

A. Frenkel

KFKI Research Institute of Particle and Nuclear Physics
Hungarian Academy of Sciences
Budapest, Hungary

A small amount of uncertainty, determined solely by constants of nature, is brought into the strucutre of classical, empty spacetime. It induces an uncertainty in the relative phases of the wave function of any isolated system. The Schrödinger evolution would then lead to the development of incoherent components (of components with completely uncertain relative phases) of the wave function. In order to prevent excessive deterioration of the coherence, random localizations of the wave function are introduced. The model gives realistic results, not only for the time evolution of microparticles and of solid bodies, but also for the decay of superpositions of traks in a cloud chamber. The formula for the spread of the relative phases is derived without the introduction of the family of auxiliary metrics used previously and without expansions in momentum space. This derivation shows clearly that the model does not contain adjustable cut-off parameters. The connection between the previous and the new derivation is given in the appendix.

Key words: uncertainty in the spacetime structure, unified dynamics for microsystems and macrosystems.

1. SPACETIME AS A POSSIBLE CARRIER OF INDETERMINACY

In the Károlyházy model of quantum mechanics a unified law of time evolution is proposed for microscopic and macroscopic systems. According to this law, the deterministic Schrödinger evolution of the quantum state of any isolated system, small or large, is repeatedly interrupted by random,

M. Ferrero and A. van der Merwe (eds.), Fundamental Problems in Quantum Physics, 89–102.

instantaneous localizations of the state to regions of appropriate sizes of the configuration space [1, 2].

The random localizations incorporate an indeterministic element into the law of time evolution of any isolated system. However, in the K model the indeterministic element has been introduced primarily not into the law of the time evolution of quantum states (like, for instance, in the model of Ghirardi, Rimini, and Weber [3]), but into the structure of classical, empty (that is matter-free) spacetime.

The idea that the concisely determined structure of classical spacetime may be an idealization pushed too far, arose from a study by Károlyházy of the quantum mechanical uncertainties of the spacetime structure. He has shown that the quantum mechanical uncertainties in the positions and momenta of material bodies induce a minimal unavoidable uncertainty in the length of time intervals (see Eq.(1) below), and he has proposed to loosen the structure of classical spacetime to the same extent. Quantum states will then propagate on a slightly hazy spacetime. Károlyházy hoped — and it turned out to be so — that the amount of uncertainty incorporated into the structure of classical spacetime is the appropriate amount needed to destroy the coherence of the quantum states of macroscopic bodies, whereas the coherence of the states of microsystems will be practically unaffected.

The aforementioned minimal quantum mechanical uncertainty $(\Delta_Q T)_{\min}$ of a time interval of length T is related to T by the formula

$$(\Delta_Q T)_{\min} \approx \left(\frac{\Lambda}{c}\right)^{2/3} T^{1/3}, \tag{1}$$

where

$$\Lambda := \sqrt{\frac{G\hbar}{c^3}} \approx 10^{-33}\ \mathrm{cm} \tag{2}$$

is the Planck length. Here and below the symbol "$\approx$" means: "equal, up to numerical constants of the order of $10^{-1} \div 10$, irrelevant in this paper."

The appearance of the constant of gravitation G in the expression for $(\Delta_Q T)_{\min}$ is due to the fact that, when evaluating $\Delta_Q T$, not only the quantum uncertainty Δx in the position of a body, but also the uncertainty $\Delta K = v\Delta p$ in the kinetic energy has to be taken into account. Indeed, ΔK contributes to the uncertainty of the structure of spacetime, and consequently to the uncertainty of the length of T. G appears in the expression of this contribution. Notice that since $\Delta x \cdot \Delta p \geq \hbar/2$, Δx and ΔK work against each other. That is why a minimum of $\Delta_Q T$ exists.

One can arrive at the expression (1) for $(\Delta_Q T)_{\min}$ in several ways ([1, 2, 4, 5]; see also [6] in these Proceedings). I shall not reproduce here any of the deductions of Eq.(1), I shall just make three remarks concerning this expression.

(i) Eq.(1) has been deduced for time intervals along world lines of slowly moving (or standing) bodies. The reference frame in which the bodies should move slowly is the one in which the 2.7° K background radiation is isotropic. Since the K model is a model of nonrelativistic quantum mechanics, this is an adequate limitation.

(ii) Eq.(1) is valid for time intervals T such that

$$T \gg T_{\text{Planck}} = \frac{\Lambda}{c} \approx 10^{-43}\,\text{sec.} \tag{3}$$

This is because, when

$$T \lesssim T_{\text{Planck}}, \tag{4}$$

the very concept of classical spacetime becomes questionable, and in that case one should not attribute physical meaning to Eq.(1). Notice that, from (1), it follows that, when (3) holds, then

$$T \gg (\Delta_Q T)_{\text{min}}. \tag{5}$$

(iii) One can see from the deduction of Eq.(1) that in order to reach the minimal uncertainty $(\Delta_Q T)_{\text{min}}$, one should allow for bodies of unrealistically high density. For realistic bodies $\Delta_Q T$ is many orders of magnitude larger than $(\Delta_Q T)_{\text{min}}$. However, in the K model only the existence of the theoretical lower bound $(\Delta_Q T)_{\text{min}}$ is exploited, there is no need to reach it in experiments. (Notice that, e.g., for $T = 1$ sec, $(\Delta_Q T)_{\text{min}}$ is only 10^{-29} sec.)

Let us consider now the problem of the introduction of an indeterministic element into the time evolution of quantum states.

First of all, it should be recalled that the existence of the quantum mechanical spreads $\Delta x, \Delta p, \Delta_Q T$, etc., of various quantities in itself does not lead to a breakdown of the deterministic Schrödinger evolution of the quantum state ψ of a closed system. In the orthodox measurement theory the departure from determinism arises due to the reduction postulate. In the GRW model, the indeterministic element is introduced in the form of the spontaneous random localizations of ψ. As mentioned already, in the K model the indeterminacy is introduced primarily not into the time evolution of ψ, but into the structure of classical spacetime. The idea of doing so is suggested by the existence of the lower bound $(\Delta_Q T)_{\text{min}}$ given in Eq.(1).

According to this equation, for a time interval of given length T, $(\Delta_Q T)_{\text{min}}$ depends solely on constants of nature. The parameters of particular bodies playing a role in the deduction of Eq.(1) dropped out from the final formula. This suggests that (1) expresses a general property of time intervals. Namely, it says that quantum mechanics and classical general relativity, taken together, produce in the length T of any time interval a spread at least as large as $(\Delta_Q T)_{\text{min}}$. Perhaps this state of affairs is of no consequences and shows only that it is vain (but permissible) to fix the structure of classical spacetime with perfect precision. However, Károlyházy has proposed to take the message of Eq.(1) more seriously and to admit that the precision of the structure of classical spacetime should not supersede the precision indicated by Eq.(1). More precisely, he proposed to associate with any time interval of length T of the classical, empty (matter-free) spacetime an inherent classical uncertainty ΔT, equal to the minimal quantal spread $(\Delta_Q T)_{\text{min}}$ of T. Therefore,

$$\Delta T \approx \left(\frac{\Lambda}{c}\right)^{2/3} T^{1/3}. \tag{6}$$

Of course, similarly to (1), Eq.(6) is applicable only to timelike worldlines with $v \ll c$, and the condition $T \gg T_{\text{Planck}}$ should also hold.

2. A MODEL OF THE HAZY SPACETIME

In order to be able to study the influence of the haziness of spacetime on the quantum states of various isolated systems, Károlyházy constructed a mathematical model of a hazy spacetime possessing the uncertainty (6). In the construction he made use of a family of appropriately chosen auxiliary metrics [2. 4]. Recently he proposed an other construction in which there is no need in those metrics [6]. Below I shall rely on the new proposal.

Let a world event, which in Minkowskian spacetime would occur at the point $(\mathbf{x}, t)$, occur at an uncertain moment

$$t_\beta(\mathbf{x}, t) := t + \tau_\beta(\mathbf{x}, t). \tag{7}$$

The τ_β's constitute a random set of $(\mathbf{x}, t)$ functions. β labels the different functions of the set. (In a relativistic generalization of the K model, presumably the spatial coordinates of the world events also should be made uncertain.)

Consider now a worldline segment $[\mathbf{x}, t; \mathbf{x}' = \mathbf{x}, t' > t]$, that is a segment of a zero velocity worldline. The time interval

$$T = t' - t \tag{8}$$

goes over into

$$T_\beta = t'_\beta - t_\beta = T + \tau_\beta(\mathbf{x}, t') - \tau_\beta(\mathbf{x}, t). \tag{9}$$

Imposing the condition

$$\langle \tau_\beta(\mathbf{x}, t) \rangle = 0, \tag{10}$$

where the symbol $\langle \cdots \rangle$ stands for an average over the set labelled by β, we find

$$\langle T_\beta \rangle = T, \tag{11}$$

that is, on the average, time intervals have their Minkowskian value. We have then

$$\Delta T_\beta := T_\beta - \langle T_\beta \rangle, \tag{12}$$

$$\left\langle (\Delta T_\beta)^2 \right\rangle = \left\langle (T_\beta - \langle T_\beta \rangle)^2 \right\rangle = \left\langle (\tau_\beta(\mathbf{x}, t') - \tau_\beta(\mathbf{x}, t))^2 \right\rangle . \tag{13}$$

Identifying $\sqrt{(\Delta T_\beta)^2}$ with ΔT in (6), we obtain the condition

$$\left\langle (\tau_\beta(\mathbf{x}, t') - \tau_\beta(\mathbf{x}, t))^2 \right\rangle \approx \left(\frac{\Lambda}{c}\right)^{4/3} |t' - t|^{2/3}. \tag{14}$$

The absolute value on the r.h.s. of (14) comes from the fact that the l.h.s. is symmetric under $t \leftrightarrow t'$.

In Eq.(14) the function

$$\left\langle (\tau_\beta(\mathbf{x}', t') - \tau_\beta(\mathbf{x}, t))^2 \right\rangle \tag{15}$$

is given along $\mathbf{x}' = \mathbf{x} =$ const worldlines. However, as we shall see presently, the influence of the uncertainty of the spacetime structure on the quantum states is expressed through function (15) taken on spacelike surfaces $t = t' =$ const. The evaluation of (15) on such surfaces can be carried out if one takes into account that (15) should be isotropic in space and homogeneous in space and time. In other words, (15) can depend only on the differences

$$r := |\mathbf{x}' - \mathbf{x}|, \tag{16}$$

$$\theta := |t' - t|. \tag{17}$$

In particular, on spacelike surfaces $t' = t =$ const, the function (15) can depend only on r. The expression which preserves the power dependence on the spacetime variables in (14) and involves the same constants of nature is

$$\left\langle (\tau_\beta(\mathbf{x}', t) - \tau_\beta(\mathbf{x}, t))^2 \right\rangle \approx \frac{\Lambda^{4/3}}{c^2} |\mathbf{x}' - \mathbf{x}|^{2/3}. \tag{18}$$

3. UNCERTAIN PHASE IN THE QUANTUM STATE OF AN ISOLATED SYSTEM

Let us consider an isolated system of N quantum particles with masses $M_1, M_2, \ldots, M_N$. In non-relativistic quantum mechanics the rest energy phase factor

$$\exp\left(-\frac{i}{\hbar}\sum_{i=1}^{N} M_i c^2 t\right) := \exp(i\phi(t)) \tag{19}$$

of the wave function $\psi(\mathbf{x}, t)$ of the system is usually omitted. Being independent of

$$x := [\mathbf{x}_1, \ldots, \mathbf{x}_N] \tag{20}$$

and of

$$p := [-i\nabla_1, \ldots, -i\nabla_N], \tag{21}$$

this factor would drop out from all observables anyway. However, with

$$t \to t + \tau_\beta(\mathbf{x}, t), \tag{22}$$

we have

$$M_i c^2 t \to M_i c^2 (t + \tau_\beta(\mathbf{x}_i, t)), \tag{23}$$

because the coordinate of the i-th particle is $\mathbf{x}_i$ in the wave function. As a result

$$\exp(i\phi(t)) \to \exp(i\phi(t)) \cdot \exp(i\phi_\beta(x, t)), \tag{24}$$

where

$$\phi_\beta(x, t) := -\frac{c^2}{\hbar}\sum_{i=1}^{N} M_i \tau_\beta(\mathbf{x}_i, t). \tag{25}$$

The rest energy phase factor can be omitted again, but the x-dependent uncertain phase factor $\exp(i\phi_\beta(x, t))$ should be kept. Therefore, in the K model a pure quantum state of an isolated system of N quantum particles is not a usual Schrödinger wave function $\psi(x, t)$, but $\phi(x, t)$ multiplied by the uncertain, x-dependent phase factor $\exp(i\phi_\beta(x, t))$.

It should be stressed that the state is not one of the functions

$$\psi_\beta(x, t) := \exp(i\phi_\beta(x, t))\psi(x, t), \tag{26}$$

with some probability p_β. This would mean that the state is a usual mixture. The uncertainty would be then not in the state, but in our knowledge of the state. In the phase of a pure K model state the uncertainty is really there, it is "objective". This can be expressed by saying that a pure K state of an isolated system is the collection

$$\{\psi_\beta(x,t)\} = \{\exp(i\phi_\beta(x,t))\psi(x,t)\}, \tag{27}$$

and not the mixture of the functions ψ_β. Only the collection (and not its individual members) contains the full information about the state. In particular, it carries the information about the amount of uncertainty of the relative phase of $\{\psi\}$ between pairs of points of the configuration space. As a matter of fact, the main difference between a pure Schrödinger state ψ and the pure $\{\psi_\beta\}$ state corresponding to it is that while a Schrödinger state is always perfectly coherent (in the sense that its relative phase between any pairs of points x, x' is uniquely determined), a $\{\psi_\beta\}$ state is never perfectly coherent because, as we shall see, the amount of uncertainty of its relative phase — in other words the spread of its relative phase — is never zero.

Before evaluating the spread of the relative phase of a K state, two remarks are in order:

(i) The uncertain phase (25) is symmetric under the permutations of the N particles, so that a K state has the same symmetry as the corresponding Schrödinger state.

(ii) The substitution $t \to t + \tau_\beta$ should have been carried out not only in the rest energy phase factor, but in the Schrödinger wave function

$$\psi(x,t) = \exp\left(-\frac{i}{\hbar}\hat{H}t\right)\psi(x,0), \tag{28}$$

too. However, the matrix elements of the Hamiltonian $\hat{H}$ in non-relativistic cases are much smaller than the rest energies of the particles. In [2] the correction has been investigated for solid bodies. In this paper we shall ignore it.

4. THE SPREAD OF THE UNCERTAIN RELATIVE PHASE

The relative phase of a member ψ_β of a K state (27) between the points x, x' is

$$\delta_\beta(x,x',t) := \phi_\beta(x',t) + \arg\psi(x',t) - \phi_\beta(x,t) - \arg\psi(x,t). \tag{29}$$

From (10), (25), and (29), one finds

$$\langle \delta_\beta(x, x', t) \rangle = \arg \psi(x', t) - \arg \psi(x, t), \tag{30}$$

$$\begin{aligned} \Delta\delta_\beta(x, x', t) &:= \delta_\beta(x, x', t) - \langle \delta_\beta(x, x', t) \rangle = \\ &= \phi_\beta(x', t) - \phi_\beta(x, t). \end{aligned} \tag{31}$$

From (31) and (25) the mean square spread of the relative phase is

$$\begin{aligned} &\left\langle (\Delta\delta_\beta(x, x', t))^2 \right\rangle = \left\langle (\phi_\beta(x', t) - \phi_\beta(x, t))^2 \right\rangle = \\ &= \frac{c^4}{\hbar^2} \sum_{i,l=1}^{N} M_i M_l \left\langle (\tau_\beta(\mathbf{x}_i', t) - \tau_\beta(\mathbf{x}_i, t))(\tau_\beta(\mathbf{x}_l' t) - \tau_\beta(\mathbf{x}_l, t)) \right\rangle. \end{aligned} \tag{32}$$

For any fixed β, i, and l, one easily finds

$$\begin{aligned} &(\tau_\beta(\mathbf{x}_i', t) - \tau_\beta(\mathbf{x}_i, t))(\tau_\beta(\mathbf{x}_l', t) - \tau_\beta(\mathbf{x}_l, t)) = \\ &= \frac{1}{2}\big[(\tau_\beta(\mathbf{x}_i', t) - \tau_\beta(\mathbf{x}_l, t))^2 + (\tau_\beta(\mathbf{x}_i, t) - \tau_\beta(\mathbf{x}_l', t))^2 - \\ &-(\tau_\beta(\mathbf{x}_i', t) - \tau_\beta(\mathbf{x}_l', t))^2 - (\tau_\beta(\mathbf{x}_i, t) - \tau_\beta(\mathbf{x}_l, t))^2\big]. \end{aligned} \tag{33}$$

Substituting (33) into (32) and taking into account (18), we obtain a time independent expression for the spread of the relative phase:

$$\Delta_\phi^2(x, x') \approx \Lambda^{4/3} \frac{c^2}{\hbar^2} \sum_{i,l=1}^{N} M_i M_l (2|\mathbf{x}_i' - \mathbf{x}_l|^{2/3} - |\mathbf{x}_i' - \mathbf{x}_l'|^{2/3} - |\mathbf{x}_i - \mathbf{x}_l|^{2/3}), \tag{34}$$

where $\Delta_\phi^2(x, x') := \langle (\Delta\delta_\beta(x, x', t)^2 \rangle$, and a factor 1/2 has been included into the symbol $\approx$.

Equation (34) can be written also in the form

$$\Delta_\phi^2(x, x') \approx \Lambda^{4/3} \frac{c^2}{\hbar^2} \int \frac{dk}{k^{11/3}} |\mu_\mathbf{k}(x') - \mu_\mathbf{k}(x)|^2, \tag{35}$$

where

$$\mu_\mathbf{k}(x) := \sum_{l=1}^{N} M_l e^{i\mathbf{k}\mathbf{x}_l} \tag{36}$$

is the Fourier transform of the mass distribution of the N particles with masses $M_1, \ldots, M_N$ in the configuration $x = [\mathbf{x}_1, \ldots, \mathbf{x}_N]$. In [2] Δ_ϕ has been given in the form (35).

5. COHERENCE PROPERTIES OF THE K STATES

As mentioned already, a K state is never perfectly coherent. Indeed, the normalized wave function in $\{\psi_\beta\}$ always occupies some finite domain of the configuration space, and between any pairs of points of this domain $\Delta_\phi > 0$, as one sees from (35). From (34) one sees also that for a given physical system Δ_ϕ increases with the separations $|\mathbf{x}'_i - \mathbf{x}_l|$. As a result, the weight of the pairs of points with large Δ_ϕ's increases with the domain occupied by ψ. In other words, the coherence of $\{\psi_\beta\}$ deteriorates if ψ expands in the configuration space.

In order to characterize the degree of coherence of a given state $\{\psi_\beta\}$ of a given physical system, let us consider two typical situations:

(i) If in $\{\psi_\beta\}$ the wave function ψ occupies only a domain Ω such that

$$\Delta_\phi(x, x') \ll \pi, \qquad \text{for all} \quad x, x' \in \Omega, \tag{37}$$

then, in good approximation, in Ω

$$\phi_\beta(x', t) = \phi_\beta(x, t) \tag{38}$$

for all β's; the ϕ_β's are practically independent of x in Ω. This means that $\{\psi_\beta\}$ is practically equivalent to the ray $\{\exp(i\phi_\beta(t)) \cdot \psi(x, t)\}$. In short, the K state is practically equivalent to the usual pure state ψ, the coherence of the K state is nearly perfect.

(ii) If ψ occupies a domain containing subdomains Ω, Ω' of considerable weights such that

$$\Delta_\phi(x, x') > \pi, \qquad \text{for all } x \in \Omega \quad \text{and for all} \quad x' \in \Omega', \tag{39}$$

then the relative phases between these subdomains are completely uncertain, the coherence between the components of $\{\psi_\beta\}$ belonging to these subdomains is lost. Now $\{\psi_\beta\}$ is far from being equivalent to a single ray. Notice however that within subdomains inside which $\Delta_\phi < \pi$ a certain degree of coherence persists, and inside sufficiently small subdomains the coherence is nearly perfect.

A (maximal) domain of the configuration space inside which $\Delta_\phi < \pi$ will be called a "coherence cell". Its extent strongly varies from system to system. Let us consider two typical systems: a microparticle and a macroscopic solid body.

For a single microparticle the configuration space is the space of the position coordinate $\mathbf{x}$ of the particle. One finds easily from (34) that the linear size a_c of a coherence cell is supra-astronomical. For example, for an electron $a_c \approx 10^{33}$ cm. In a domain occupied by any realistic wave function of a microparticle $\Delta_\phi \lll \pi$. Therefore the K state of an isolated microparticle is always practically equivalent to a usual pure quantum state. The same result holds for isolated microsystems composed of not too many microparticles, bound or not bound to each other.

In the case of an isolated macroscopic solid body, after a short calculation based on Eq.(35) one finds [2] (see also [4]) that the value of Δ_ϕ is determined by the extent of the wave function ψ in the center of mass coordinate subspace of the configuration space. The relative coordinates of the microscopic constituents of the body do not contribute appreciably to Δ_ϕ. The linear size of a coherence cell in the c.m. subspace is microscopic. For example, for a body of 1 g (and of usual terrestial density $\varrho \approx 1$ g/cm^3), $a_c \approx 10^{-16}$ cm [2, 4, 5]. A K state of the isolated body in which ψ extends with considerable weight over a domain with linear size $2a_c$ in the c.m. subspace, already contains incoherent components of considerable weights.

THE LAW OF TIME EVOLUTION OF THE K STATE

The Schrödinger equation is beyond doubt applicable to isolated microsystems. Accordingly, the unified law of time evolution for microsystems and macrosystems should include as an essential element the deterministic Schrödinger evolution of the wave function $\psi(x,t)$ in the K state (27) of any isolated system. The question is how to add an indeterministic element which leads to an appropriate departure from the Schrödinger evolution.

The Schrödinger wave function ψ of an isolated system expands in configuration space, at least in the 3 degrees of freedom of the center of mass. Consequently, the coherence of the $\{\psi_\beta\}$ state deteriorates during the Schrödinger evolution. The indeterministic element is then brought into the law of time evolution through the assumption that the loss of the coherence is prevented by random localizations of ψ onto such domains of the configuration space occupied by ψ, inside which the coherence still persists [1, 2].

In [2] (see also [4, 5]) the law has been given in terms of the following algorithm: In the K state (27) of any isolated system $\psi(x,t)$ obeys the Schrödinger equation as long as $\{\psi_\beta\}$ does not develop incoherent components of considerable weights in non-overlapping coherence cells. When such components appear, a random localization of ψ (and with it of $\{\psi_\beta\}$, too,) to one of these cells takes place. After the localization ψ expands again until a second localization occurs, and so on. The repeated localization-expansion cycles prevent the expansion of $\{\psi_\beta\}$ to domains much larger than a coherence cell.

Since there is no free parameter in the law of time evolution of the K model, the behavior of the various physical systems is, in principle, fully specified by this law. Let us illustrate on three generic cases that the K model leads to realistic results.

(i) As we have seen above, in the case of an isolated microparticle a coherence cell is so large that Δ_ϕ remains always much smaller than π. Therefore no random localization of ψ occurs and the Schrödinger equation holds without restriction: an isolated microparticle behaves quantum mechanically.

(ii) On the other hand, the wave function of an isolated macroscopic solid body can expand to a domain larger than a coherence cell. If at some moment of time $\psi(x,t)$ occupies in the center of mass coordinate subspace a single coherence cell of linear dimension a_c, then after a time

$$\tau_c \approx \frac{M a_c^2}{\hbar} \tag{40}$$

it will occupy a domain of linear size $2a_c$. Then $\{\psi_\beta\}$ will already have components of considerable weights in distinct coherence cells, and a localization to one of them takes place. From there ψ expands again, and so on. For a body of 1 g $a_c \approx 10^{-16}$ cm, so that $\tau_c \approx 10^{-5}$ sec. The localization-expansion cycles, at a rate $\tau_c^{-1} \approx 10^5$ localizations/sec, keep the quantum mechanical spread of the position of the center of mass of the body within the range of 10^{-16} cm. The localization of the center of mass of a macroscopic body is thus practically pointlike. In this respect the body behaves classically.

For solid bodies the mathematical description of the random localizations [2, 4] is similar to that of the GRW model for finite cycle periods [3]. The transition between microscopic and macroscopic behavior, the influence of the surroundings on the localization-expansion cycles of solid bodies and the anomalous Brownian motion of solid bodies in the framework of the K model have been discussed in [2, 4, 5, 7]. A detailed comparison of the GRW model with the K model is given in [7].

(iii) Let us now turn to the application of the K model to gases. As an example, we shall consider a cloud chamber filled with vapor. The walls of the chamber are taken into account only as boundary conditions. We shall look at the fate of a K state in which $\psi(x,t)$ is a superposition with two components of comparable weights. In each component there is a water droplet surrounded by vapor molecules. The components differ in the location of the droplet.

Imagine first a superposition with the droplet in the two branches, but without the surrounding vapor. Then, if the droplet is massive enough, the two components of the superposition will belong to different coherence cells and a random localization will choose one of the components. If the droplet is not massive enough, no localization occurs, the state remains a superposition.

Now, as shown in [2], the thermal motion of the vapor molecules leads to a decay of the superposition even if the droplet is not massive enough to trigger its own localization-expansion cycles. Namely, due to the difference in the location of the droplet in the two branches of the superposition, in the configuration space of the vapor molecules such subdomains develop between which $\Delta_\phi > \pi$, but inside which $\Delta_\phi < \pi$. The ensuing random localization to one of these subdomains does not select at once one of the branches of our superposition. Instead, it modifies randomly the coefficients c_i of the superposition by a small, calculable amount. The repetition of this procedure amounts to the well known win or lose game between the coefficients. The i-th coefficient wins the game with probability $|c_i|^2$, in agreement with the prediction of orthodox quantum mechanics. The smaller the droplet, the longer the game lasts. A rough estimate shows [2] that with a droplet composed of 10^6 molecules in each component of our superposition, the game — that is the decay of the superposition — takes place in 10^{-7} seconds if initially $|c_1| = |c_2|$, and if the separation between the positions of the droplet in the two branches is macroscopic (e.g. 1 cm) (see also [6, 8]). The win or lose game as a mechanism for the decay of superpositions has been also investigated by Pearle [9] and Gisin [10].

Let us conclude the discussion of the time evolution of the K state with the following remark. We have seen that only the time independent spread $\Delta_\phi(x, x')$ of the relative phases $\phi_\beta(x, t)$ plays a role in this time evolution, the time dependence of the ϕ_β's themselves can be left unspecified. On the other hand, in the spacetime model constructed with the help of a family of auxiliary metrics [2, 4, 5], the time dependence of the ϕ_β's has been fixed. The comparison of that model of spacetime with the one presented here is carried out in the appendix.

6. COMMENT ON THE PREFERRED BASIS PROBLEM

There is no a priori chosen preferred basis in the K model. One can speak only of more or less suitable variables for the characterization of the domains $\Delta_\phi < \pi$ of the configuration space of a given physical system. In case of a microparticle this is the position variable of the particle, because then Δ_ϕ has a simple dependence on $|\mathbf{x}' - \mathbf{x}|$ (see (34)). In the case of a solid body the suitable variable is the c.m. coordinate of the body. In the case of water droplets surrounded by vapor molecules in a cloud chamber, the suitable variables are the c.m. coordinates of the droplets and the Fourier transforms (36) of the mass distribution of the droplets and of the vapor molecules, because Δ_ϕ can be conveniently expressed in terms of these variables [2].

7. THE CONTINUOUS LOCALIZATION LIMIT IN THE K MODEL

As we have seen, the cycle period τ_c is the time during which a K state expands from a single coherence cell to several non-overlapping cells. This time is of course always finite. In order to arrive at a continuous localization version of the K model, one has to assume that the random localization intervenes as soon as the coherence of the K state begins to deteriorate in the course of the Schrödinger evolution. On this basis a law of time evolution with continuous localization, in principle applicable to any isolated system, has been proposed recently by Károlyházy. The application of this law to systems in which Δ_ϕ depends on the extent of the wave function only in a single degree of freedom has been presented in [11]. The continuous localization treatment of the decay of a superposition in a cloud chamber is given in [6]. Various aspects of the continuous localization formulation of the K model are under investigation.

ACKNOWLEDGEMENTS

The author is indebted to F. Károlyházy for numerous illuminating discussions and to L. Diósi for critical remarks, in particular for pointing out a formal mistake in the manuscript, corrected in the present version.

APPENDIX

In Eq. (3.4.13) of Ref. [2] an expression with explicit (x,t) dependence has been given for the random phases $\phi_\beta(x,t)$. Taking into account the simple relation (25) between ϕ_β and τ_β in the case $N = 1$, after a short calculation one finds the expression of the function (15) in the model of the hazy space-time proposed in [2]:

$$\left\langle (\tau_\beta(\mathbf{x}',t') - \tau_\beta(\mathbf{x},t))^2 \right\rangle \approx \frac{\Lambda^{4/3}}{c^2} f(r,\theta), \tag{41}$$

where

$$f(r,\theta) := \frac{1}{r}\left[|r + c\theta|^{5/3}\mathrm{sign}(r + c\theta) + |r - c\theta|^{5/3}\mathrm{sign}(r - c\theta) \right], \tag{42}$$

with

$$r = |\mathbf{x}' - \mathbf{x}|, \qquad \theta = |t' - t|. \tag{43}$$

In particular,

$$f(r,\theta) = \begin{cases} \frac{10}{3}(c\theta)^{2/3} & \text{if } r = 0, \quad (44a) \\ 2r^{2/3} & \text{if } \theta = 0, \quad (44b) \\ 2(2r)^{2/3} & \text{if } c\theta = r. \quad (44c) \end{cases}$$

Equations (44a) and (44b) show that (41) is compatible with our equations (14) and (18). The irrelevant numerical coefficients 10/3 and 2 in (44a,b) come from the particular choice of the family of auxiliary metrics in [2]. The correlation function of a stochastic gravitational potential $\phi(\mathbf{x}, t)$ which corresponds to the time derivative of our $\tau_\beta(\mathbf{x}, t)$ has been given in [12].

(44c) gives the correlation function along the path of a light signal. In the model of spacetime presented in this paper one arrives at this expression if one assumes that the uncertainty in the travel time of a light signal from a point $\mathbf{x}$ to $\mathbf{x}'$ and back to $\mathbf{x}$ has to be of the same order of magnitude as the uncertainty of the time elapsed at $\mathbf{x}$ between the moments of departure and arrival of the signal. In other words, one assumes that the uncertainty of a time interval T along a world line with $v = 0$ cannot be reduced by sending light signals from and to the points of this world line.

REFERENCES

1. F. Károlyházy, *Nuovo Cimento* **42**, 390 (1966).
2. F. Károlyházy, "Gravitation and Quantum Mechanics of Macroscopic Bodies", *Magyar Fizikai Folyóirat* **22**, 23 (1974) (Thesis, in Hungarian).
3. G. C. Ghirardi, A. Rimini, and T. Weber, *Phys. Rev. D* **34**, 470 (1986).
4. F. Károlyházy, A. Frenkel, and B. Lukács, in *Physics as Natural Philosophy*, A. Shimony and H. Feschbach editors (MIT Press, Cambridge, 1982), p. 204.
5. F. Károlyházy, A. Frenkel, and B. Lukács, in *Quantum Concepts of Space and Time*, R. Penrose and C. J. Isham, editors (Clarendon Press, Oxford, 1986), p. 106.
6. F. Károlyházy, "State vector reduction via spacetime imprecision", in these Proceedings.
7. A. Frenkel, *Found. Phys.* **20**, 159 (1990).
8. A. Frenkel and F. Károlyházy, in *Bell's Theorem and the Foundations of Modern Physics*, A. van der Merwe, F. Selleri, and G. Tarozzi editors (World Scientific, Singapore, 1992), p. 208.
9. Ph. Pearle, *Phys. Rev. D* **13**, 857 (1976); *Int. J. of Theor. Phys.* **18**, 489 (1979); *Found. Phys.* **12**, 249 (1982).
10. N. Gisin, *Phys. Rev. Lett.* **52**, 1657 (1984); *Helv. Phys. Acta* **62**, 363 (1989).

CAN THE EXPERIMENTS BASED ON PARAMETRIC-DOWN CONVERSION DISPROVE EINSTEIN LOCALITY?

Augusto Garuccio

Dipartimento di Fisica, Università di Bari
Via Amendola 173, I-70126 Bari, Italy

The experiments on Einstein locality based on parametric-down-conversion source are discussed. The theoretical analysis shows that this technique fails to discriminate between quantum mechanics and local realism. A deeper analysis of the standard approach shows that the Clauser-Horne factorizability of joint detection probability is not so general as to describe this type of source. A more adequate description of these photon pairs leads to define a new Bell-type inequality which is always satisfied by quantum mechanical predictions.

Key words: Einstein locality, parametric down-conversion, factorizability, photon correlation.

1. INTRODUCTION

In last years a new technique for producing pairs of correlated photons has been proposed as a suitable source for testing Einstein locality via Bell-type inequalities.

In this technique a laser beam produces a pair of down-converted photons in a non-linear crystal of potassium dihydrogen phosphate. When the condition for phase matching is satisfied, the two down-converted photons emerge with the same linear polarization along different paths. One of the two photons passes through a 90° polarization rotator, while the other traverses a compensating glass plate. At this stage the two photons are first reflected from two mirrors and then , from opposite sides, they impinge onto a beam splitter (Fig. 1). If the beams emerging from the beam splitter pass through linear dichotomic polarizers, set at angles $\theta 1$ and $\theta 2$, respectively, the correlation function between the results of the measurement of polarization is

$$E(\theta_1;\theta_2) = P(\theta_{1+},\theta_{2+})-P(\theta_{1+},\theta_{2-})-P(\theta_{1-},\theta_{2+})+P(\theta_{1-},\theta_{2-}), \quad (1)$$

M. Ferrero and A. van der Merwe (eds.), Fundamental Problems in Quantum Physics, 103–112.

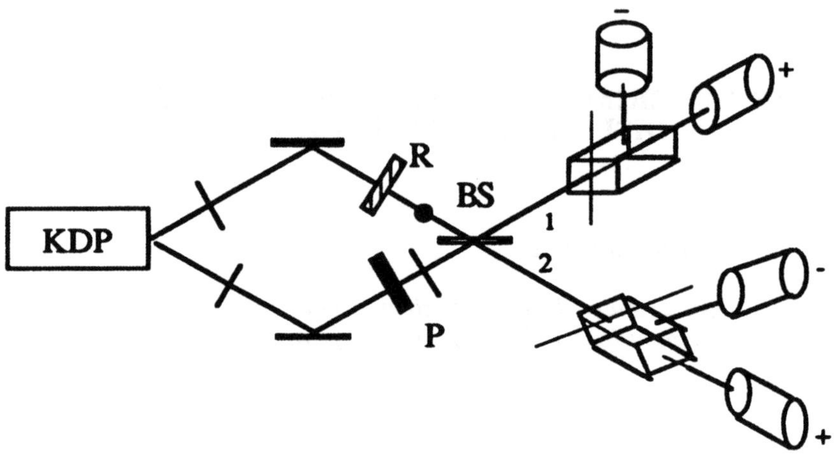

Fig1 Set-up of an experiment of correlation in polarization with parametric down conversion source

where $P(\theta_{1i}, \theta_{2j})$ are the probabilities that the photon that travels along the channel 1 is transmitted through polarizer 1 along i-channel (i=+ for the ordinary channel, i = - for the extraordinary one) and the photon which travels along channel 2 is transmitted through polarizer 2 along the j-channel.

In 1964 Bell[1] proved that for a realistic and local theory the expectation value (1) must obey the inequality

$$|E(\theta_1;\theta_2) - E(\theta_1;\theta'_2)| + |E(\theta'_1;\theta_2) + E(\theta'_1;\theta'_2)| \leq 2, \qquad (2)$$

where $\theta_1,\theta'_1;\theta_2,\theta'_2$ are four different directions of the polarizers.

In the experiments performed only one of four joint probabilities $P(\theta_{1i},\theta_{2j})$ (i,j = +,-) have been measured, consequently a different inequality, containing only experimentally measurable quantities , has been used.

Two different approaches have been developed to deduce this new inequality. The firs due to Clauser, Horne, Shimony and Holt[2] (CHSH) in 1969 and the second to Clauser and Horne[3] (CH) in 1974. Both the approaches start from Einstein locality and, with some supplementary assumptions, arrive at the inequality.

$$-1 \leq B(\theta_1,\theta'_1;\theta_2,\theta'_2) \leq 0, \qquad (3)$$

with $B(\theta_1,\theta'_1;\theta_2,\theta'_2) = P(\theta_1;\theta_2)-P(\theta_1;\theta'_2)+P(\theta'_1;\theta_2)+P(\theta'_1;\theta'_2)-P(\theta'_1;\infty)-P(\infty;\theta_2)$, where $P(\theta_1;\theta_2)$ is the joint detection probability of a photon pair, when the beams emerging from the beam splitter pass through linear polarizers, set at angles θ_1 and θ_2, respectively, $P(\theta'_1;\infty)$ and $P(\infty;\theta_2)$ are the corresponding probabilities when either one of the linear polarizer is removed.

We will prove in the first part of this talk that the CHSH approach leads to an inequality different from (2) if it is correctly applied to parametric down-conversion source[4]. In the second part, we will show that the CH approach leads in this case to the previous inequality, but it is not applicable *in principle* to this class of experiment[5].

2. CHSH APPROACH

For every choice of polarizer orientations θ_1 and θ_2 CHSH. have assumed a dichotomic choice between *transmission and absorption* in the polarizers. Thus, for instance, $P(\theta_{1+},\theta_{2-})$ represents the probability that the photon which travels along the channel 1 is transmitted through polarizer 1 and the photon which travels along channel 2 is absorbed by polarizer 2. The correlation function is every expressed by formula (1) and the following relations hold:

$$P(\theta_{1+}, \theta_{2+})+P(\theta_{1+}, \theta_{2-})+P(\theta_{1-}, \theta_{2+})+P(\theta_{1-}, \theta_{2-}) = 1, \quad (4a)$$

$$P(\theta_{1+}, \theta_{2+})+P(\theta_{1+}, \theta_{2-}) = P(\theta_{1+}, \infty_+), \quad (4b)$$

(4)

$$P(\theta_{1+}, \theta_{2+})+P(\theta_{1-}, \theta_{2+}) = P(\infty_+, \theta_{2+}), \quad (4c)$$

$$P(\infty_+, \infty_+) = 1, \quad (4d)$$

where $P(\theta_{1+}, \infty_+)$, $P(\infty_+, \theta_{2+})$ and $P(\infty_+, \infty_+)$ are the corresponding probabilities with one, the other or both linear polarizers removed, respectively. Using Eqs. (4), Eq. (1) becomes:

$$E(\theta_1;\theta_2) = 4P(\theta_{1+}, \theta_{2+})-2P(\theta_{1+}, \infty_+)-2P(\infty_+, \theta_{2+})+1. \quad (5)$$

In the above equation only cases of double transmission, which can be experimentally detected, appear. This allows us to transform Bell's inequality into inequality (3).

In quantum mechanics the state of the emerging pair is given by:

$$|\psi\rangle = \sqrt{T_x T_y}\,|x_1\rangle|y_2\rangle + \sqrt{R_x R_y}\,|y_1\rangle|x_2\rangle$$

$$- i\sqrt{R_y T_x}\,|x_1\rangle|y_1\rangle + i\sqrt{R_x T_y}\,|x_2\rangle|y_2\rangle, \quad (6)$$

where: R_x, R_y and T_x, T_y are the beam-splitter reflectivities and transmissivities, respectively, with $R_x+T_x=R_y+T_y=1$; $|x_i\rangle$ $[|y_i\rangle]$ is the polarization state along the x-direction [y-direction] for the photon in the i-th beam-splitter output channel.

But, from Eq. (1) it follows that some photon pairs can travel along the same channel, reaching only one of the two polarizers. Thus, this quantum state does not satisfy all Eqs.(4). In fact, Eqs. (4-a) and (4-d) must be replaced by

$$P(\theta_{1+}, \theta_{2+})+P(\theta_{1+}, \theta_{2-})+P(\theta_{1-}, \theta_{2+})+P(\theta_{1-}, \theta_{2-}) = 1/2\,, \quad (7a)$$

$$P(\infty_+, \infty_+) = 1/2\,, \quad (7b)$$

which causes Eqs. (5) to be replaced by:

$$E(\theta_1;\theta_2) = 4P(\theta_{1+}, \theta_{2+})-2P(\theta_{1+}, \infty_+)-2P(\infty_+, \theta_{2+})+1/2 \quad (8)$$

and, consequently inequality (3) to be written as

$$-3/4 \leq B(\theta_1,\theta'_1;\theta_2,\theta'_2) \leq 1/4. \quad (9)$$

The inequality given by Eq. (9) cannot be violated by the quantum-mechanic joint transmission probabilities for the correlated photon pairs described by Eq. (6). Indeed, for ideal polarizers and detector these probabilities are given by:

$$P(\theta_1;\theta_2)=[\sqrt{T_xT_y}\cos\theta_1\sin\theta_2+\sqrt{R_xR_y}\sin\theta_1\cos\theta_2]^2$$
$$\cong 1/4\sin^2(\theta_1+\theta_2), \quad (10a)$$

$$P(\theta_1;\infty) = T_xT_y\sin^2\theta_1+R_xR_y\cos^2\theta_1 \cong 1/4, \quad (10b)$$

if we assume, as done in the performed experiments, $R_x/T_x=R_y/T_y \cong 1$. In fact, for example, the maximum value of observable $B(\theta_1,\theta'_1;\theta_2,\theta'_2)$ according to the quantum-mechanical predictions expressed by Eqs. (10) is

$$B_{QM} = \frac{1}{4}(\sqrt{2}-1) < \frac{1}{4},$$

and, therefore, the parametric down-conversion photon source cannot give quantum states with which to test quantum mechanics versus local realism not even in the case of ideal behavior of polarizers and detectors.

Many authors proposed and used[6-9] this parametric down-conversion source to test Einstein locality and in his papers they claim that the experimental results violate of several standard deviation the upper limit zero of Eq. (3) and therefore they draw the conclusion that these experiment prove the violation of the Einstein locality. We have now proved that this conclusion is not correct and that in this case the locality is expressed by inequality (9). This result is in complete agreement with theoretical results on the subject. In fact, state (6) can be written in the factorized form

$$|\psi> = (\sqrt{T_x}\,|x_1> + i\sqrt{R_x}\,|x_2>)\,(\sqrt{T_y}\,|y_2> - i\sqrt{R_y}\,|y_1>, \qquad (11)$$

and it has been proved that factorable states always satisfy Einstein locality or, equivalently, Bell-type inequalities [10]. This theoretical consideration seems to be decisive in proving that the discussed result is reliable.

The problem is then to analyse the approach of the authors of the proposals of these experiment in order to identify which assumptions fails in representing this source.

In most of the experiments carried out to test the locality versus quantum mechanics, both with atomic sources[11-16] and parametric down-conversion sources, the so-called Clauser-Horne inequality has been used. This inequality has been deduced by starting from the assumption of locality and separability and adding a suitable supplementary hypothesis. This hypothesis, usually called the no-enhancement hypothesis, plays a fundamental role in deducing an inequality containing experimentally measurable quantities. The violation of this inequality in the performed experiments has gave rise to a long and intense debate on the validity of Clauser-Horne and similar hypotheses[17].

The mathematical formulation of causality and separability has also been subject to a long, although less intense, debate. As well known, the basic assumptions for the Clauser and Horne (CH) deduction of a Bell's type inequality are the following:

(a) There exists a (hidden) variable λ which represents the physical state of a pair of correlated quantum objects (α,β);
(b) $p(a,\lambda)$ is the probability that the object α in state λ crosses the analyser with parameter 'a' and is subsequently detected;
(c) $q(b,\lambda)$ is the similar probability for particle β ;
(d) $\omega(a,b,\lambda)$ is the probability that both α and β cross the analysers and are both detected;
(e) locality is expressed by
(e.1) the factorizability of joint detection probability (CH factorizability)

$$\omega(a,b,\lambda) = p(a,\lambda)\, q(b,\lambda) ;$$

(e.2) both probability density $\rho(\lambda)$ and set Λ of the values of λ are independent of the orientation of the two analysers a and b.

In many papers it has been pointed out that the last two definitions do not include all the possible local situations, consequently it has been

suggested and that a more general definition where assumption (e.1) remains valid, but (e.2) must be substituted with a different condition, can be considered[18].
We will discuss assumption (e.1) and will prove that does not hold in the case of parametric down-conversion.

Let us now consider the Clauser-Horne approach to the problem to determine an experimentally testable Bell's-type inequality.

If $p(\theta_1,\lambda)$ and $p(\infty,\lambda)$ are the photon detection probabilities along path 1 when the polarizer is either set at angle θ_1, or removed, and $q(\theta_2,\lambda)$ and $q(\infty,\lambda)$ are similar for the photon in channel 2, then we can assume that

$$0 \leq p(\theta_1,\lambda), p(\theta'_1,\lambda) \leq p(\infty,\lambda) \tag{12a}$$

$$0 \leq q(\theta_2,\lambda), q(\theta'_2,\lambda) \leq q(\infty,\lambda) \tag{12b}$$

Using the well known algebraic inequality

$$-XY \leq xy - xy' + x'y + x'y' - Yx' - Xy \leq 0$$

in which x, y, x', y', X, Y are any real numbers satisfying the conditions $0 \leq x, x' \leq X$ and $0 \leq y, y' \leq Y$, we can write

$$-p(\infty,\lambda)q(\infty,\lambda) \leq p(\theta_1,\lambda)q(\theta_2,\lambda) - p(\theta_1,\lambda)q(\theta'_2,\lambda) +.p(\theta'_1,\lambda)q(\theta_2,\lambda)$$
$$+p(\theta'_1,\lambda)q(\theta'_2,\lambda) - p(\theta'_1,\lambda)q(\infty,\lambda) - p(\infty,\lambda)q(\theta_2,\lambda) \leq 0.$$

If, by using the CH factorizability (e.1) we first identify the products $p(i,\lambda)q(j,\lambda)$ as joint detection probabilities and then integrate over the set Λ of the values of the hidden variables, we obtain

$$-P(\infty;\infty) \leq B(\theta_1,\theta'_1;\theta_2,\theta'_2) \leq 0. \tag{13}$$

The upper limit of Eq. (13) is different from that of Eq.(9) deduced in the case of parametric down-conversion and the comparison of the experimental results with the upper limit of (13) leads to incorrect conclusion about the validity of Einstein locality, as we discussed above.

Let us note that in deducing Eq. (13) we used Eqs. (12) which expresses the so called no-enhancement hypothesis, but the use of this assumption is not the cause of the discrepancy between the two results. Even if we do not assume (7) and if we put value one as the upper bound of the single detection probabilities, the right term of (8) does not change, and this is violated in the actual experiments.

It is the assumption of the factorizability of the detection probability (e.1) that does not hold in this case.

This can be proved by considering a pair of photons with hidden variable λ. Let us trying to write the simplest expression for the joint detection probability. A photon joint detection occurs when:

-the photon with polarization x is transmitted in channel 1, it passes through polarizer θ_1 and then is detected in photomultiplier 1; contemporaneously the photon with polarization y is transmitted in channel 2, it passes through polarizer θ_2, and is detected in photomultiplier 2;
-the photon with polarization x is reflected in channel 2 , it passes through polarizer $\theta 2$, and it is detected in photomultiplier 2; viceversa, the photon y is reflected in channel 1, passes through polarizer θ_1 and is detected in photomultiplier 1.

In other words, the mathematical expression of the joint detection probability is

$$\begin{aligned} w(\theta_1,\theta_2,\lambda) &= P(1|x,\lambda)P(\theta_1|x,\lambda)P(2|y,\lambda)P(\theta_2|y,\lambda) \\ &\quad + P(2|x,\lambda)P(\theta_2|x,\lambda)P(1|y,\lambda)P(\theta_1|y,\lambda) \end{aligned} \tag{14}$$

where:

- $P(i|a,\lambda)$ is the conditional probability to have the photon in channel i when it impinges on the beam-splitter with polarization a and hidden variable λ;
- $P(\theta i|a,\lambda)$ is the conditional probability for a photon in channel i of polarization a and hidden variable λ to be transmitted through polarizer θ_i and detected.

Equation (14) cannot be written in a factorized form and this proves that the CH assumption of the factorizability of the detection probabilities is not so general as to include the analysed case.

On the other hand the probability of single detection $p(\theta,\lambda)$, introduced in the CH assumptions , integrated over all λ, gives , by definition, the single detection probability. However, in the case of the parametric-source experiments two physically different situations contribute to the single detection. The first is concerned with the experimental situation with the photons of the pair travelling along both channels (50% of pairs); the second with the photons travelling along the same channel (25% of pairs). It is obvious that only the first case contributes to the coincidence rate, while the second contributes only to the single rate, but not in a negligible way.

The condition of the Clauser-Horne factorizability is then not satisfied in the case of parametric down-conversion and the inequality deduced applying the CH approach does not express the Einstein locality.

3. SOURCES OF PHOTON COUPLES BASED ON TYPE II PARAMETRIC DOWN-CONVERSION.

In order to avoid the previous criticisms, we are studying, in collaboration with the group of Prof Mandel in Rochester, a new type of source of correlated photon pairs. Let us describe briefly this new source.

The previous experiment have been performed using a particular process of parametric down-conversion called "type I". In this process a photon beam with energy ω and wave number $\mathbf{k}$ impinge into a non linear crystal. The impinging photon are polarized along extraordinary direction (i.e. in the plane containing the direction of propagation of impinging beams and the optical axis of the crystal). Inside the crystal the pump photon is split into two photons, called signal and idler, which satisfy the phase matching condition $\omega=\omega_s + \omega_i$ and $\mathbf{k} = \mathbf{k}_s + \mathbf{k}_i$. The emerging photons are polarized along extraordinary direction (i.e. in the plane containing the direction of propagation of impinging beams and the optical axis of the crystal). In general, $\mathbf{k}_s$ and $\mathbf{k}_i$ make an angle θ_s and θ_i with the direction of impinging beam and are distributed along two coaxial cones and are polarized along ordinary direction. The condition $\mathbf{k} = \mathbf{k}_s + \mathbf{k}_i$ implies that the two photons are strictly correlated in momentum, i.e. if we detect a signal photon along the direction defined by $\mathbf{k}_s$, we are sure that the idler photon travels, along $\mathbf{k}_i = \mathbf{k} - \mathbf{k}_s$, in the plane containing the direction of the impinging beam and of the signal photon , at angle θ_i with photon axis and by opposite side with respect $\mathbf{k}_s$.

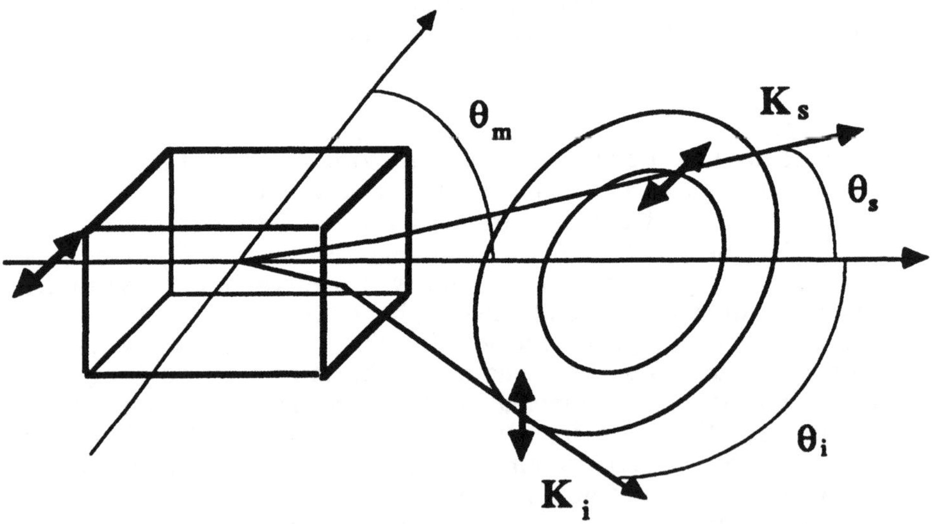

Fig.2 Type-II parametric down conversion. The two photons are orthogonally polarized. If $\theta_i = \theta_s$, the photon pair is decribed by an entanled state.

With a suitable choice of the angle between impinging beam and the optical axis, it can be possible obtain the condition $\omega_s = \omega_i$ in this case $\theta_s = \theta_i$ and the two cones overlap.

In type II phase-matching the pump beam has ordinary polarization and the output photon have orthogonal polarization. More precisely the signal photons travel along the cone with angle θ_s and ordinary polarization, while the idlers travel along the cone with angular aperture θ_i and extraordinary polarization.

If the phase-matching angle is choosen so that the two cones overlap, the state of the pairs traveling in the same plane (for example, the plane a of figure 2) is the superposition of signal photon in direction $\mathbf{k}_1$ and idler in direction $\mathbf{k}_2$ *plus* idler photon in direction $\mathbf{k}_1$ and signal photon in direction $\mathbf{k}_2$, or equivalently

$$|\psi\rangle = \frac{1}{\sqrt{2}}\{|s\rangle_1|i\rangle_2 + |i\rangle_1|s\rangle_2\}.$$

Unfortunately, in type II parametric down conversion, if we impose the overlapping of the cones, we cannot verify the condition $\theta_s = \theta_i$. If it will be possible to find a commercial crystal in which the energy difference $\omega_s - \omega_i$ is so small to lie inside the bandwidth of the filters usually used in these experiments, the two photon will be indistinguishable and the source will produce pairs totally correlated in momentum.

REFERENCES

[1] J.S. Bell, *Physics* **1**, 195 (1965).
[2] J. F. Clauser, M.A. Horne, A. Shimony, and R.A. Holt, *Phys. Rev. Lett.* **23**, 880 (1969).
[3] J. F. Clauser and M.A. Horne, *Phys. Rev.* **D23**, 526 (1974).
[4] L. De Caro and A. Garuccio, *Phys. Rev. A*, in press.
[5] A. Garuccio, "On the validity of Clauser and Horne factorizability", preprint 1993.
[6] Z. Y. Ou and L. Mandel, *Phys. Rev. Lett.* **61**, 50 (1988).
[7] Y. H. Shish and C. O. Alley, *Phy. Rev. Lett.* **61**, 2921 (1988).
[8] Z. Y. Ou, C. K. Hong, and L. Mandel, *Optics Commun.* **67**, 159 (1988).
[9] S. M. Tan and D. F. Walls, *Optics Commun.* **71**, 235 (1989).
[10] V. Capasso, D. Fortunato, and F. Selleri, *Int. J. Theor. Phys.* **7**, 319 (1973).
[11] S.J. Freedman and J.F.Clauser, *Phys. Rev. Lett.* **28**, 938 (1972).
[12] J.F.Clauser, *Phys. Rev. Lett.* **36**, 1223 (1976).
[13] E.S. Fry and R.C. Thompson, *Phys. Rev. Lett.* **37**, 465 (1972).

[14] A. Aspect, P.Grangier and G. Roger, *Phys. Rev. Lett.* **47**, 460 (1981);49, 91 (1982).
[15] A. Aspect, J. Dalibard, and G. Roger, *Phys. Rev. Lett.* **49**, 1804 (1982).
[16] A.J. Duncan, W. Perrie, H.J. Beyer and H. Kleinpoppen, in *Fundamental Processes in aAomic Collision Physics* (Plenum, New York, 1985), p. 555..
[17] *Quantum Mechanics versus Local Ralism*, F. Selleri ,ed. (Plenum, New York, 1991).
[18] A. Garuccio, V.L. Lepore, and F. Selleri, *Found. Phys.* **20**, 173 (1990).

QUANTUM-MECHANICAL HISTORIES AND THE UNCERTAINTY PRINCIPLE

J. J. Halliwell[1]

Theory Group, Blackett Laboratory
Imperial College
London SW7 2BZ, United Kingdom

This paper is a brief summary of earlier papers on the question of how the uncertainty principle arises in formulations of quantum mechanics whose central goal is the assignment of probabilities to histories. Two possible anwers to this question are offered. Firstly, and most generally, the uncertainty principle arises as a lower bound on the Shannon information of the probabilities for histories. I show that the lower bound on the information has the universal form $\ln(V_H/V_S)$, where V_H is a "volume element of history space" and V_S is the volume of that space probed by the string of projection operators characterizing the quantum-mechanical histories. A second expression of the uncertainty principle is specific to histories characterized by sequences of position samplings. These histories are used to contruct a distribution on the value of the field equations, $F = m\ddot{x} + V'(x)$. I show that there is a peak about $F = 0$, and that the uncertainty principle arises as a lower bound on the width of the peak ΔF. For the case of Gaussian samplings continuous in time, it has the form $\sigma \Delta F \geq \hbar$, where σ is the width of the position samplings. I also consider the modification of these results due to the presence of thermal fluctuations in open quantum systems.

Key words: information theory, uncertainty principle, decoherence.

1. INTRODUCTION

This paper is concerned with exploring the properties of the formula for the probability of a quantum-mechanical history:

$$p(\alpha_1, \alpha_2, \cdots, \alpha_n) = \mathrm{Tr}\left(P^n_{\alpha_n}(t_n) \cdots P^1_{\alpha_1}(t_1) \rho P^1_{\alpha_1}(t_1) \cdots P^n_{\alpha_n}(t_n)\right) \qquad (1.1)$$

Here, ρ is the initial density matrix of the system, and the histories of the system are characterized by strings of projection operators $P^1_{\alpha_1}, \cdots, P^n_{\alpha_n}$ at

M. Ferrero and A. van der Merwe (eds.), Fundamental Problems in Quantum Physics, 113–121.

times $t_1, \cdots, t_n$. The projections effect a partition of the possible alternatives α_k of the system at time t_k. They are positive hermitian operators that are both exclusive and exhaustive:

$$P_\alpha P_\beta = \delta_{\alpha\beta}\, P_\alpha, \qquad \sum_\alpha P_\alpha = 1 \tag{1.2}$$

Evolution between each projection is described by the unitary evolution operator $e^{-\frac{i}{\hbar}Ht}$, and the time-dependent projections $P^k_{\alpha_k}(t)$ appearing in (1.1) are given in terms of the time-independent ones $P^k_{\alpha_k}$ by

$$P^k_{\alpha_k}(t_k) = e^{\frac{i}{\hbar}H(t_k - t_0)} P^k_{\alpha_k} e^{-\frac{i}{\hbar}H(t_k - t_0)} \tag{1.3}$$

The formula (1.1) arises in two different contexts. Firstly, it is central to the decoherent histories approach to the quantum mechanics of closed systems. [1,2,3,4]. Secondly, it is a concise summary of the Copenhagen approach to the quantum mechanics of measured subsystems, incorporating both the unitary evolution of states together with the "collapse of the wave function" incurred as a result of measurement by an external agency, modeled by the projection operators [5].

Irrespective of which interpretational scheme one is concerned with, it is an interesting problem to understand how the various familiar features of quantum theory are expressed in the language of histories. The purpose of this paper is to discuss the particular question of how the uncertainty principle arises in formulations of quantum mechanics based on Eq.(1.1). I shall show that in a formulation of quantum mechanics based on histories, the uncertainty principle arises as a limitation on the degree to which the probability (1.1) may be peaked about a particular history. Two different forms of the uncertainty principle for histories will be obtained. One is a lower bound on the Shannon information of the probability distribution (1.1), and is described in Sec. 2. The second, considered in Sec. 3, is specific to histories characterized by sequences of position samplings. Sequences of position samplings will be used to construct a probability distribution on the value of the field equations, $F = m\ddot{x} + V'(x)$, and it will be shown that it is peaked about $F = 0$. I will then show that the uncertainty principle arises as a lower bound on the width ΔF of the peak. Finally, I will discuss the modifications of these versions of the uncertainty principle due to thermal fluctuations in open systems. This paper is largely a summary of Refs.[6,7,8].

2. INFORMATION-THEORETIC UNCERTAINTY RELATIONS

Given a probability distribution $p(\alpha)$, where α may be either a discrete or continuous variable, its Shannon information is defined to be

$$I = -\sum_{\alpha} p(\alpha)\ln p(\alpha). \tag{2.1}$$

The Shannon information (2.1) is most properly thought of as a measure of the uncertainty expressed in the probability distribution $p(\alpha)$. It is large for very spread out distributions, whilst for very concentrated distributions it tends to zero (in the discrete case) or minus infinity (in the continuous case). For continuous distributions $p(x)$, it is connected with the variance Δx (the more usual measure of uncertainty), by the inequality,

$$I \leq \ln\left(2\pi e(\Delta x)^2\right)^{\frac{1}{2}}, \tag{2.2}$$

with equality if and only if $p(x)$ is a Gaussian. These and other results on information theory are described in, for example, Refs. 9 and 10. As stated, one expects the uncertainty principle for histories to arise as a limitation on the degree of peaking about a given history. The uncertainty principle should therefore arise as a lower bound on the Shannon information of the probability for a set of histories. In what follows, I will derive explicit forms of such lower bounds in a variety of situations.

2A. Phase Space Samplings

As our first example of an information-theoretic uncertainty relation, consider a history characterized by a phase space sampling at a single moment of time:

$$p(k,x) = \mathrm{Tr}\left(P_z\rho\right). \tag{2.3}$$

Here, $P_z = |k,x\rangle\langle k,x|$ is a coherent state projector (actually only an approximate projection operator), and the coherent states $|k,x\rangle$ are defined by $|k,x\rangle = U(k,x)|0\rangle$ where $U(k,x)$ is the usual unitary Weyl operator and $|0 \rightarrow$ is a Gaussian. The coherent state projectors probe a region of phase space of fixed size of order $2\pi\hbar$.

As a measure of the degree of spread, or uncertainty, in the distribution (2.3) I use the Shannon information

$$I(K,X) = -\int \frac{dkdx}{2\pi\hbar}\, p(k,x)\ln p(k,x), \tag{2.4}$$

where the factor of $2\pi\hbar$ is included for convenience. This quantity is sometimes called Wehrl entropy [11]. The question is now whether (2.4) has a lower bound, over all states ρ. It turns out that it does. As shown by Lieb, the information (2.4) satisfies the inequality,

$$I(K,X) \geq 1, \tag{2.5}$$

with equality if and only if ρ is a coherent state [12]. The proof of this result is non-trivial and uses a number of inequalities from Fourier analysis.

A simple but important generalization of this result was noted by Grabowksi [13]. This is that the inequality (2.6) continues to hold for projections constructed from a class of *generalized* coherent states, of the form $|\psi_{kx}\rangle = U(k,x)|\psi\rangle$, where $|\psi\rangle$ is an arbitrary state. This generalized result, together with the result (2.2), may be shown to imply the usual uncertainty relation. The reverse implication may not be made, however, and thus these information-theoretic expressions of the uncertainty principle are more general than the usual one.

Numerous papers containing similar results exist. See Ref. 6 for an extensive list of references.

2B. Position and Momentum Samplings

I now consider the expansion of the above results to more general types of samplings and more general histories. See Ref. 6 for further details on the remainder of this section.

Approximate position samplings may be effected using the projection operator

$$P_\alpha^x = \int dx\ \Upsilon(x - \bar{x}_\alpha)\ |x\rangle\langle x|, \tag{2.6}$$

where $\Upsilon(x - \bar{x}_\alpha)$ is a sampling function. Υ may be taken to be a window function – a function which is equal to 1 in an interval of width σ_x centred around $\bar{x}_\alpha$, and zero outside that interval. Here, $\bar{x}_\alpha = \alpha\sigma_x$ is the sampled position and α is an integer. It is also often of interest to take Υ to be a Gaussian of width σ_x, in which case α is a continuous label, and the projections are only approximately exclusive. In a similar manner, one may construct projections for samplings of momentum,

$$P_\beta^k = \int dk\ \Gamma(k - \bar{k}_\beta)\ |k\rangle\langle k|, \tag{2.7}$$

for some sampling function $\Gamma(k - \bar{k}_\beta)$, of width σ_k, where $\bar{k}_\beta = \beta\sigma_k$.

Using these projection operators, more general samplings of phase space may be carried out. There are two ways of doing this. The first is to use a position sampling immediately followed by a momentum sampling:

$$p(\alpha,\beta) = \mathrm{Tr}\left(P_\beta^k P_\alpha^x \rho P_\alpha^x\right). \tag{2.8}$$

The second is to use two position samplings and perform a time-of-flight estimation of momentum,

$$p(\alpha,\beta) = \mathrm{Tr}\left(P_{\alpha'}^x e^{-\frac{i}{\hbar}Ht} P_\alpha^x \rho P_\alpha^x e^{\frac{i}{\hbar}Ht}\right), \tag{2.9}$$

where t is taken to be small. Here, $\beta = \alpha' - \alpha$, and the effective width of the momentum sampling is $\sigma_k = m\sigma_x/t$. Both methods in fact yield approximately the same result. The advantage of this type of phase space sampling, over Eq.(2.3), is that phase space regions of arbitrary size may be sampled. The Shannon information of each of these distribution is

$$I(K,X) = -\sum_{\alpha\beta} p(\alpha,\beta)\ln p(\alpha,\beta). \tag{2.10}$$

Now there are two cases to consider in obtaining lower bounds on (2.10). First, when the sampling functions Υ and Γ are Gaussians, the samplings (2.8) and (2.9) may be related to the sampling by generalized coherent state projectors described above. The Lieb inequality may then be used to derive the result

$$I(K,X) \geq 1 + \ln\left(\frac{2\pi\hbar}{\sigma_x\sigma_k}\right). \tag{2.11}$$

The second case is when the sampling functions are exact window functions and the probability distributions are discrete. One may then use the fact that the probabilities $p(\alpha,\beta)$ have an upper bound, $p_{\max}$, when maximized over all α, β and states ρ. The Shannon information (2.10) then has a lower bound $\ln(1/p_{\max})$. The maximum probability may be computed by studying the spectrum of the operator traced with ρ in (2.8), (2.9) (*e.g.*, the operator $P^x_\alpha P^k_\beta P^x_\alpha$ in the case of (2.8)). In the limit of small and large sampling volumes, one thus obtains the approximate result,

$$I(K,X) \geq \ln\left(\frac{1}{p_{\max}}\right) \approx \begin{cases} 0, & \text{if } \sigma_k\sigma_x >> 2\pi\hbar; \\ \ln\left(\frac{2\pi\hbar}{\sigma_k\sigma_x}\right), & \text{if } \sigma_k\sigma_x << 2\pi\hbar. \end{cases} \tag{2.12}$$

Improved versions of this and related results have been given by Meister [14].

The results (2.11) and (2.12) are intuitively appealing. For samplings of phase space whose volume $\sigma_x\sigma_k$ is much larger than the fundamental phase space volume $2\pi\hbar$, there is essentially no restriction on the uncertainty, the lower bound approaching zero in the discrete case, and approaching minus infinity in the continuous case. However, as is to be expected, the bound is non-trivial when the sampling volume $\sigma_x\sigma_k$ is comparable to the size of the fundamental cell. In that case, the bound itself is then the logarithm of the number of samplings required to sample an entire fundamental cell.

2C. More General Histories

The above inequalities may be used to obtain the lower bound on the Shannon information for a history consisting of position samplings at two moments of time, using (2.9). One obtains, in the interesting regime of small sampling width σ_x,

$$I(X_1, X_2) \geq \ln\left(\frac{2\pi\hbar t}{m\sigma_x^2}\right). \tag{2.13}$$

This result does in fact extend considerably, to histories consisting of an arbitrary number of samplings of any continuous variable. The probability for a history of n samplings at times $t_1, \cdots, t_n$. is given by (1.1). Let the times to be closely spaced. Take the widths of the samplings (of position, momentum, *etc.*) to be $\sigma_1, \cdots, \sigma_n$. Then the lower bound on the information of (1.1), in the interesting regime of small σ_i, is

$$I(A_1, A_2, \cdots A_n) \geq I_{\min} \approx \ln\left(\frac{V_H}{\sigma_1\sigma_2^2 \cdots \sigma_{n-1}^2\sigma_n}\right). \tag{2.14}$$

Here, the quantity V_H is given by

$$V_H = \prod_{k=1}^{n-1} |\Delta(t_{k+1}, t_k)|^{-2}, \tag{2.15}$$

where $\Delta(t_{k+1}, t_k)$ is the short-time limit of the propagator prefactor between the sampled quantities at times t_k and t_{k+1}. For example, for the case of histories characterized by position samplings, $\Delta(t, 0) = (m/2\pi i\hbar t)^{\frac{1}{2}}$, and Eq.(2.14) reduces to (2.13) for a two-time history. Similarly, (2.12) readily follows from (2.14) for a two-time history consisting of a position and momentum sampling.

It is of interest to understand the significance of the factor V_H in Eq.(2.14). Take for example the case of sequences of position samplings. The quantity $|\Delta|^2$ has dimension $(\text{length})^{-2}$ and obeys a continuity equation. It thus has the interpretation as a "density of paths." V_H is a product of the inverse of quantities of this form, and might therefore reasonably be thought of as a fundamental volume element of the "history space", to which the histories belong. A similar interpretation plausibly holds for other types of samplings also. With this understanding, the argument of the logarithm of the lower bound (2.14) is then the ratio of a fundamental history space volume element to the volume of history space sampled, and thus it has precisely the same form as the lower bounds on phase space samplings (2.11) and (2.12). One thus finds that in the interesting regime of small sampling volume, and for short time separations between samplings, all lower bounds on the Shannon information of the probability for a set of histories have the form

$$I \geq I_{\min} \approx \ln\left(\begin{array}{c}\#\text{ of samplings required to}\\ \text{cover the fundamental volume}\end{array}\right) \tag{2.16}$$

This is the main result of this section. For further details see Ref. 6.

3. FLUCTUATIONS ABOUT CLASSICAL PREDICTABILITY

I now focus on the special case of histories characterized by strings of approximate position samplings. I am therefore considering the probability for histories (1.1), with the projection operators given by (2.6), and I denote the probablity $p(\bar{x}_1, \bar{x}_2, \cdots \bar{x}_n)$, where $\bar{x}$ denotes the sampled position. Given such a probability distribution, there are then two questions one can ask. First, to what extent does this distribution indicate classical predictability, *i.e.*, to what extent are successive position samplings correlated according to classical laws? Secondly, to what extent is classical predictability restricted by the uncertainty principle?

3A. Histories of Position Samplings

To answer both of these questions, consider the following change of variables in the probability distribution $p(\bar{x}_1, \bar{x}_2 \cdots \bar{x}_n)$ from the $\bar{x}_j$'s, to $\bar{x}_1$, $\bar{k}_1$, $\bar{F}_2, \cdots \bar{F}_{n-1}$, where $k_1 = m(\bar{x}_2 - \bar{x}_1)/t$ is the initial momentum, and

$$\bar{F}_j = \frac{m}{t^2}(\bar{x}_{j+1} - 2\bar{x}_j + \bar{x}_{j-1}) + V'(\bar{x}_j), \quad j = 2, \cdots, n-1, \tag{3.1}$$

are discrete approximations to the value of the field equations $F = m\ddot{x} + V'(x)$, at $n-2$ times (where t is the time separation between projections). By integrating out the initial data, one thus obtains a distribution on the $\bar{F}_j$'s:

$$p(\bar{F}_2, \cdots \bar{F}_{n-1}) = \int d\bar{x}_1 d\bar{k}_1 \; p(\bar{x}_1, \bar{k}_1, \bar{F}_2, \cdots \bar{F}_{n-1}). \tag{3.2}$$

The expression (3.2) was evaluated in Ref. 7 for a variety of different situations. In brief, the main result are as follows:

1. The distribution (3.2) is peaked about $\bar{F}_i = 0$. Classical predictability is thus exhibited.
2. The width ΔF of the peak about $\bar{F}_i = 0$ is largely independent of the initial state.
3. The width has a lower bound, which in the case of Gaussian samplings continuous in time, takes the form,

$$\sigma \Delta F \geq \hbar, \tag{3.3}$$

where σ is the Gaussian width of the sampling, and has dimension (length) $\times$ (time)$^{\frac{1}{2}}$. This is therefore a second form of the uncertainty principle for histories, and confirms an earlier heuristic analysis of Mensky [15].

3B. Thermal Fluctuations in a Quantum Brownian Motion Model

A situation the results of this and the previous section extend to is the quantum Brownian motion model, in which one has a particle coupled to an environment consisting of a large number of harmonic oscillators in a thermal state (see Ref. 16, for example). Models of this type are studied extensively in connection with decoherence. I now briefly outline the results of Ref. 7.

In the presence of fluctuations due to the presence of an environment, one would still expect to obtain classical predictability, but the width of the fluctuations about it should be broadened due to thermal effects. Repeating the above analysis for a thermal environment characterized by a temperature T and dissipation coefficient γ, one finds first of all, that the probability distribution on the field equations is again peaked about $\bar{F} = 0$, but now $\bar{F} = m\ddot{x} + m\gamma\dot{x} + V'(x)$, *i.e.*, it is modified by a dissipative term. Secondly, the width of the peak about $\bar{F} = 0$ is

$$(\Delta F)^2 \approx \frac{\hbar^2}{2\sigma^2} + 4M\gamma kT. \tag{3.4}$$

One may thus see that there are clearly identifiable contributions to the width from quantum and thermal fluctuations. In particular, one may see that the thermal fluctuations dominate the quantum ones when $M\gamma kT\sigma^2/\hbar^2 >> 1$. This is in fact the condition for decoherence in quantum Brownian motion models [17]. It follows that decoherence is unavoidably accompanied by thermal fluctuations.

Thermal modifications of the information-theoretic results (2.5) may also be obtained, and lead to essentially the same conclusions [8]. Also, similar results have previously been obtained by Hu and Zhang [18].

ACKNOWLEDGEMENTS

I would like to thank the organizers for inviting me to such an interesting meeting. I am also grateful to numerous colleagues for useful conversations, including Arlen Anderson, Carl Caves, Murray Gell-Mann, Jim Hartle, Salman Habib, Bei-Lok Hu, Chris Isham, Raymond Laflamme, Seth Lloyd, Warner Miller, Roland Omnes, Juan Pablo Paz, and Wojciech Zurek. This work was supported by a University Research Fellowship from the Royal Society.

REFERENCES

1. M.Gell-Mann and J.B.Hartle, *Phys.Rev.* **D47**, 3345 (1993).
2. R.Griffiths, *J. Stat. Phys.* **36**, 219 (1984).
3. R.Omnès, *Rev. Mod. Phys.* **64**, 339 (1992); *Ann. Phys. (N.Y.)* **201**, 354 (1990); see also his contribution in this volume.
4. H.F.Dowker and J.J.Halliwell, *Phys. Rev.* **D46**, 1580 (1992).
5. See, for example, E.P.Wigner, in *Quantum Theory and Measurement*, edited by J.A.Wheeler and W.H.Zurek (Princeton University Press, Princeton, 1983).
6. J.J.Halliwell, *Phys. Rev. D* **48**, 2739 (1993).
7. J.J.Halliwell, *Phys. Rev. D* **48**, 4785 (1993).
8. A.Anderson and J.J.Halliwell, *Phys.Rev. D* **48**, 2753 (1993).
9. C.E.Shannon and W.W.Weaver, *The Mathematical Theory of Communication* (University of Illinois Press, Urbana, 1949).
10. T.M.Cover and J.A.Thomas, *Elements of Information Theory* (Wiley, New York, 1991).
11. A.Wehrl, *Rep.Math.Phys.* **16**, 353 (1979).
12. E.H.Lieb, *Comm. Math. Phys.* **62**, 35 (1978).
13. M. Grabowski, *Rep.Math.Phys.* **20**, 153 (1984).
14. B.Meister, Imperial College preprint, 1993.
15. M.Mensky, *Phys. Lett.* **A155**, 229 (1991).
16. A.O.Caldeira and A.J.Leggett, *Physica* **121A**, 587 (1983).
17. W.H.Zurek, in *Frontiers of Non-Equilibrium Statistical Mechanics*, G.T. Moore and M.O.Scully, eds. (Plenum, New York, 1986).
18. B.L.Hu and Y.Zhang, "Uncertainty relation at finite temperature," University of Maryland Preprint, 1992.

NOTE

1. E-mail address: j_halliwell@vax1.physics.imperial.ac.uk

EXPERIMENTS WITH COHERENT ELECTRON WAVE PACKETS

Franz Hasselbach

Institut für Angewandte Physik der Universität Tübingen
Auf der Morgenstelle 12
D-72076 Tübingen, Germany

In an electron biprism interferometer the coherent wave packets are travelling on laterally separated paths from the source to the interference plane. They may experience on their paths different electromagnetic and inertial fields, be deflected and pick up quantum mechanical phase shifts due to the corresponding potentials. In order to study the influence of electromagnetic fields, a Wien filter (i.e., crossed magnetic and electric fields perpendicular to the beam path, matched in such a way that the electric force on the electrons is just canceled by the Lorentz force) inserted into the beam path has proved to be a very versatile tool: It introduces a longitudinal shift of the wave packets and allows to measure the coherence lengths, the spectral content of the wave packet, to optimize longitudinal coherence and to observe Aharonov-Bohm phase shifts. It turns out that longitudinal coherence is not destroyed easily but sometimes "hidden" only, if the longitudinal shifts caused by electric fields exceed the coherence length. Longitudinal coherence and fringe visibility turns out to be a local feature of the electron wave field. We will see that "which-path information" and fringe visibility exclude one another. Quantum phase shifts induced by inertial fields and vector potentials are accessible due to the possiblity to put the miniaturized electron interferometer on a turntable.

Key words: Dispersive propagation, coherence length and Fourier spectroscopy of matter wave packets, "which-path information", influence of electromagnetic and gravito-inertial potentials and fields on the phase of matter waves, Aharonov-Bohm effects.

1. INTRODUCTION

The electron biprism interferometer, invented by Möllenstedt and realized by Düker 40 years ago [1], has proved to be a very successful tool to investigate the fundamentals of quantum mechanics. Immediately after its

M. Ferrero and A. van der Merwe (eds.), Fundamental Problems in Quantum Physics, 123–139.

invention it has been used to verify the phase shift caused by the magnetic vector potential which had been predicted by Ehrenberg and Siday [2] in the context of electron optics in 1949 already. However, the physicists at that time did not comprehend the far reaching consequences of these ideas. Therefore only ten years later the seminal paper of Aharonov and Bohm (A-B) on the importance of potentials in quantum mechanics [3] triggered the interest in the subject. But then it was an electron biprism interferometer, that led to the first experimental proof [4] immediately after the A-B paper had appeared. As early as 1962 the most beautiful experiment of Bayh [5] was performed in Tübingen. The ongoing discussion has been settled only recently by an experiment of Tonomura et al. [6].

In 1980 the possibilities for investigating the fundamentals of quantum mechanics electron interferometrically were augmented firstly by a new type of electron interferometer [7], which was insensitive to environmental perturbations, namely magnetic fields caused by alternating currents and mechanical vibrations and, secondly, Möllenstedt and Wohland discovered in a Wien filter a device [8], that allowed to shift two electron wave packets longitudinally which propagate in the two legs of an interferometer. While the insensitivity of the new interferometer to vibrations opened the possibility to apply inertial potentials and forces to matter wave packets and study their effects, the Wien filter gives answers to questions on the influence of electromagnetic fields and potentials on the propagation of the charged matter wave packets and the phase structure inside these wave packets.

The paper is organized as follows: Its first section is devoted to the dispersive propagation of matter wave packets and non-dispersive propagation of coherence length packets. Then the influence of electric and magnetic fields and of their combination (Wien filter) on the wave packets is discussed extensively. The last chapter is devoted to the influence of inertial forces and potentials on the phase of electron waves. The formal mathematical analogy between magnetic forces and potentials with the Coriolis force resp. the rotational vector potential which holds in nonrelativistic quantum mechanics, makes evident the close relationship between electrical and mechanical phenomena.

2. DISPERSIVE PROPAGATION OF MATTER WAVE PACKETS; COHERENCE LENGTH IS A CONSTANT OF MOTION

The coherence length of matter waves is not identical with the spatial extend of the wave packet. This is due to the intrinsically dispersive propagation of matter waves. The linear extend Δq of a matter wave packet increases in time according to:

$$\Delta q = \Delta q_0 \left[1 + \tfrac{1}{16\pi}\left(\tfrac{D\lambda}{(\Delta q_0)^2}\right)^2\right]^{\frac{1}{2}},$$

where D is the distance covered by the wave packet during the time t and λ is the average de Broglie wavelength. The wave packets spread in time in a rather complex way: The plane waves with the shorter wavelengths correspond to the particles with higher kinetic energy and propagate with a higher velocity than those corresponding to the particles with lower kinetic energy. When we could observe the arrival of electron wave packets downstream the interferometer we would at first observe plane wave components with the shorter wavelength and only at the end of the packet those with the longest wavelengths.

It is obvious that no interference is to be observed, if the path difference exceeds the spatial extend of the wave packets. Clearly, the the question arises, does the coherence length of a wave packet increase simultaneously with its spreading? Then, we could always enforce longitudinal coherence by increasing the length of the interferometer. Klein, Opat, and Hamilton [9] showed, that the coherence length of a de Broglie wave packet remains unchanged, even when its length Δq increases upon propagation. That is, the "coherence length packet" is time invariant, while the spatial extend of the wave packet is not. A stationary interference pattern is observed in the interference plane only when the path length difference is smaller than the coherence length irrespective of the distance travelled by the wave packet.

In Tübingen coherence length measurements were performed downstream about [10] D=7 cm and [8] D=100 cm from the electron source in two different interferometers. Although the experimental parameters as well as the accuracy differed largely in both experiments it is possible to conclude that there is no spreading in time of the coherence length, which is given by $\lambda_c = \frac{\lambda^2}{\Delta\lambda} = 2\frac{\lambda U}{\Delta U_H}$. The coherence length is independent of the distance travelled by the wave packet, i.e., of the time for which it has evolved.

$E[keV]$	$\Delta E[eV]$	$\lambda_c[nm]$	$\lambda_c{}^{exp}[nm]$	$D[cm]$	Δq
4,0	0,36	430	280	7	1,15 q_0
5,0	0,5	350	170	100	11,5 q_0

In order to measure coherence lengths, an artificial variable time delay must be inserted in one leg of the interferometer. In the interference plane the wave packet is recombined with a replica of itself after introducing different longitudinal shifts Δx. Subsequently the fringe visibility as a function of Δx is measured. Before 1980 no variable time delay was at hand for matter waves. Then, Möllenstedt and Wohland discovered that a Wien filter in its compensated state is a component that makes the introduction of variable time delays into the legs of an electron interferometer possible by varying its excitation [8]. Möllenstedt and Wohland calculated, from the strength of the electric field in the Wien filter for vanishing fringe contrast, the coherence length λ_c^{exp} for 5keV electrons (2nd line in the table) with rather high error margins of 40%; λ_c^{exp} for 4keV electrons was obtained using the highly accurate method described later in Sec. 4.1. The definition for the coherence length used in this latter experiment was the longitudinal shift which corresponds to a decrease of the fringe contrast to 1/e.

Originally, the Wien filter was invented as a spectrometer for charged particles. It consists of crossed electric and magnetic fields perpendicular to each other and to the beam path. Their strengths are usually chosen in such a way that no force acts on charged particles of a certain energy which are traversing these fields. Charged particles different in their energy are deflected and form a spectrum behind the Wien filter. Our application of a Wien filter in an electron interferometer has, at least at first sight, nothing in common with its conventional application as a spectrometer.

In order to understand the action of the Wien filter on electron wave packets let's consider beforehand the influence of electric and magnetic fields on electron waves separately.

3. INFLUENCE OF ELECTROMAGNETIC POTENTIALS AND FIELDS ON THE PHASE OF MATTER WAVES

3.1. Deflection in a Homogeneous Electric Field

In the following chapters we discuss the deflection of a primary interference pattern (Fig. 1) by homogeneous electromagnetic fields. The reason why we consider an interference pattern and not an arbitrary picture is a pedagogical one: By its very nature an interference pattern reveals local phase shifts in the plane wave fronts, introduced, e.g., by electromagnetic fields and /or path length differences.

Imagine two plane wave fronts (AA', BB' in Fig. 1), e.g., coming from an electron optical biprism. In the region of overlap interference fringes are formed. In Fig. 1 these fringes are depicted only in a small region before and after the homogeneous electric deflection field. The electric field deflects the fringe field say, by an angle α to the right hand side. When we increase the angle of deflection slowly from zero deflection, we observe experimentally (Fig. 1) always the same interference field irrespective of the deflection angle α. It does not change its appearance although an extra, continuously increasing path length difference $\Delta = \alpha d$ — represented by the bold line on the left-hand side boundary of one of the plane waves (Fig. 1) — is introduced with increasing deflection α; d represents the distance from the right hand side boundary of the plane waves. The fact that we observe *no* continuous change from bright to dark of a single fringe of the pattern with increasing deflection angle means that the varying path length Δ and the corresponding phase shift Δ/λ is compensated in some way. This compensation is due to the scalar Aharonov-Bohm effect. In other words the A-B phase shift keeps exact pace with the path length difference caused by the deflection. Consequently, the fringe system does not change its appearance with increasing deflection angle.

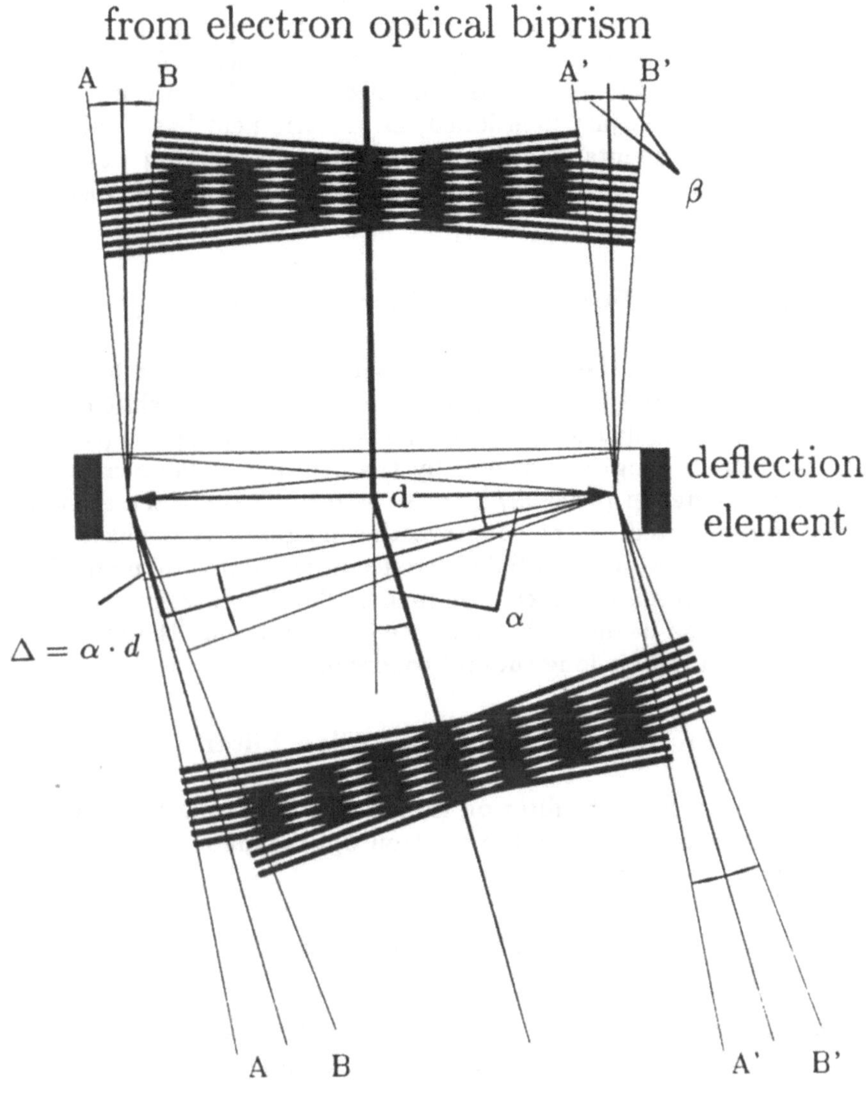

Fig. 1. Deflection of two plane waves AA' and BB' by homogeneous electric or magnetic fields. The angle of incidence is $\pm\beta$. The electric resp. magnetic field deflects the wave fronts by an angle α. The path length difference $\Delta = \alpha \cdot d$ introduced between the two coherent beams by the deflection is marked by a bold line. In the case of an electric field, the wavelets making up the plane wave fronts propagate inside the deflection element on different electric potentials (depending on d). The corresponding Aharonov-Bohm phase shift compensates the path length difference Δ. In the case of a magnetic field, the field direction is perpendicular to the plane of the paper. Here, difference in magnetic flux enclosed between the coherent beams compensates the geometric phase shift Δ/λ for every wavelet making up the wave fronts.

This analysis demonstrates that not even a simple electron optical deflection element would work properly without A-B quantum phase shifts. Ehrenberg and Siday realized already in 1949 that these phase shifts exist and that they are absolutely essential for electron optics to work.

Besides the effects just mentioned, an electric field has a third, in the context of our experiments most important effect on electron wave packets, namely, a longitudinal shift of the wave packets. This will be discussed in the next chapters.

3.2. Deflection in a Homogeneous Magnetic Field

In a magnetic field, the electron wave packets are deflected by the action of the Lorentz force and acquire an Aharonov-Bohm phase shift due to the difference in magnetic flux enclosed by the two coherent beams (see Fig. 1). The additional path length Δ caused by the bending of the beams and the corresponding change in phase Δ/λ are — as in the case of the electric field — exactly compensated by the counteracting magnetic A-B phase shift.

However, there is one essential difference in the action of magnetic fields on wave packets compared to electric fields: The Lorentz force is directed perpendicular to the beam path, therefore it cannot shift the wave packets longitudinally and impair longitudinal coherence.

3.3. Combination of Both Fields, the Wien Filter

The action of the Wien filter on the electron wave packets is easiest to understand, when we introduce the electron optical index of refraction n in order to describe the phase shifts.

$$n(\mathbf{r}, \tfrac{\mathbf{v}}{v}) = \sqrt{\tfrac{\Phi(\mathbf{r})}{\Phi_0}} - \sqrt{\tfrac{e}{2mU_B}}\mathbf{A}(\mathbf{r})\tfrac{\mathbf{v}}{v}$$

It contains an isotropic term depending on the electric potential $\Phi(\mathbf{r})$ and an anisotropic term depending on the vector potential $\mathbf{A}(\mathbf{r})$

$$\Phi_0 = -U_B, \qquad \Phi(\mathbf{r}) = E \cdot x - U_B,$$

where U_B is the accelerating voltage of the electrons and E the value of the electric field in the Wien-filter:

$$\sqrt{\tfrac{\Phi(\mathbf{r})}{U_B}} = \sqrt{1 - \tfrac{E \cdot x}{U_B}} = 1 - \tfrac{E \cdot x}{2U_B} + O(x^2), \qquad \mathbf{A} = (0, B \cdot x, 0)$$

This gauge (phase shifting field) [8] must be chosen in order to make the tangential component of $\mathbf{A}$ continuous at the entrance and exit planes of the

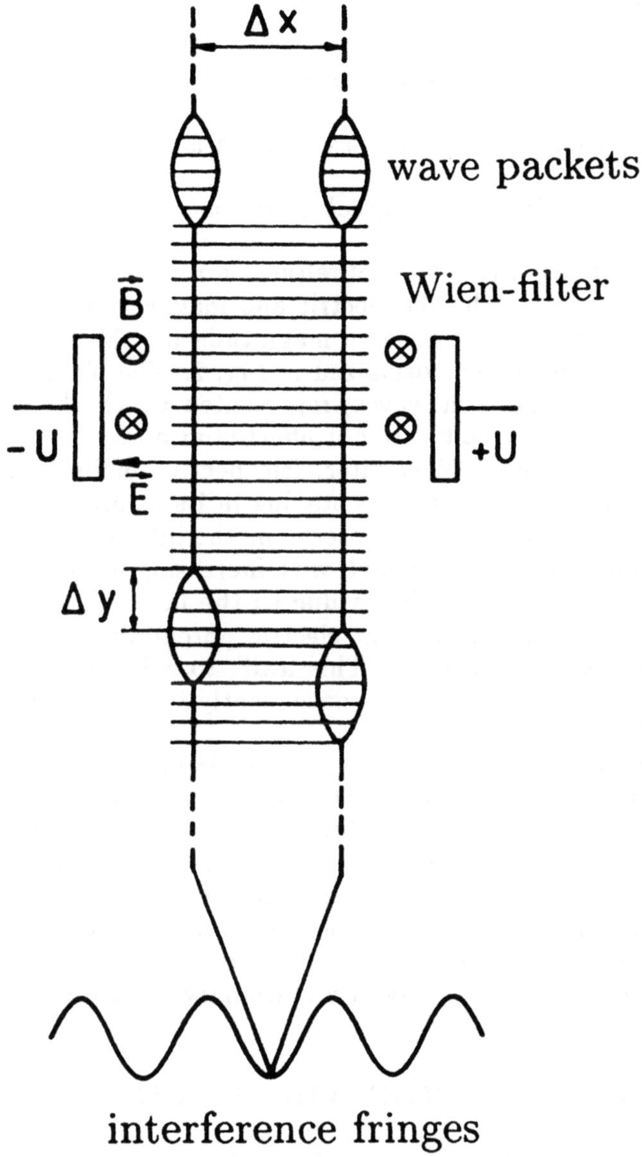

Fig. 2. Influence of a Wien filter in its compensated state on two spatially separated electron wave packets. The wave packets travel inside the Wien filter with different group velocities and leave it shifted longitudinally while the phase velocity is not affected, irrespective of the excitation of the Wien filter. Therefore, the positions of the horizontal lines, which symbolize the crests of the waves are not affected in the Wien filter. Note: The electric potential difference on the two paths increases with increasing excitation of the Wien filter. The acceleration and deceleration of the wave packets to the value inside the Wien filter occurs in the electric fringing fields of the Wien filter condensor.

Wien filter. Then

$$n = 1 - \frac{Ex}{2U_B} - \sqrt{\frac{e}{2mU_B}}Bx + O(x^2).$$

In the compensated state, the Wien condition is fullfilled:

$$E = -Bv, \qquad v = \sqrt{2eU_B/m} \Longrightarrow n = 1 + O(x^2).$$

The first-order terms of the index of refraction n vanish due to the fact, that the phase shifts excerted by the electric and magnetic fields are opposite to each other and of exactly the same magnitude for the compensated Wien filter. The consequences are visualized in Fig. 2. The planes of equal phase (e.g., crests) of the electron waves, represented by the horizontal lines in Fig. 2, and the phase velocity are not affected at all by the electromagnetic fields of the *compensated* Wien filter, irrespective of its excitation. That means, when we increase — always staying in its compensated state — the excitation of the Wien filter, we observe in the interference plane a stationary field of interference fringes, *but*, with increasing excitation a decreasing fringe contrast is observed. This is due to the fact that the electron wave packets travel on paths of different electric potentials with different group velocities inside the Wien filter. This leads to a longitudinal shift of the wave packets at the exit plane and consequently to a reduced contrast of the interference fringes.

The magnetic A-B effect has been proved in many experiments [4],[5],[6] while the electric effect in its pure form is not. The fact that the phase shift caused by the enclosed magnetic flux in the Wien filter (Fig. 2) is compensated by the phase shift caused by the difference of the scalar electric potential, is an indirect proof of the existence of the scalar A-B effect.

It is important to direct the readers attention explicitly to the fact that the *Wien filter in its compensated state is not a phase shifter.* The wave packets are shifted longitudinally in a stationary "phase wave sea."

4. APPLICATIONS OF THE WIEN FILTER

4.1. Measurement of Coherence Lengths

In order to measure coherence lengths, the wave packets and their identical replicas, which are produced by wavefront splitting by an electron biprism, are superimposed with different longitudinal shifts until the contrast of the interference fringes is reduced to, e.g., 1/e or vanishes. This longitudinal shift can be measured with a precision of a small fraction of a wavelength when the Wien filter is aligned to its compensated state in the following two step process [10]: At first, the magnetic and electric fields are zero (Fig. 3a). We postulate for the moment that maximum fringe contrast is observed.

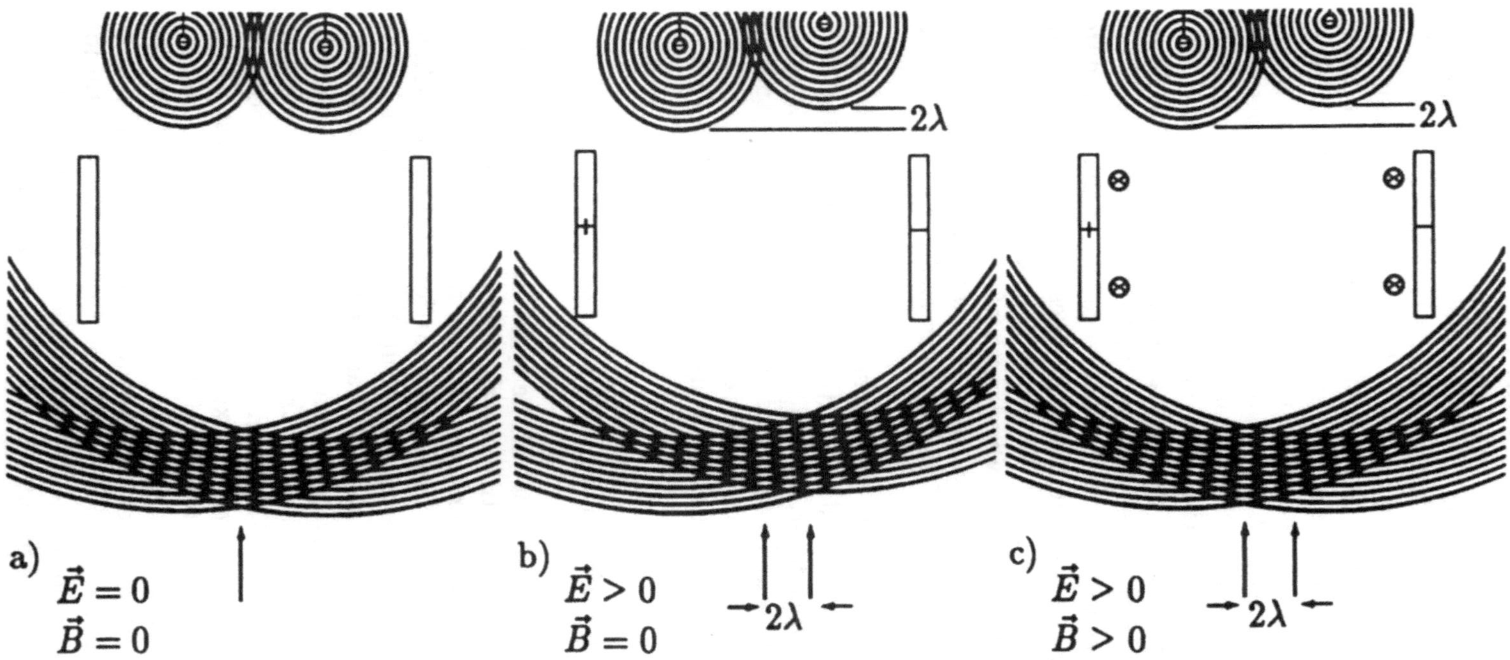

Fig. 3. The two step process to reach the compensated state of the Wien filter: Electron wave packets emerging from two coherent sources traverse the Wien filter. (a) The Wien filter is switched-off. A fringe pattern symmetrical to the axis is observed. (b) As a first step the electric field is increased. Now, inside the Wien filter the packet on the right hand side propagates on a more negative potential with a lower group velocity compared to that on the left hand side. This is simulated in this figure by moving backwards the right hand source by 2 wavelengths. As a consequence the central maximum of the pattern is shifted by exactly 2 fringe widths to the right in Fig. 3b. (c) Then, the magnetic field is increased until the shift of the pattern is compensated. The wave packets leaving the now again compensated Wien filter are shifted longitudinally by exactly 2 wavelengths.

Now we increase in a first step the electric field only. The Wien condensor works as a deflection element. The interference fringes are deflected, e.g., by 2 fringe widths to the right on the fluorescent screen of the interferometer (Fig. 3b). This is due to the fact, that the wave packet travelling in the more negative region is slower and arrives in the interference plane with a delay. Fig. 3b demonstrates that a delay of $2\lambda/v$ where v is the phase velocity, corresponds to a shift of the interference fringes of exactly 2 fringe widths. We now increase the magnetic field until the deflection due to the electric field is just compensated. This state of the now again compensated Wien filter corresponds to the following physical situation (Fig. 3c): Both beams travel rectilinearly through the Wien filter but the right hand side wave packet is shifted longitudinally by two wavelengths in the Wien filter. The overlap of the wave packets is reduced and in turn the contrast of the interference fringes. In order to measure the coherence length this procedure is repeated while counting the total number of fringes until the contrast in the fringe field has decreased to 1/e. We define the coherence length by twice this number of fringes times the wavelength of the electrons. With this definition for field emitted electrons of 4 keV with an energy spread of about 0.36 eV a coherence length of 280 nm was measured (see table in Sec. 2).

It has to be noted here that this measurement method works without any knowledge about the field strengths and (in-)homogenities of the electromagnetic fields in the Wien filter, not to mention that of the fringing fields. It is extremely precise and enables us to measure the longitudinal shift excerted by the Wien filter with a precision of the order of 1% of a wavelength or less. This high precision was a prerequisite to realize Fourier spectroscopy for matter waves.

4.2. Fourier Spectroscopy of Electron Waves

For an interferometric measurement of the spectrum it is not sufficient just to count the number of fringes until a certain decrease of the contrast is reached. Here the contrast of the interference fringes and the periodicity of the fringes must be recorded quantitatively as a function of the longitudinal shift in the whole interference field consisting of up to 20 000 fringes in our low voltage interferometer. Provisionally, this was done for the contrast only, restricting the technique temporarily to symmetric spectra. The pioneering work on this subject was done at the end of the last century and the beginning of this one by A.A. Michelson [11] and H. Rubens and R.W. Wood [12] for electromagnetic waves. In the present experiment, the contrast of the interference fringes was recorded in sets of 10 fringes with a CCD-camera densitometer successively. The digitized data sets were corrected for the cylinder lens action of the Wien filter, put together with matched phases in a personal computer and subsequently transfered to a VAX computer for Fourier analysis. In the first experiment [13] a resolution of about 0.5 eV was obtained. State of the art now is about 50 meV [14]. This resolving power surpasses that of electron spectrometers which are used in todays analytical electron microscopes by about an order of magnitude. Our Fourier spectrometer is the *first spectrometer for particles that fully relies on quantum mechanics, i.e., the wave properties of matter.* It seems remarkable that this experiment proves that the *quantum mechanical probability waves* exhibit the same features as the "real" waves, e.g., those of an electromagnetic field. All *conventional* spectrometers for charged particles are based on the dispersion of particle trajectories in electromagnetic fields.

5. LONGITUDINAL COHERENCE, A LOCAL FEATURE OF AN ELECTRON WAVE FIELD

In the course of the development of the miniaturized low voltage interferometer [7],[15],[16] with which all experiments described here were performed, our attention was attracted by a — at first sight unaccountable and up to that time never observed — strong dependence of the fringe contrast on the alignment of the electron beam to the optical axis, more exactly on the excitation of the electrostatic alignment system. The cause of this dependence is, as we have already seen in Sec. 3, the inevitable longitudinal shift that electron wave packets experience in electrostatic deflection elements. Figure 4 gives a sketch of the wave packet shifting action of electrostatic deflection elements and its compensation by a suitably excited Wien filter.

In the region between electron source and the first deflection condensor the wave packets propagate in single file. Longitudinal coherence is fullfilled. The electrostatic deflectors introduce a longitudinal shift of the wave packets and a corresponding decrease of longitudinal coherence and fringe contrast. When working with very low electron energies, for example a few hundred eV only, this longitudinal shift in the well aligned interferometer often exceeds the coherence length. Then fringes become visible only after compensating this time delay by exciting the Wien filter suitably. This is demonstrated in Fig. 5. Fringe contrast is absent for the unexcited Wien filter, appears, goes through its optimum, and disappears again when the Wien filter is overexcited.

Longitudinal coherence is a feature of an electron wave field that is not necessarily present in the whole interferometer but locally only. The exchange of energy of the wave packets within the electrostatic fields gives rise to this locality. In areas of the interferometer where a superposition of the plane waves does not lead to interference fringes, the partial waves may not really be incoherent, but coherence can be "hidden" only. In this situation full overlap of the coherence length packets and in turn fringe visibility can be reestablished by introducing a compensating delay of the wave packets. These facts do not corroborate the prevailing idea that coherence is very fragile and destroyed easily. On the contrary, this experiment demonstrates the robustness of quantum mechanical coherence. This has been confirmed recently in a conceptually similar experiment for neutrons [17].

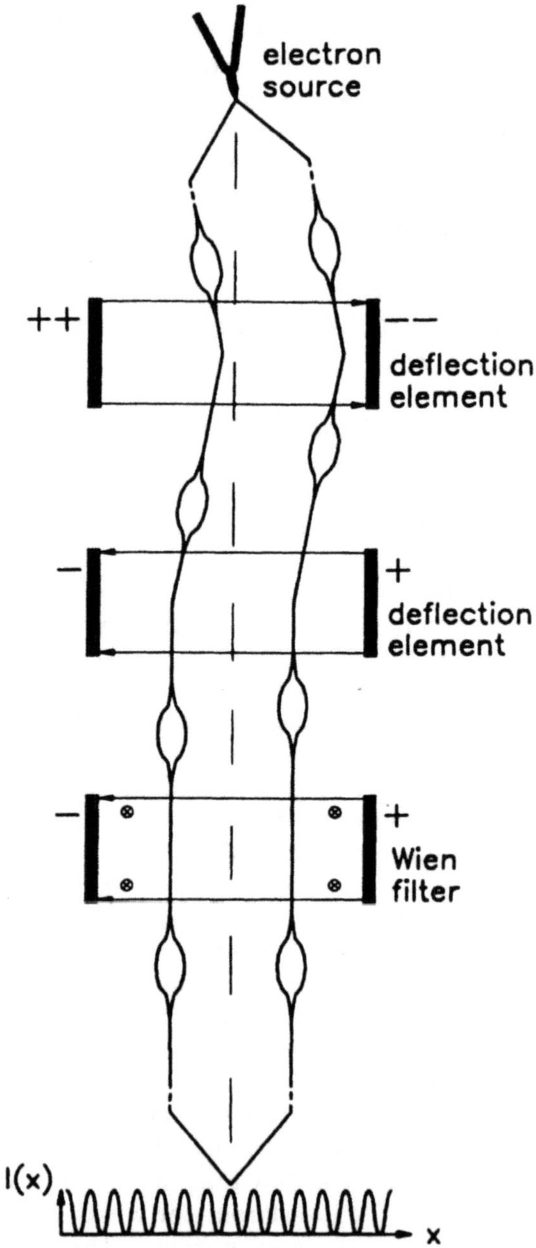

Fig. 4. Combined wave packet shifting effect of two deflection elements and a Wien filter. As an example, the electron source is slightly misaligned (the angular deviation is exaggerated in the drawing). The first deflection element bends the beams back to the optical axis, and the second one makes them parallel again with the optical axis. Because the first deflection is approx. twice as large as the second one, the longitudinal shifts do not cancel completely. The remaining net delay is eliminated by the Wien filter.

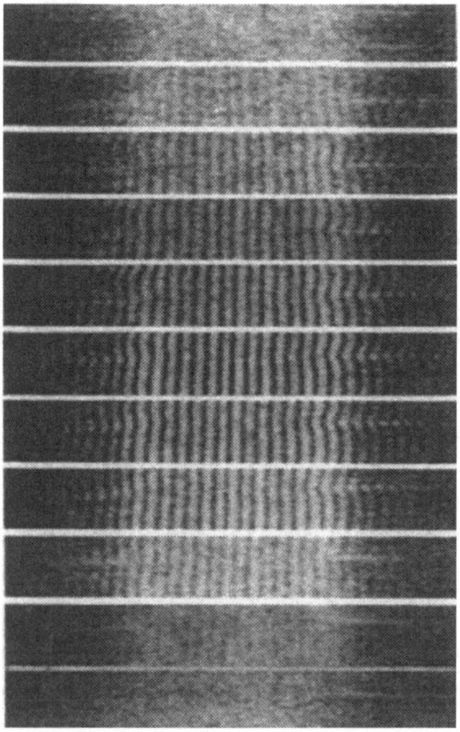

Fig. 5. Reestablishing longitudinal coherence by a Wien filter. With switched-off Wien filter (top of figure) no fringes are visible although the paths of the wave packets are aligned to the optical axis of the interferometer. With increasing excitation interference fringes appear, their contrast reaches its maximum (middle) and disappears (bottom) when the longitudinal shift produced by the Wien filter exceeds that caused by the electrostatic alignment elements plus one coherence length.

6. LONGITUDINAL COHERENCE AND "WHICH-PATH INFORMATION"

When optimum fringe contrast is observed, the transit time of the two wave packets on the spatially separated paths to the interference plane is exactly equal for both packets. No "which-path information" is available by considering the transit time. The transit time differs by an amount equal to the coherence time when the Wien filter (see, e.g., Fig. 2) is excited to that point that fringe contrast just vanishes. If we now measured the transit times of many electrons from the electron source to the interference plane, we would find that it is centered around two discrete values. The larger transit time corresponds to the path in the Wien filter traversing the area with lower electric potential and the shorter one to that in the higher, i.e., for a transit time difference just in excess of the coherence time, we know in principle the path of the electron. The knowledge of its path is exclusively the reason for the disappearance of the interferences. This can be shown easily by placing a second Wien filter behind the first one in Fig. 2. We can excite this Wien filter in such a way that the energy exchange in the first Wien filter is exactly compensated. The interference fringes reappear, "which-path information" is ereased by the second Wien filter. This proves that no uncontrolled scattering events or energy transfers were present in the whole experiment, on the contrary, the energy exchange of the electron wave field with the environment is very well defined.

Essentially, this experiment has already been performed in 1987 [16],[15], only the interpretation presented here is new. Another experimental proof that the availability of "which-path information" and nothing else is the fundamental cause for invisibility resp. visibility of interference fringes was recently performed by Scully et al. [18] and involved a micromaser/atomic beam combination.

7. INFLUENCE OF INERTIAL FORCES AND POTENTIALS ON THE ELECTRON PHASE, SAGNAC EFFECT

The effect of a magnetic field on the phase of an electron wave has been discussed extensively. A comparison between the relations of a massive particle subjected to the Coriolis force and a charged particle subjected to a magnetic field makes evident the mathematical equivalence of a rotation $\boldsymbol{\Omega}$ and a magnetic field $\mathbf{B}$:

$$\mathbf{F}_{coriolis} = 2m\mathbf{v} \times \boldsymbol{\Omega}, \qquad \mathbf{F}_{lorentz} = (e/c)\mathbf{v} \times \mathbf{B}.$$

The relation of the canonical momentum in a rotating frame of reference,

$$\mathbf{p} = m\mathbf{v} + m\boldsymbol{\Omega} \times \mathbf{x},$$

and that of the canonical momentum of a charged particle,

$$\mathbf{p} = m\mathbf{v} + (e/c)\mathbf{A} = m\mathbf{v} + (e/2c)\mathbf{B} \times \mathbf{x},$$

adds additional evidence to the correlation between a rotating frame of reference and a magnetic field.

In a rotating two beam interferometer where the coherent beams enclose a finite area A, a Sagnac phase shift due to rotation is detected:

Sagnac phase shift	Aharonov-Bohm phase shift
$\Delta\phi = \frac{2m}{\hbar} \int_A \boldsymbol{\Omega} d\Sigma$	$\Delta\phi = \frac{e}{\hbar \cdot c} \oint \mathbf{A} d\mathbf{s} = \frac{e}{\hbar \cdot c} \int_A \mathbf{B} d\Sigma$

Where the surface integrals are over the oriented enclosed area A. Evidently, the Sagnac effect is the mechanical analog of the Aharonov-Bohm effect. The following quantities correspond to each other: $2m\mathbf{\Omega} \Longleftrightarrow (e/c)\mathbf{B}$. In their recent paper [19] Hendriks and Nienhius substantiate this by showing that a rotation has the same effect on the Schrödinger equation as does a magnetic field described by a vector potential **A** or by the corresponding magnetic flux. Due to the formal mathematical analogy both effects are manifestations of a geometric- or Berry phase shift [20].

In the present experiment, the whole Sagnac interferometer was placed on a turntable. An enclosed $\mathbf{\Omega}$-flux and an $\mathbf{\Omega}$ dependent Coriolis force shifted the phase of the electrons resp. the fringe pattern. The effect of this flux and of the Coriolis force is equivalent to the already discussed deflection of electrons in a magnetic field. Our results [15] agree within the error limits of about 30% with the theoretically expected values. The difference of our Sagnac experiment to the already mentioned A-B experiments, is that it is not so purely topological as the A-B experiments since in addition to the non-local $\mathbf{\Omega}$-flux dependent phase shift the inertial Coriolis force acts on the electrons in the Sagnac experiment. It is equivalent to the Lorentz force in a homogeneous magnetic field. In complete analogy to the deflection of electron interference fringes by a magnetic field, the fringe pattern as a whole is shifted in our Sagnac experiment. While the Sagnac effect with matter waves was predicted by C.V. Heer [21] in 1961, the investigation of the Aharonov-Bohm effect for an inertial vector potential was discussed in the form of thought experiments by Aharonov and Carmi [22] in 1973. When reviewing these papers J.H. Harris and M.D. Semon [23] came across the experimental paper of J.E. Zimmermann and J.E. Mercereau [24] on the "Compton wavelength of superconducting electrons." In this experiment Zimmermann and Mercereau rotated a Josephson junction interferometer (a SQUID). In a rotating superconducting ring magnetic and electric fields are induced which compensate exactly the Coriolis- and centrigugal-forces on the charge carriers [25],[26]. Additionally, no magnetic field is induced by the rotation in the area enclosed by the superconductor [27]. Thus, in the enclosed area a pure $\mathbf{\Omega}-$flux exists which gives rise to an uncanceled inertial vector potential in the force free superconducting region. The rotationally induced phase shift is given by the A-B relation with the substitution given above. The current in the SQUID becomes a periodic function of $\mathbf{\Omega}$. Taking the Compton wavelength as given, this experiment is so far the only verification of the Aharonov-Bohm effect for an inertial field.

ACKNOWLEDGEMENT

Appreciation is given to my colleagues Dr. Marc Nicklaus, Andreas Schäfer, Uwe Maier, Harald Wachendorfer and Matthias Bauer for many helpful discussions.

REFERENCES

[1] G. Möllenstedt, H. Düker, "Fresnelscher Interferenzversuch mit einem Biprisma für Elektronenwellen," *Naturwiss.* **42**(1954), 41; G. Möllenstedt and H. Düker, "Beobachtungen und Messungen an Biprisma-Interferenzen mit Elektronenwellen," *Z. Phys.* **145**(1956), 377-397.
[2] W.E. Ehrenberg and R.E. Siday, "The refractive index in electron optics and the principles of dynamics," *Proc. Phys. Soc. (London)* **B62**(1949), 8-21.
[3] Y. Aharonov and D. Bohm, "Significance of electromagnetic potentials in the quantum theory," *Phys. Rev.* **115**(1959), 485-491.
[4] R.G. Chambers, "Shift of an electron interference pattern by enclosed magnetic flux," *Phys. Rev. Lett.* **5**(1960), 3-5.
[5] W. Bayh, "Messung der kontinuierlichen Phasenschiebung von Elektronenwellen im kraftfeldfreien Raum durch das magnetische Vektorpotential einer Wolfram-Wendel," *Z. Phys.* **169**(1962), 492-510.
[6] See, e.g., M. Peshkin and A. Tonomura, *The Aharonov-Bohm Effect* (Lecture Notes in Physics Vol. 340) (Springer, Berlin Heidelberg New York, 1989).
[7] F. Hasselbach, "A ruggedized miniature UHV electron biprism interferometer for new fundamental experiments and applications," *Z. Phys. B* (Condensed Matter) **71**(1988), 443-449; F. Hasselbach, 19. Tagung der Deutschen Gesellschaft für Elektronenmikroskopie, Tübingen, 1979, Abstract 7L1, p. 90.
[8] G. Möllenstedt and G. Wohland, "Direct interferometric measurement of the coherence length of an electron wave packet using a Wien filter," Electron Microscopy 1980, edited by P. Bredoro and G. Boom (Seventh European Congress on Electron Microscopy Foundation, Leiden, 1980), Vol.1, p. 28-29; G. Wohland, "Messung der Kohärenzlänge von Elektronen im Elektroneninterferometer mit Wien-Filter," PhD Thesis, Universität Tübingen, 1981.
[9] A.G.Klein, G.I. Opat and W.A. Hamilton, "Longitudinal coherence in neutron interferometry," *Phys. Rev. Lett.* **50**(1983), 563-565; H. Kaiser, S.A. Werner and E.A. George, "Direct measurement of the longitudinal coherence length of a thermal neutron beam," *Phys. Rev. Lett.* **50**(1983), 560-563; G. Comsa, Comment on "Direct measurement of the longitudinal coherence of a thermal neutron beam," *Phys. Rev. Lett.* **51**(1983), 1105-1106.
[10] I. Daberkow, H. Gauch and F. Hasselbach, "Measurement of the longitudinal coherence of electrons from a fieldemission source," Joint Meeting on Electron Microscopy, Antwerp 1983, Programm and Abstract book, p. 100.
[11] A. A. Michelson, "Visibility of interference fringes in the focus of a telescope," *Phil. Mag.* **34**(1890), 256-259; A. A. Michelson, "On the application of interference-methods to spectroscopic measurements I, " *Phil. Mag.* **31**(1891), 338-346; A. A. Michelson, "On the application of interference-methods to spectroscopic measurements II, " *Phil. Mag.* **34**(1892), 280-299.

[12] H. Rubens and R.W.Wood, "Focal isolation of long heat-waves," *Phil. Mag. Ser. 6* **21**(1911), 249-261.
[13] F. Hasselbach and A. Schäfer, "Interferometric (Fourier-spectroscopic) measurement of electron energy distributions," Proc., 12th. Int. Congress for Electron Microscopy, Seattle 1990, L.D. Peachey, D.B. Williams, editors (San Francisco Press, Inc., Box 6800, San Francisco), Vol. **2**, p. 110-111.
[14] F. Hasselbach, A. Schäfer and H. Wachendorfer, "State of the art of a novel spectroscopic method: Fourier-spectroscopy of charged particle waves," Proceedings of the International Conference and Exhibition on Instrumental Analysis (Peking University Press, 1993), Vol. **A**, Electron Optics, p. A1.
[15] F. Hasselbach and M. Nicklaus, "An electron optical Sagnac experiment," *Physica B* **151**(1988), 230-234.
[16] M. Nicklaus and F. Hasselbach, "Wien filter: A wave packet shifting device for restoring longitudinal coherence in charged matter wave interferometers," *Phys. Rev. A* **48**(1993), 152-160.
[17] R. Clothier, H. Kaiser, S.A. Werner, H. Rauch and H. Wöllwitsch, "Neutron phase echo," *Phys. Rev. A.* **44**(1991), 5357.
[18] M. O. Scully, B.-G. Englert and H. Walther, "Quantum optical tests of complementarity," *Nature* **351**(1993), 111-116.
[19] B.H.W.Hendriks and G. Nienhuis, "Sagnac effect as viewed by a corotating observer," *Quant. Opt.* **2**(1990), 13-21.
[20] M.V. Berry, "Quantal phase factors accompanying adiabatic changes," *Proc. Roy. Soc. (London) A* **392**(1984), 45-57; H.J. Bernstein and A.V. Phillips, "Fiber bundles and quantum mechanics," *Scientific American* **245** (1) (1981), 95-109.
[21] C.V. Heer, "Interference of electromagnetic and matter waves in a nonpermanent gravitational field," *Bull. Am. Phys. Soc.* **6**(1961), 58.
[22] Y. Aharonov and G. Carmi, "Quantum aspects of the equivalence principle," *Found. Phys.* **3**(1973), 493-498.
[23] J.H. Harris and M.D. Semon, "A review of the Aharonov-Carmi thought experiment concerning the inertial and electromagnetic vector potentials," *Found. Phys.* **10**(1980), 151-156; M.D. Semon, "Experimental verification of an Aharanov-Bohm effect in rotating reference frames," *Found. Phys.* **12**(1982), 49-57.
[24] J.E. Zimmermann and J.E. Mercereau, "Compton wavelength of superconducting electrons," *Phys. Rev. Lett.* **14**(1965), 887.
[25] R. Becker, F. Sauter and C. Haller, "Über die Stromverteilung in einer supraleitenden Kugel," *Z. Phys.* **85**(1933), 772.
[26] R.G. Rystepanick, "Electromagnetic fields in rotating superconductors," *Am. J. Phys.* **44**(1976), 647.
[27] A.F. Hildebrand and M.M. Saffren, *Proceedings. of the 9th. International Conference on Low Temperature Physics*, J.E. Daunt et al., eds. (Plenum, New York, 1965). p. 459.

THE ONTOLOGICAL INTERPRETATION OF QUANTUM FIELD THEORY APPLIED IN A COSMOLOGICAL CONTEXT

B. J. Hiley and A. H. Aziz Muft

Physics Department, Birkbeck College
University of London, Malet Street
London WC1E 7HX, United Kingdom

In this paper it will be shown how the ontological (Bohm) interpretation of quantum field theory can be applied to the Guth-Pi model of the inflationary universe in a de Sitter spacetime. The Bohm interpretation has the advantage of providing a simple way of discussing the transition between the quantum and classical levels and thus we are able to demonstrate in a straight forward manner how the classical world emerges from the quantum origins of this specific model universe.

Keywords: Quantum field theory, Bohm interpretation, classical limit.

1. INTRODUCTION

By now the Bohm approach to quantum particle mechanics in the non-relativistic domain is well understood. [See for example, Bohm and Hiley (1993) and Holland (1993)]. It is internally consistent and provides a clear intuitive picture of a possible ontology underlying quantum processes. There are no sustainable objections to this approach in spite of claims to the contrary. In the last few years the Bohm approach has been extended to field theory in Minkowski space-time (See for example, Bohm, Hiley and Kaloyerou (1987)). Our own work has been confined to a consideration of the scalar field in order to keep the mathematical complexity to a minimum and thus give a clearer insight into the physical principles underlying such an approach. The purpose of this paper is to outline how this approach can be extended to curved space-times and, as a consequence, we are able to show how a classical world will emerge from a quantum universe in the inflationary scenario considered by Guth and Pi (1985). Our approach offers an extremely simple way to demonstrate how this transition arises.

M. Ferrero and A. van der Merwe (eds.), Fundamental Problems in Quantum Physics, 141–156.

2. SUMMARY OF THE MAIN POINTS OF THE BOHM INTERPRETATION FOR NON-RELATIVISTIC QUANTUM MECHANICS

Let us begin by outlining the Bohm interpretation. [See Bohm and Hiley (1993) and Holland (1993)]. We begin by writing the wave function in polar form $\psi(\mathbf{r},t) = R(\mathbf{r},t)\exp[iS(\mathbf{r},t)/\hbar]$ and substitute this form into Schrödinger's equation

$$\hbar\frac{\partial\psi}{\partial t} = -\frac{\hbar^2}{2m}\nabla^2\psi + V\psi, \tag{1}$$

where V is the classical potential. This gives rise to two equations

$$\frac{\partial S}{\partial t} + \frac{(\nabla S)^2}{2m} + V + Q = 0, \tag{2}$$

where

$$Q = -\frac{\hbar}{2m}\frac{\nabla^2 R}{R} \tag{3}$$

and

$$\frac{\partial R^2}{\partial t} + \nabla.(R^2\frac{\nabla S}{m}) = 0. \tag{4}$$

If Q is zero, then eq. (2) would be the classical Hamiltonian-Jacobi equation representing a particle with momentum

$$\mathbf{p} = \nabla S, \tag{5}$$

which moves normal to the wave front S = const. As is well known the Hamilton-Jacobi theory actually defines an ensemble of classical particle trajectories.

Suppose now Q were to increase slowly then we could solve the corresponding equations and find a new set of trajectories. These trajectories would deviate more and more from the classical trajectories as Q increased in a continuous way. At no point is it possible to argue that we must give up the concept of a trajectory. We will therefore assume that eqs. (2) and (5) define an ensemble of quantum trajectories. These trajectories differ from their classical counterparts because of the presence of an additional potential Q acting on the particle to produce the quantum behaviour. For this reason Q is called the quantum potential.

If we interpret $R^2 = |\psi(\mathbf{r},t)|^2$ as the probability of a particle being at position $\mathbf{r}$ at time t, then eq. (4) can be regarded as an equation for the conservation of probability in an ensemble of quantum particles, all moving normal to the same wave front with momentum given by eq. (5). Thus we have an interpretation in which all particles have well-defined properties, but not all these properties can be known to us simultaneously.

In summary then the key points of the Bohm approach are:

1. The electron actually is a particle with simultaneously well defined position **r**(t) and momentum **p**(t). To distinguish these variables from the "observables" in the conventional approach, we follow Bell and call them "beables". It is these beables that enable us to define trajectories that are casually determined.

2. The particle is never separate from a new type of quantum field that fundamentally affects it. This field is given by R and S or alternatively by $\psi = R\exp(iS/\hbar)$. ψ then satisfies Schrödinger's equation (rather than, for example, Maxwell's equation), so that it too changes continuously and is causally determined.

3. The particle has an equation of motion

$$m\frac{dv}{dt} = -\nabla(V) - \nabla(Q). \qquad (6)$$

This means that the forces acting on it are not only classical forces, $-\nabla V$, but also there is an additional quantum force, $-\nabla Q$

4. The particle momentum is restricted to $\mathbf{p} = \nabla S$. This is also known as the guidance condition.

5. In a statistical ensemble of particles, selected so that all have the same quantum field ψ, the probability density is $P = R^2$. It should be noted that if $P = R^2$ holds initially, then the conservation eq. (4) guarantees that it will hold for all time.

Until now there has been no way to demonstrate in a specific example how the classical trajectories are limiting features of quantum trajectories. The model of the inflationary universe considered by Guth and Pi (1985) provides such an example and one of the purposes of this paper is to demonstrate this feature. This model actually involves using quantum field theory, but we will show that in a suitable limit, the field Hamiltonian becomes equivalent to an upside-down harmonic oscillator (UDHO) and it is in this context that we can show the continuous evolution of quantum trajectories into classical trajectories as the effect of the quantum potential decreases with time. In this way we see that the particle is the same object in both the quantum and classical domains. It merely behaves in a different way in each domain simply because of the presence or absence of the quantum potential. Thus in the ontological interpretation we assume that there is only one "reality" for both the quantum and classical domains; see Leggett and Garg (1985).

The second purpose of this paper will be to demonstrate how the classical world emerges in the inflationary scenario considered by Guth and Pi (1985). To discuss this it will be necessary to generalise the Bohm interpretation so that it can be applied to quantum field theory. Fortunately this extension has already been suggested by Bohm (1952). A more detail account of the relevant material may be found in Bohm, Hiley and Kaloyerou (1978). In this paper we will sketch the Bohm approach to quantum field theory in Minkowski space-time and apply these techniques to the inflationary model of Guth and Pi (1985).

3. THE ONTOLOGICAL INTERPRETATION OF QUANTUM FIELD THEORY

Fortunately the ontological approach to quantum field theory is fairly straight forward. In this case the beables are the field configurations, $\phi(\mathbf{r},t)$, and their conjugate momenta, $\pi(\mathbf{r},t)$. The particle trajectory is replaced by a well-defined time development of the field $\phi(\mathbf{r},t)$. In order to describe the quantum properties of the field, a wave functional $\Psi(\phi(\mathbf{r},t))$ is introduced. This wave functional, sometimes called the super-wave function, is assumed to satisfy the super-Schrödinger equation

$$i\frac{\partial\Psi(\phi(\mathbf{x},t),t)}{\partial t} = H\Psi,\phi(\mathbf{x},t),t) \tag{7}$$

where H is the Hamiltonian of the field. (We put $c = \hbar=1$). In the case of a massless scalar field

$$H = \frac{1}{2}\int[-\frac{\delta^2}{(\delta\phi)^2}+(\nabla\phi)^2]\,d^3\mathbf{r}, \tag{8}$$

where $\frac{\delta}{\delta\phi}$ is the functional derivative.

If we now write the wave functional as

$$\Psi(\phi(\mathbf{r},t)) = R(\phi(\mathbf{r},t))\exp[iS(\phi(\mathbf{r},t))], \tag{9}$$

where R and S are real functionals of $\phi(\mathbf{r},t)$ and substitute it into the super-Schrödinger eq. (7), we find once again, from the real and imaginary parts, that

$$\frac{\partial S}{\partial t} + \frac{1}{2}\int[(\frac{\delta S}{\delta\phi})^2 + (\nabla\phi)^2]\,d^3\mathbf{r} + Q = 0, \tag{10}$$

with Q , the super-quantum potential, given by

$$Q = -\frac{1}{2}\int\left[\frac{\left(\frac{\delta^2 R}{(\delta\phi)}\right)}{R}\right]d^3\mathbf{r}, \tag{11}$$

and that

$$\frac{\partial P}{\partial t} + \int \frac{\delta}{\delta\phi} [P(\frac{\delta S}{\delta\phi})] \, d^3\mathbf{r} = 0. \tag{12}$$

Comparing eqs. (10) and (12) with (2) and (4) for the particle, we see that eq. (10) is a generalised Hamilton-Jacobi equation for the field functional $S(\phi(\mathbf{r},t))$, while eq. (12) is a conservation of probability equation, if we note that the field velocity is

$$\frac{\partial\phi(\mathbf{r},t)}{\partial t} = \frac{\delta S(\phi(\mathbf{r},t))}{\delta t}, \tag{13}$$

and the the probability of the field configuration being $\phi(\mathbf{r},t)$ is given by

$$P(\phi(\mathbf{r},t)) = |\Psi(\phi(\mathbf{r},t))|^2 = R^2(\phi(\mathbf{r},t)). \tag{14}$$

The particle guidance condition eq. (5) is replaced by the field guidance condition

$$\pi(\phi(\mathbf{r},t)) = \frac{\delta S}{\delta\phi}. \tag{15}$$

In order to obtain the quantum field equations for the scalar field, we use the Hamiltonian density

$$H' = \pi^2 + (\nabla\phi)^2, \tag{16}$$

so that Hamilton's equations of motion give

$$\frac{\partial\phi}{\partial t} = \frac{\delta H'}{\delta\pi} = \pi, \tag{17}$$

$$\frac{\partial\pi}{\partial t} = -\frac{\delta H'}{\delta\phi} = \nabla^2\phi - \frac{\delta Q}{\delta\phi}. \tag{18}$$

These can be combined to give the quantum field equation in the form

$$\frac{\partial^2\phi}{\partial t^2} = \nabla^2\phi - \frac{\delta Q}{\delta\phi}. \tag{19}$$

It is the presence of the last term that turns this equation into one describing the evolution of the quantum field. This corresponds in the case of the quantum particle to eq. (6).

4. THE INFLATIONARY MODEL OF GUTH AND PI

We now turn to the main point of this paper and show how these techniques can be applied to a specific example, namely, the inflationary universe model proposed by Guth and Pi (1985). In this case the universe evolves according to the Lagrangian density

$$L = \sqrt{-g}\,[\,-\frac{1}{2}g^{\mu\nu}\partial_\mu\phi\partial_\nu\phi - V(\phi)], \qquad (20)$$

where the effective potential $V(\phi)$ is given by

$$V(\phi) = -\frac{1}{2}\,[m^2 - \frac{\lambda T^2}{4}]\,\phi^2. \qquad (21)$$

Here m is the mass of the scalar field and λ is the self-acting coupling constant, T being the temperature of the universe at time t.

As the universe cools, the scalar field becomes trapped in the false vacuum at $\phi \approx 0$ with energy density $\rho_0 = \frac{m^4}{4\lambda}$. Once the temperature falls below a critical value $T_C = 2m\sqrt{\frac{1}{\lambda}}$, the field then slowly "rolls over" into the solution $\phi_C = \frac{m}{\sqrt{\lambda}}$; see Fig. 1.

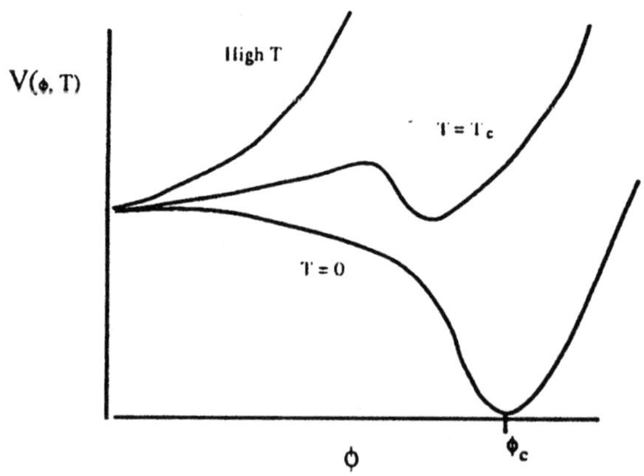

Fig. 1. Effective potential, ϕ, as a function of field and temperature, T, in the inflationary scenario of Guth and Pi.

We will assume that during this process the space-time structure is defined by the de Sitter metric

$$ds^2 = -dt^2 + \exp[2\chi t]\mathbf{dr}.\mathbf{dr}, \tag{22}$$

where χ is the expansion rate defined by $\chi = \sqrt{\frac{8\pi}{3}G\rho_0}$. If we also assume that the temperature drops exponentially, viz.,

$$T = T_0 \exp(-\chi t),$$

the Lagrangrian density finally takes the form

$$L = \frac{1}{2}[\dot{\phi}^2 - \exp(-2\chi t)[(\nabla\phi)^2 + \gamma^2\phi^2] + m^2\phi^2]\exp(3\chi t), \tag{23}$$

where $\gamma^2 = \dfrac{\lambda T_0^{\,2}}{4}$.

This leads to the Hamiltonian density

$$H = \sqrt{-g}\,\frac{1}{2}[\dot{\phi}^2 + \exp(-2\chi t)[(\nabla\phi)^2 + \gamma^2\phi^2] - m^2\phi^2 + \text{const.}]. \tag{24}$$

To allow sufficient time for adequate inflation to take place, χ must be large ($\chi \sim 10^{10}$ light years), then, as $\chi t \rightarrow \infty$, the Hamiltonian density becomes

$$H \rightarrow \sqrt{-g}\,\frac{1}{2}[\dot{\phi}^2 - m^2\phi^2 + \text{const.}] \tag{25}$$

which is just the Hamiltonian of the upside-down harmonic oscillator with unit mass and with m playing the role of the force constant. Thus the inflationary period of the universe can be approximated by the upside-down harmonic potential as shown in fig. 2.

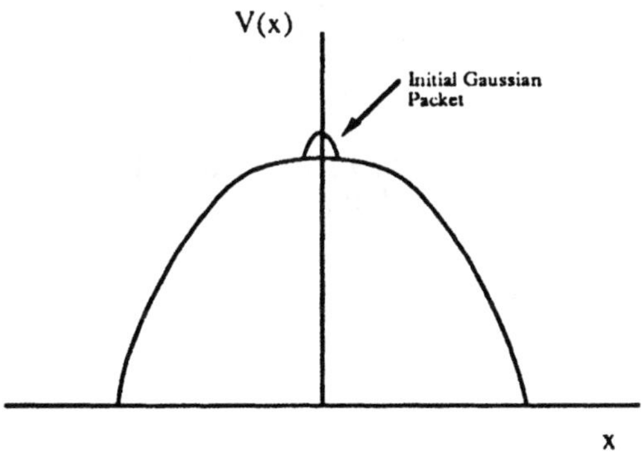

Fig. 2. Up-side Down Harmonic Oscillator (UDHO) potential with initial Gaussian wave packet.

5. THE UPSIDE-DOWN HARMONIC OSCILLATOR

In order to obtain a clearer idea of how the inflationary field $\phi(\mathbf{r},t)$ behaves with time, let us first consider what happens to a particle moving in an upside-down harmonic potential. Let us first consider the classical situation. The Hamiltonian is

$$H = \frac{1}{2}\frac{p^2}{m} - \frac{1}{2}kx^2 \qquad (k > 0). \qquad (26)$$

From this we find that the Hamilton-Jacobi equation is

$$\frac{\partial S}{\partial t} + \frac{(\nabla S)^2}{2m} - \frac{1}{2}kx^2 = 0, \qquad (27)$$

which gives the solutions

$$S(x,t) = \pm \int \sqrt{2m(\beta + \frac{1}{2}kx^2)}\, dx - \beta t, \qquad (28)$$

where β is the energy of the particle. From this we can easily obtain the equation of motion for the particle

$$x = \sqrt{\frac{2\beta}{k}} \sinh(\omega t + \phi), \qquad \beta > 0, \qquad (29)$$

$$x = c\exp[\omega t + \phi], \qquad \beta = 0. \qquad (30)$$

Here $\omega = \sqrt{\frac{k}{m}}$ and ϕ is a constant.

Now we consider the standard quantum mechanical treatment; see Barton (1968). The Schrödinger equation is simply

$$i\hbar\frac{\partial \psi}{\partial t} = -\frac{\hbar^2}{2m}\frac{\partial^2 \psi}{\partial x^2} - \frac{1}{2}kx^2\psi. \qquad (31)$$

We will asume that the wave function of the particle is initially a Gaussian. The wave function at a later time can be written in the form

$$\psi(x,t) = A(t)\exp[-B(t)x^2], \qquad (32)$$

with

$$A(t) = \frac{1}{(2\pi)^{1/4}}[\,b\cos(\phi - i\omega t)]^{-1/2} \qquad (33)$$

and

$$B(t) = \frac{1}{2a^2}\tan(\phi - i\omega t), \tag{34}$$

where

$$a^2 = \frac{\hbar}{\sqrt{mk}},\ b = a(\sin 2\phi)^{-1/2} \quad \text{and} \quad \sin 2\phi > 0.$$

In order to compare this with the classical behaviour we need to know the mean energy of the particle described by the wave function (32). This can easily be shown to be

$$\langle E\rangle = \frac{1}{4}\hbar\omega\left(\tan\phi - \frac{1}{\tan\phi}\right). \tag{35}$$

Now let us consider how this problem can be treated using the Bohm method. The modified Hamilton-Jacobi equation in this case is

$$\frac{\partial S}{\partial t} + \frac{(\nabla S)^2}{2m} - \frac{1}{2}kx^2 + Q = 0. \tag{36}$$

Form the wave function (32), we have

$$R = \frac{b^{1/2}}{(2\pi)^{1/4}}\left[(\cos\phi\cosh\omega t)^2 + (\sin\phi\sinh\omega t)^2\right]^{-1/4}\exp\left[-\frac{1}{2a^2}\frac{x^2\sin 2\phi}{\cos 2\phi + \cosh 2\omega t}\right]$$

and

$$S = \frac{1}{2a^2}\frac{x^2\sinh 2\omega t}{\cos 2\phi + \cosh 2\omega t} - \frac{1}{2}\tan^{-1}[\tan\phi\tanh\omega t].$$

Using

$$p = \frac{\partial S}{\partial x} = \frac{1}{a^2}\frac{\sinh 2\omega t}{\cos 2\phi + \cosh 2\omega t}x, \tag{37}$$

we find the quantum particle trajectory is

$$x_q = C\sqrt{\alpha + \cosh 2\omega t}, \tag{38}$$

where $\alpha = \cos 2\phi$ and C = constant. By comparing this with the classical trajectories we see that they are different. However, if we consider the case as $t \rightarrow \infty$, eq. (38) becomes

$$x_q \approx c\exp(\omega t + \phi_1), \tag{39}$$

which is now identical in form to the classical trajectories. In fact by suitable choice of boundary conditions we can show how a quantum trajectory merges into a classical trajectory (see fig. 3).

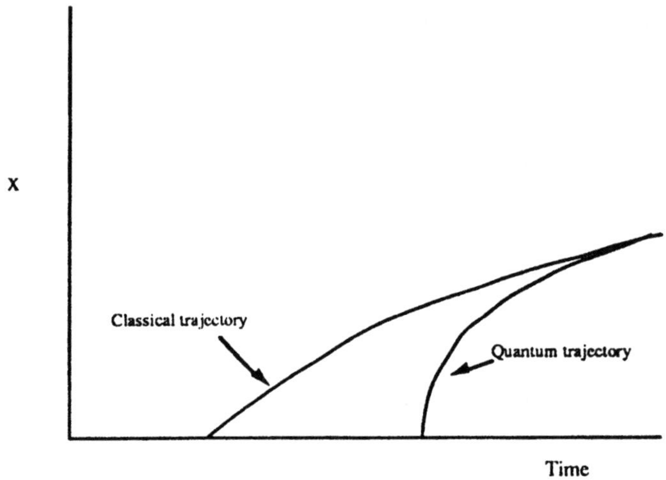

Fig. 3. The merging of classical and quantum trajectories in the UDHO model.

The reason for this merging can be seen by examining the expression for the quantum potential. This takes the form

$$Q = \frac{\hbar}{2m} D(t)[\, 1 - 2D(t)x^2] \,, \tag{40}$$

with

$$D(t) = \frac{\sin 2\phi}{a^2(\cos 2\phi + \cosh 2\omega t)} .$$

Thus, when $t \rightarrow \infty$, $Q \rightarrow$ const. $\exp[\, -2\omega t]$, which clearly becomes negligible with time.

In this example we see one of the advantages of the Bohm interpretation, namely, it provides a clear way of investigating the transition between the quantum and classical domains. Indeed, the UDHO model shows that there can, in fact, be a smooth transition from the quantum trajectories to the classical trajectories. This feature is clearly seen in fig. 3. This result confirms the proposals made by Bohm and Hiley (1986), namely, that there is no need to assume two different types of "reality", one for the quantum level and one for the classical level. Nor is there any need to assume that the classical level emerges from some statistical features as was argued by Guth and Pi (1985).

6. THE CLASSICAL SCALAR FIELD IN THE INFLATIONARY UNIVERSE

Now let us return to consider the emergence of the classical level from the quantum level in the inflationary model. Here we have to study how the field $\phi(\mathbf{r},t)$, which is quantum in origin, transforms into one that is classical in its long time behaviour. Starting with the Lagrangian density (23), we find that the classical field equation is

$$\ddot{\phi} + 3\chi\dot{\phi} - \exp[-2\chi t]\{\nabla^2\phi + \gamma^2\phi\} + m^2\phi = 0. \tag{41}$$

Before proceeding to the quantum behaviour of this field it is necessary to expand it into its normal modes. We then write

$$\phi(\mathbf{r},t) = \frac{1}{(2\pi)^{3/2}}\left(\frac{2\pi}{L}\right)^{3/2}\left[\sqrt{2}q_0(t) + \sum_{k\neq 0}\left[q_+(\mathbf{k},t)\cos\mathbf{k.r} + q_-(\mathbf{k},t)\sin\mathbf{k.r}\right]\right] \tag{42}$$

where we have box normalised so that

$$k = \frac{2\pi n}{L}, \qquad n = 0, \pm 1, \pm 2, \ldots$$

The corresponding expansion for the conjugate momentum is

$$\pi(\mathbf{r},t) = \frac{1}{(2\pi)^{3/2}}\left(\frac{2\pi}{L}\right)^{3/2}\left[\sqrt{2}\pi_0(t) + \sum_{k\neq 0}\left[\pi_+(\mathbf{k},t)\cos\mathbf{k.r} + \pi_-(\mathbf{k},t)\sin\mathbf{k.r}\right]\right]. \tag{43}$$

The Hamiltonian can be written in terms of the variables $q_\alpha(\mathbf{k},t)$ and $\pi_\alpha(\mathbf{k},t)$, where $\alpha = 0,+,$ or $-$, so that

$$H = \sum\left\{\frac{1}{4}\left(\frac{L}{2\pi}\right)^3\right\}\exp(-3\chi t)\pi_{\alpha c}{}^2 + \ldots$$
$$+\left(\frac{2\pi}{L}\right)^3\exp(3\chi t)\left[\exp[-2\chi t](k_\alpha{}^2 + \gamma^2) - \mu^2\right]q_{\alpha c}{}^2). \tag{44}$$

The classical equation of motion for $q_{\alpha c}(\mathbf{k},t)$ is then

$$\ddot{q}_{\alpha c} + 3\chi\dot{q}_{\alpha c} = [m^2 - \exp[-2\chi t](k_\alpha{}^2 + \gamma^2)]q_{\alpha c}, \tag{45}$$

and this is identical with eq. (3.18) of the paper of Guth and Pi (1985). They show that the general solution of eq. (45) can be written in terms of Bessel functions as

$$q_{\alpha c}(\mathbf{k}_\alpha,t) = \left(\frac{\pi}{\chi}\right)^{1/2} \exp(-3\chi t/2)\{a(\mathbf{k})J_p(z) + b(\mathbf{k})N_p(z)\}. \tag{46}$$

Here $p = \left(\frac{9}{4}+\frac{m^2}{\chi^2}\right)^{1/2}$ and $z = \frac{(k_\alpha{}^2 + g^2)^{1/2}}{\chi}\exp[-\chi t]$ and $a(\mathbf{k})$ and $b(\mathbf{k})$ are arbitrary real functions.

For our purposes it will not be necessary to fix the integration constants. We are interested in $q_{\alpha c}$ as $t \to \infty$. Now as $t \to \infty$, $z \to 0$. Therefore we will be interested in the behaviour of $J_p(z)$ and $N_p(z)$ as $z \to 0$. In this limit $J_p(z) \to 0$, while

$$N_p(z) \approx -\, A(p,\mathbf{k}_\alpha)\, z^{-p}[1 + o(z^2)] + O(z^p). \tag{47}$$

Thus, as $t \to \infty$,

$$q_{\alpha c} \approx B(p,\mathbf{k}_\alpha)\exp(\omega t), \quad p > \frac{3}{2}. \tag{48}$$

For the universe considered by Guth and Pi (1985), one finds that $\omega = (p - \frac{3}{2})\chi$. In natural units $\frac{m}{\chi} \gg \frac{3}{2}$ so that $\omega = m$. Then eq. (48) becomes identical to the result we obtained for the UDHO eq. (39).

7. THE QUANTUM SCALAR FIELD IN THE INFLATIONARY UNIVERSE

Let us now turn to consider the quantum evolution of the scalar field $\phi(\mathbf{r},t)$. We will discuss this in terms of the techniques developed in Bohm, Hiley and Kaloyerou (1987). In order to proceed we examine the behaviour of the normal modes of the field. We write the wave functional in the form

$$\Psi = \Psi(\ldots,q_\alpha,\ldots,t) = \prod_\alpha \Psi_\alpha(q_\alpha,t), \tag{49}$$

where the $\Psi_\alpha(q_\alpha,t)$ are the wave functions of each degree of freedom, while q_α are the quantum field co-ordinates. Since the Hamiltonian is given by eq. (44), these fields must satisfy the Schrödinger equation

$$i\frac{\partial\Psi_\alpha}{\partial t} = H_\alpha\Psi_\alpha = \left\{-\frac{1}{4}\left(\frac{L}{2\pi}\right)^3 \exp(-3\chi t)\frac{\partial^2}{\partial q_\alpha{}^2} + \ldots \right.$$
$$\left. + \left(\frac{2\pi}{L}\right)^3 \exp(3\chi t)\left[\exp(-2\chi t)(k_\alpha{}^2 + \gamma^2) - m^2\right]q_\alpha{}^2\right\}\Psi_\alpha. \tag{50}$$

Here the q_α play a role analogous to the position co-ordinate x in the case of the UDHO. Again exploiting this analogy we will assume the initial distribution of q_α to be Gaussian, so that the exact solution at a later time t will be given by

$$\psi_\alpha(q_\alpha,t) = A(\mathbf{k}_\alpha,t)\exp\left[-2\left(\frac{2\pi}{L}\right)^3 B(\mathbf{k}_\alpha,t)\, q_\alpha{}^2\right]. \tag{51}$$

Then, by adopting a suitable normalisation convention, it is straight-forward to show that

$$B(\mathbf{k}_\alpha,t) = \frac{1}{4|\Phi|^2}\,[\,1 - i\,\exp(3\chi t)\frac{\partial}{\partial t}|\Phi(\mathbf{k}_\alpha,t)|^2] \tag{52}$$

and

$$A(\mathbf{k}_\alpha,t) = \left[\left(\frac{L}{2\pi}\right)^3 \pi|\Phi(\mathbf{k}_\alpha,t)|^2\right]^{--1/4}, \tag{53}$$

where $\Phi(\mathbf{k}_\alpha,t)$ is the general solution of the classical field eq. (45) [i.e., we follow Guth and Pi and write $\Phi(\mathbf{k}_\alpha,t)$ instead of $q_{\alpha c}(\mathbf{k}_\alpha,t)$]. It can then be shown that in order to get agreement with the flat space calculations it is necessary to put $C_1 = 1$ and $C_2 = 0$, so that

$$\Phi(\mathbf{k}_\alpha,t) = \frac{1}{2}\left(\frac{\pi\hbar}{\chi}\right)^{1/2}\exp(-\,\frac{3}{2}\,\chi t)H_p{}^{(1)}(z), \tag{54}$$

which is eq. (3.38) in the Guth-Pi paper. $H_p{}^{(1)}(z)$ is the Hankel function of the first kind.

8. THE BOHM APPROACH TO THE INFLATIONARY UNIVERSE

In order to see how the normal modes evolve in this case, we must first substitute

$$\psi_\alpha(q_\alpha,t) = R_\alpha(q_\alpha,t)\exp[iS_\alpha(q_\alpha,t)/\hbar] \tag{55}$$

into eq. (50). After some work, we find the quantum Hamilton-Jacobi equation to be

$$\frac{\partial S}{\partial t} + \frac{1}{4}\left(\frac{L}{2\pi}\right)^3\exp(-3\chi t)\left(\frac{\partial S}{\partial q_\alpha}\right)^2 + V_o q_\alpha{}^2 + Q = 0, \tag{56}$$

where

$$V_0 = \left(\frac{2\pi}{L}\right)^3 \exp[3\chi t]\{\exp(-2\chi t)[k_\alpha{}^2 + \gamma^2] - \mu^2\}, \tag{57}$$

and the quantum potential Q is

$$Q = -\frac{1}{4}\left(\frac{L}{2\pi}\right)^3 \hbar^2 \exp(-3\chi t)\frac{\partial^2 R}{\partial q_\alpha{}^2}\left(\frac{1}{R}\right). \tag{58}$$

The solution of (56) is

$$S(q_\alpha,t) = \frac{1}{2}\left(\frac{2\pi}{L}\right)^3 \frac{1}{|\Phi|^2} \exp(3\chi t)\frac{\partial}{\partial t}|\Phi(k_\alpha t)|^2 q_\alpha. \tag{59}$$

Then

$$\pi_\alpha = \frac{\partial S}{\partial q_\alpha} = \left(\frac{2\pi}{L}\right)^3 \frac{1}{|\Phi|^2} \exp(3\chi t)\frac{\partial}{\partial t}|\Phi(k_\alpha,t)|^2 q_\alpha. \tag{60}$$

Since

$$\pi_\alpha = 2\left(\frac{2\pi}{L}\right)^3 \exp(3\chi t) q_\alpha \tag{61}$$

so that

$$2\frac{\partial q_\alpha}{\partial t} = \frac{1}{|\Phi|^2}\frac{\partial}{\partial t}|\Phi(k_\alpha,t)|^2 q_\alpha \tag{62}$$

which gives

$$q_{\alpha q} = A\exp\left(-\frac{3\chi t}{2}\right)|H_p{}^{(1)}(z)|. \tag{63}$$

This gives a significantly different behaviour from the classical field described by eq. (46). However it is clear that in the limit $t \to \infty$ both the classical $q_{\alpha c}$ and the quantum $q_{\alpha q}$ become the same since as $z \to 0$, $|H_p{}^{(1)}(z)| \approx N_p(z)$. Hence the quantum field evolves into a classical field in the long time limit. This is exactly what is to be expected from our considerations of the UDHO model.

Once again we find that the reason for this agreement at large t is that, in this limit, the super-quantum potential becomes negligible. In order to show this we first note that from eqs. (51) and (52), we see that

$$R(q_\alpha,t) = A(k_\alpha,t)\exp\left[-\left(\frac{2\pi}{L}\right)^3 \frac{q_\alpha^{\ 2}}{2|\Phi|^2}\right]. \tag{64}$$

By using eq. (54) we find that the quantum potential can be written in the form

$$Q = \frac{1}{|H_p^{(1)}(z)|^2}\left[1 - 2\left(\frac{2\pi}{L}\right)^3 \frac{\chi}{\pi\hbar}\frac{\exp(3\chi t)q_\alpha^{\ 2}}{|H_p^{(1)}(z)|^2}\right]. \tag{65}$$

If one examines the limiting form of this, one finds

$$Q \approx \exp[-2(p - \tfrac{3}{2})\chi t] \approx \exp(-2mt) \tag{66}$$

if we use the fact that $\frac{m}{\chi} \gg \frac{3}{2}$ as discussed earlier. Since m is positive, we see that the quantum effects disappear in the long time limit and a classical world can emerge from its quantum origins.

9. CONCLUSIONS

What we have shown here is that the Bohm interpretation provides a very simple criterion to test whether a quantum system will develop a classical behaviour in some suitable limit. This criterion involves examining the quantum potential and seeing under what conditions it becomes negligible. The examples we have used here illustrate that there is no need to have two types of "reality", one for the quantum domain and one for the classical world. Rather there is one "reality" but the qualitatively different behaviours arise in a well defined way which involves the quantum potential playing a central role.

Our application of these ideas to the inflationary scenario considered by Guth and Pi illustrates the simplicity of our approach compared with the conventional approach used by these authors themselves. However it must not be concluded from the work presented here that we now understand how, in general, the classical world emerges from the quantum domain. The questions involved are quite subtle and great care is needed in this analysis. A detailed discussion of this more general problem will be found in chap. 8 of Bohm and Hiley (1993).

REFERENCES

Barton , G, Ann. Phys. **166**, 322 (1968).
Bohm, D., Phys. Rev. **85**, 166 (1952).
Bohm, D., and Hiley, B. J., Phys. Rev. Lett., **55**, 2511 (1985).
Bohm, D., Hiley, B. J. and Kaloyerou, P. N., Phys. Rep. **144**, 349 (1987).

Bohm, D., and Hiley, B. J., *The Undivided Universe: an Ontological Interpretation of Quantum Mechanics*, (Routledge, London, 1993).
Guth, A. H., and Pi, S., Phys. Rev. **32D**, 1899 (1985).
Holland, P.,*The Quantum Theory of Motion, An Account of the deBroglie-Bohm Causal Interpretation of Quantum Mechanics*, (Cambridge University Press, Cambridge, 1993).
Leggett, A. J., and Garg, A., Phys. Rev. Lett. **54**, 857 (1985).

STATE VECTOR REDUCTION VIA SPACETIME IMPRECISION

F. Karolyhazy

Institute for Theoretical Physics
R. Eötvös University, Budapest, Hungary

A slight stochastic modification of the Schrödinger equation is discussed. The time evolution of the state vector can be best visualized as going through (infinitesimal) expansion – reduction cycles, much in the spirit of the GRW model. However, the model presented here doesn't contain any free parameter. First, we argue that the metric structure of spacetime necessarily includes a certain amount of imprecision. Then we propose to show that this imprecision is a reliable guide in formulating the reduction postulates quite generally, applicable to any (nonrelativistic) system. In the microworld, traditional results are re-established. In the free or nearly free motion of macroscopic bodies the model predicts tiny deviations, apparently amenable to experimental check making use of present day technology.

Key words: reduction, measurement, localization.

1. TAKING SIDES

However well known it is, let me begin with the picture of an electron right behind the slits (and in front of a photographic plate) in a two-slit interferometer.

According to the simplest version of "textbook principles" of quantum mechanics, the system under investigation in this case is the electron, its wave function (with obvious notation) is the superposition

$$\varphi(x) = c_1\varphi_1(x) + c_2\varphi(x) \tag{1}$$

the photographic plate playing the role of a measuring apparatus.

The superposition "has to be taken seriously": one could not write a - sign in place of the + sign in (1) without possibly getting in conflict with subsequent

M. Ferrero and A. van der Merwe (eds.), Fundamental Problems in Quantum Physics, 157–167.

events. I shall risk the statement that in this case wave function and "reality" (the "actual state" of the system) are in a one to one correspondence with each other, however sweatening it would be to describe more closely, what reality means.

The photographic plate can be included into the system. The initial state of the total system electron + plate will then be described by

$$\Psi_0(X) = (c_1\varphi_1(x) + c_2\varphi_2(x))\Phi(\xi) = c_1\Psi_{10} + c_2\Psi_{20} \ , \tag{2}$$

with $X=(x,\xi)$, and $\Phi(\xi)$ corresponding to the initial state of the plate.This wave function will soon develop into the superposition

$$\Psi = c_1\Psi_1 + c_2\Psi_2 \ , \tag{3}$$

following the Schrödinger equation for the total system, where Ψ_1 corresponds to the state with a spot in the upper part of the plate, Ψ_2 to the state with a spot in the lower part. (For simplicity, writing down (3), we have assumed that the plate is close to the slits.)

Now, the + sign has to be taken seriously in (2), the same as in (1). Nobody takes it seriously in (3). Why? "Well, the members of the superposition in (2) are only microscopically different, whereas in (3) they differ macroscopically, as befitting a measurement process."

Yet our intuition fails completely if we are asked about the border line (or transition region) between microscopic and macroscopic. This will be even more apparent if we think of a Wilson chamber, instead of a photographic plate. Imagine a situation in which a Wilson chamber, at the beginning, contains merely a single ion (around which a droplet can develop), only, the chamber "doesn't know", whether the ion is at a certain place in the upper part of the chamber or at a certain place in the lower part. Fig. 1*a* represents this initial superposition. The wave function of the chamber will soon develop, following Schrödinger's equation, into what is symbolized by Fig. 1*b*, with nice fat droplets. But the character of the process (diffusion of vapor molecules) doesn't seem to change much as time passes. No clue seems to exist to tell apart superpositions with microscopically different and with macroscopically different members.

Perhaps, we ought not talk about the wave function of the Wilson chamber at all. The chamber cannot be a closed system, it is exposed to the bombardment of gas molecules from the outside, to heat radiation,etc. True, but the thermal motion inside the chamber is quite erratic even if we imagine the chamber as isolated from the surrounding. (This is a remarkable situation indeed. We may wish to include the laboratory or the whole Galaxy into the system – initially the Wilson chamber. The bigger the total system, the more ridiculous it is to regard it closed and write down its wave function. Yet, at the same time, the more perplexing the idea seems that the border between microscopic and macroscopic,

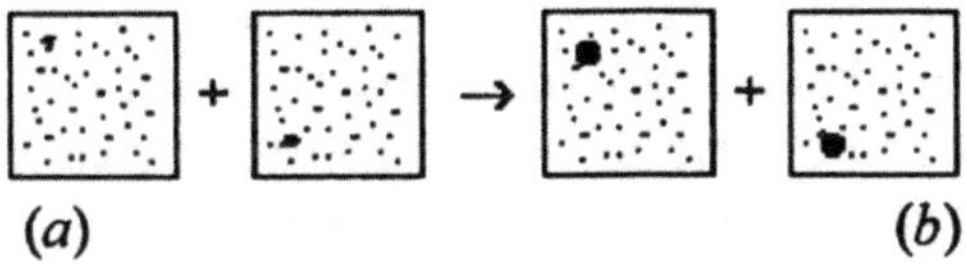

Fig. 1. Superposed state of a Wilson chamber,
(*a*) initially, (*b*) appreciably later.

during the evolution of the droplet, is determined by heat radiation, etc., coming from outside the Galaxy.)

To say that the superposition of Fig. 1*b* *does stand* in a one to one correspondence with reality, until an interaction with some *other* macroscopic body takes place, seems too easy a way out to me.

Our conclusion is the following:

Somewhere along the way from Fig. 1*a* to Fig. 1*b* (or, more generally, on the way toward a superposition with "very" different members), the one to one correspondence between wave function (obeying Schrödinger equation) and reality gets lost, without the excuse that "from this point on – not before – we may talk about a measurement".

But how? Where? Who is the culprit?

On the one hand side, during the last two decades, many experiments have been performed, involving microscopic systems (consisting of a few elementary particles) but macroscopic distances. (See, for example, Clauser and Shimony, 1978; Aspect et al., 1982.) These experiments confirmed the validity of quantum mechanics and, thus, the validity of the superposition principle.

On the other hand, many physicists are of the opinion that the whereabouts of the breakdown of the superposition principle is irrelevant, if only it occurs deep enough in the region where the components of the superposition are already macroscopically different.

In fact, one can take the view, as was first boldly done by H. Everett, that no reduction ever occurs, the different branches (eventually of the whole universe) all possess their own reality, unaware of each other.

A more recent attitude, similar in spirit, is that it is a matter of taste whether we believe or not that reductions actually occur in nature: because of the continuous, inevitable microscopic interactions with the surrounding: macroscopic (or even submacroscopic) systems that start out to develop "funny superpositions", decohere very quickly into ("classical") states that become dynamically irrevocably independent of each other (see, for example, Joos and Zeh, 1985; Zeh, 1993).

Our purpose is to turn a matter of taste into a matter of experiment.

As a matter of taste, we join those who abhor the idea that Schrödinger's cat can be dead and alive simultaneously. However, we shall argue that the *principal impossibility* of states corresponding to superpositions with macroscopically different members may lead to directly observable effects, not merely to the principal exclusion of certain extravagant interference phenomena, which, for practical reasons, would remain unobservable anyway.

Once we adopt the view that coexistence of macroscopically different branches in a superposition is not only latent – because of the enormous complexity of the branches – but principally impossible, then it is no longer necessary to regard the system open. Indeed, there is nothing in the present formalism of quantum mechanics that forbids us to apply it to an imaginary situation in which a Wilson chamber (or any other complex system) is ideally isolated. In our view, therefore, the message of the contrast between the stochastic nature of the outcome of observations and the deterministic emergence of unrealistic superpositions is that we must not assign unlimited validity to the Schrödinger equation. We need, instead, a slight modification in the law of the wave propagation, in the spirit of Ghirardi, Rimini, and Weber (1986), which:

(*a*) prevents the wave function from developing unacceptable superpositions;

(*b*) makes the deterministic and stochastic aspect of the wave propagation go hand in hand generally, deleting the distinguished status of "measuring processes"; and, thereby,

(*c*) keeps the close correspondence between wave function and actual state.

2. IMPRECISION IN SPACETIME STRUCTURE AND THE BREAKDOWN OF THE SUPERPOSITION PRINCIPLE

The danger of arbitrariness in seeking a slight modification of the Schrödinger equation is diminished if we look for some other unsolved problem in physics and try to find a connection. Our candidate, already in some early work on the subject (Karolyhazy, 1966, 1974) was the resistance of general relativity against quantization.

We shall try to interpret this resistance as a sign that nature "tries to keep" spacetime structure classical as much as possible. However, we must concede at least some minimal quantum mechanical uncertainty in position (and energy, etc.) of the macroscopic bodies. Otherwise, quantum mechanics would be completely rejected. Spacetime with a perfectly sharp structure then is an impermissible idealization.

On the other hand, the empirical fact that massive bodies never develop large uncertainties in position, suggests that spacetime, with a little blur in it, is apt to narrow the validity of the Schrödinger equation. [The idea that if spacetime structure has some "smear" in it, then the relative phase between distant parts of a wave function may gradually, with increasing distance, become indefinite, had already – without elaboration – occurred to Feynman (1962). For other discussions, making reference to spacetime fluctuations, see, e.g., Sanchez Gomez (1992), and Diosi (1987).]

In the model presented here the basic tasks were: (*a*) to make an estimate about the degree of imprecision which "should be put" into spacetime, and (*b*) to create the rules to describe wave propagation on such a spacetime.

Let us briefly review them in turn.

2.1. Imprecision in Spacetime Structure

Suppose we want to "realize" (measure out) a world line segment of length s in a flat spacetime domain. Trying to keep the basic features of both quantum mechanics and general relativity we get the following estimate on the absolutely maximal precision with which we can achieve this goal. The absolutely unavoidable uncertainty Δs in the length s of any world line segment is given by

$$\Delta s^2 = \Lambda^{4/3} s^{2/3} , \tag{4}$$

Λ is Planck's length.

The argument leading to (4) is admittedly heuristic. Its main ingredients are, first, the unavoidable broadening of the position uncertainty of any "clock," as time passes, because of the Heisenberg uncertainty relations, second, the uncertainty of the length s connected with the Schwarzschild metric created by the mass of the "pointer," the uncertainty here arising from the fact that the mass of the pointer is slightly uncertain because its energy is (quantum mechanically) uncertain. Choosing the optimal combination of the various uncertainties we obtain (4).

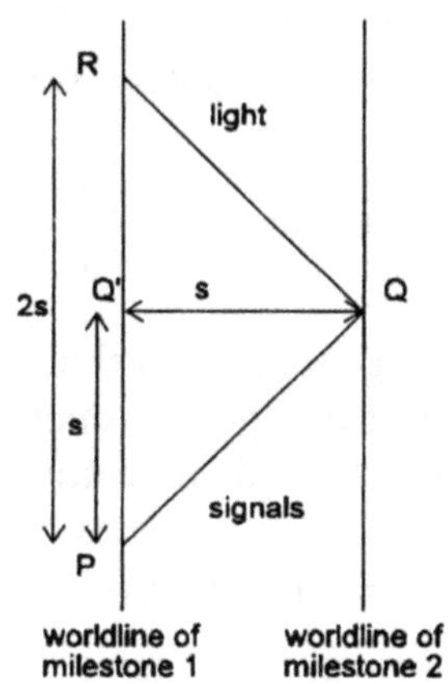

Fig. 2. Establishing simultaneity in an inertial frame of reference.

While deriving (4), we were, of course, not interested in practical possibilities. We disregarded any restrictions on our "devices" other than those posed by the very basic postulates of quantum mechanics and classical general relativity. The value of Δs obtained this way may prove too small. However, in that case, the consequences to be expected might be even more pronounced.

2.2. Wave Propagation

We shall transform the imprecision expressed by (4) into a "relative phase uncertainty " concept.

First, we try to visualize an inertial frame of reference in a flat spacetime domain, in which we wish to write down the wave function of the system under investigation. Taking a look at Fig.2, we immediately see that the simultaneity of the points Q and Q', separated by the spatial distance s in space, can only be checked by measuring the time interval PQ'. In view of (4), we recognize that simultaneity at a *spatial distance s* cannot be established with an uncertainty smaller than expressed by the same formula

$$\Delta s^2 = \Lambda^{4/3} s^{2/3} , \tag{5}$$

How to proceed to a relative phase uncertainty concept? Answer:

(*a*) Replace the simple hyperplane $t = t_0$ by a set of hypersurfaces

$$t = t_0 + \tau_\beta(\mathbf{x}) , \tag{6}$$

close to $t = t_0$, β labelling the members of the set.

If we choose the set $\{ \tau_\beta(\mathbf{x}) \}$ properly, we easily get the desired relation

$$\left\langle \left[c\tau_\beta(\mathbf{x}) - c\tau_\beta(\mathbf{x}+\mathbf{s}) \right]^2 \right\rangle = \Lambda^{4/3} s^{2/3} , \tag{7}$$

where the brackets mean average taken over the set.

(*b*) For a single, slow, say, spinless microscopic particle of mass m the connection between the relativistic and nonrelativistic wave function is

$$\varphi(\mathbf{x},t) = \psi(\mathbf{x},t) \exp(-imc^2 t/\hbar) . \tag{8}$$

Replacing t by $t + \tau_\beta(\mathbf{x})$ before omitting the factor $\exp(-imc^2 t/\hbar)$, we get, to a good approximation,

$$\psi_\beta(\mathbf{x},t) = \psi(\mathbf{x},t) \exp\left[i\Phi_\beta(\mathbf{x}) \right] , \tag{9}$$

with

$$\Phi_\beta(\mathbf{x}) = -(mc^2/\hbar)\,\tau_\beta(\mathbf{x})\ . \tag{10}$$

This extra phase can immediately be generalized to the case of any number of particles, gases, solids (of nonrelativistic motion). The many particle wave function $\Psi(X) = \Psi(\mathbf{x}_i)$, corresponding to t=const., will go over into the set

$$\left\{\Psi_\beta(X)\right\} = \left\{\Psi(X)\exp\left[i\Phi_\beta(X)\right]\right\}\ , \tag{11}$$

with

$$\Phi_\beta(X) = -\sum_i (m_i c^2/\hbar)\,\tau_\beta(\mathbf{x}_i)\ \ ; \tag{12}$$

X denotes the coordinate in configuration space, and the sum is taken over the particles.

We shall regard the set (11) as the mathematical representative of a single *pure state* in blurred spacetime. Also, we may say that the set embodies a wave function with a relative phase uncertainty. The simplest way, however, to see the content of the pure state (11) is to form, by definition, the density matrix

$$\rho(X,X') = \Psi(X)\Psi^*(X')\left\langle \exp\left[i\Phi_\beta(X) - i\Phi_\beta(X')\right]\right\rangle\ , \tag{13}$$

where the brackets again mean average over β. The quantity

$$D(X,X') = \left\langle \left[\Phi_\beta(X) - \Phi_\beta(X')\right]^2 \right\rangle \tag{14}$$

measures the relative phase uncertainty between the points X and X' in configuration space.

We have $Tr\,\rho = 1$, but

$$Tr\rho^2 = \int dX dX' |\Psi(X)|^2 |\Psi(X')|^2 \exp[-D(X,X')] <\ . \tag{15}$$

The value of $Tr\rho^2$ reflects the "quality of the coherence" of the pure state (11). The smaller the extension of $\Psi(X)$ in configuration space, the smaller (in average) $D(X,X')$, the closer the value of $Tr\rho^2$ to 1.

2.3. The Reduction Postulate

Up to now, no mention has been made of any deviation in the time evolution of $\Psi(X)$ from the Schrödinger equation.However, as $\Psi(X)$ expands with time, $Tr\rho^2$ diminishes, the coherence of the wavefunction deteriorates.To counteract this, we now introduce the postulate of (infinitesimal) expansion-reduction cycles in the following way:

From some time t to $t+r$ (r infinitesimal) the Schrödinger equation be valid.

$$\Delta Tr\rho^2 = Tr\rho^2(t+r) - Tr\rho^2(t) \tag{16}$$

is then negative. At the moment $t+r$ a stochastic reduction $\Psi(t+r) \rightarrow \Psi_i$ should occur, with probability P_i, to some more coherent Ψ_i, such that

$$Tr\rho_i^2 > Tr\rho^2 \tag{17}$$

We specify the set (P_i, Ψ_i) by three demands:

1. $$|\Psi(X)|^2(t+r) = \sum_i P_i |\Psi_i(X)|^2, \quad \text{for all } X \tag{18}$$

(postulate of no bias). The second demand concerns the "degree of intrusion" upon the deterministic time evolution.

Through the actual transition $\Psi \rightarrow$ some definite Ψ_i the coherence quality improves, but at the expense of damaging in an other way the "ideal" time evolution: through the unpredictability of the outcome of the reduction. The measure of the latter can be best expressed with the help of the auxiliary quantity

$$\rho^{aux} = \sum_i P_i \rho_i \ , \tag{19}$$

by forming

$$Tr(\rho^{aux})^2 - Tr\rho^2(t+r) \tag{20}$$

(this quantity is also negative). We also demand

2. $$Tr(\rho^{aux})^2 - Tr\rho^2(t+r) = \Delta Tr\rho^2 \tag{21}$$

(postulate of tolerance). That is the only reasonable way to accept the hint of deteriorating coherence to take steps to prevent it. Otherwise, we would either be negligent in saving the coherence, or do more harm than good to the time evolution.

The third demand reflects the assumption that there is a tendency in nature for the reductions to be effective:

3. $$\sum_i P_i Tr \rho_i^2 \rightarrow \max!$$ (22)

(postulate of efficiency). These prescriptions determine the possible paths of evolution for the system under investigation.

As we see, in our model spacetime imprecision in itself does not automatically trigger stochastic reductions. They have to be postulated. But the concept of the "quality of the coherence" for a pure state helps us to do so "smoothly," with rather little arbitrariness. Deterministic and stochastic aspect of the time evolution of any system appear to be both inherent, they go hand in hand.

3. APPLICATIONS

We briefly mention two simple cases, where the calculations could be carried out explicitly.

(*a*) *Wilson chamber model.* Making use of our postulates, we can easily predict the fate of the initial superposition

$$\Psi(0) = c_1(0)\Psi_{up} + c_2(0)\Psi_{down} \tag{23}$$

in the Wilson chamber model of Fig.1, if we make some further simplifying assumptions (e.g. , that the chamber is at rest, or, that apart from the growth of the droplet, the density of the gas in the chamber remains uniform).

In this (somewhat artificially) simple case, the prediction for any later moment t involves only two possible states,

$$\Psi_A(t) = c_{1A}(t)\Psi_{up} + c_{2A}(t)\Psi_{down} \tag{24}$$

and

$$\Psi_B(t) = c_{1B}(t)\Psi_{up} + c_{2B}(t)\Psi_{down} \quad , \tag{25}$$

with the corresponding probabilities P_A and P_B . Ψ_{up} and Ψ_{down} describe, of course, the states "droplet above" and "droplet below," respectively. (They are not constant in time, because the mass $M(t)$ of the droplet increases with time.)

The time dependence of the c's and P's are displayed in Fig.3. We see that the actual state of the system (either Ψ_A or Ψ_B), not too long after the initial moment t=0, still will be a superposition in the droplet position, but as time passes, Ψ_A tends to Ψ_{up} and Ψ_B tends to Ψ_{down} . The limiting values of P_A and P_B are $c_1^2(0)$ and $c_2^2(0)$ respectively, in agreement with standard measurement

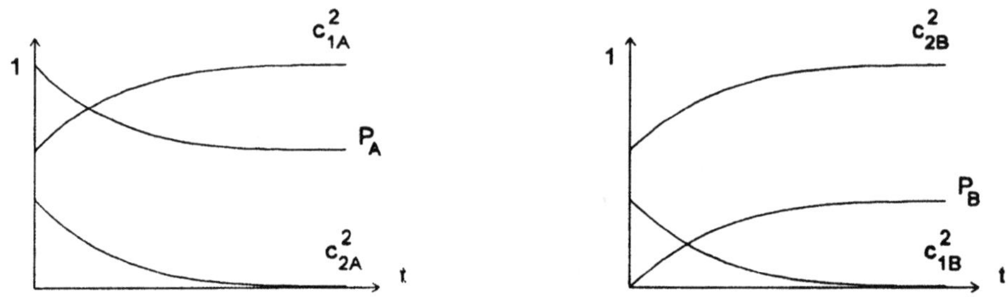

Fig. 3. Decay of the superposition in a simple Wilson chamber model.

theory. (Thus, this type of result, though important theoretically, offers nothing new to the experimentalist.)

The time dependence of the c's and P's is determined solely by the time dependence of the mass $M(t)$ of the droplet and the distance a of the two different positions of the droplet. The relevant quantity is

$$D(t) = (G/\hbar^2)^{2/3} M(t) a^{2/3} \quad , \tag{26}$$

deriving from $D(X,X')$ in (14). G is the constant of gravity.

It turns out that $c_{1A}(t)$ and $c_{2B}(t)$ are equal:

$$c_{1A}(t) = c_{2B}(t) = c_1(t) \ , \tag{27}$$

and similarly,

$$c_{2A}(t) = c_{1B}(t) = c_2(t) \ . \tag{28}$$

Furthermore, with

$$f = (c_1^2 - c_2^2)^2 \ , \tag{29}$$

we have

$$1 - f = (1 - f_0)\exp[D(0) - D(t)] \tag{30}$$

and

$$\ln(P_A - P_B) = -\frac{1}{2}\ln\frac{1-(1-f_0)\exp[D(0)-D(t)]}{f_0} \ . \tag{31}$$

If, for example, $a \approx 1$ cm, then the superposition practically decays by the time the mass of the droplet reaches the value $M \approx 10^{-15}\ g$.

(*b*) *Free motion of a body.* The experimental possibilities are connected with the motion of solid bodies. For a ball of radius R and mass M the function $D(X,X')$ in (14) turns out to depend only on these quantities and the difference in the center of mass coordinate. Any spread larger than

$$a \approx (\hbar^2 / G)^{1/3} R^{2/3} / M \quad (32)$$

in the center of mass coordinate will be killed by the reductions. With $R \approx 1$ cm and ordinary density, we would have $a \approx 10^{-16}$ cm. Since every (infinitesimal) stochastic reduction gives rise to an infinitesimal jump in the momentum and energy of the free or nearly free mass, the tiny anomaly – sort of an anomalous Brownian motion – in the behaviour of the testbodies could serve, if observed, as a confirmation of the breakdown of the superposition principle. (The slight violation of the conservation laws in our model can be shown to be harmless in ordinary observations.)

ACKNOWLEDGEMENT

This work was partially supported by the National Research Foundation of Hungary (OTKA), Grant No. I/3, 2193.

REFERENCES

Aspect, A., Dalibard, J., and Roger, G. (1982), *Phys. Rev. Lett.* **49**, 1804.
Clauser, J. F., and Shimony, A. (1978), *Rep. Prog. Phys.* **41**, 1881.
Diosi, L. (1989), *Phys. Rev. A* **40**, 1165.
Feynman, R. P. (1962), Lecture on Gravitation, California Institute of Technology.
Ghirardi, G. C., Rimini, A., and Weber, T. (1986), *Phys. Rev. D* **34**, 470.
Joos, E., and Zeh, H. D. (1984), *Z. Phys. B* **59**, 223.
Karolyhazy, F. (1966), *Nuovo Cimento* **42**, 390.
Karolyhazy, F. (1974), Magyar Fizikai Folyoirat **22**, 23.
Karolyhazy, F., Frenkel, A., and Lukacs, B. (1982), in *Physics as Natural Philosophy: Essays in Honor of Laszlo Tisza on his Seventy-Fifth Birthday*, Shimony, A., and Feshbach, H., eds. (MIT Press, Cambridge, Massachusetts).
Karolyhazy, F. (1990), in *Sixty two years of uncertainty*, Miller, A. I., ed. (Plenum, New York).
Unturbe, J., and Sanchez - Gomez, J. L. (1992), *Nuovo Cimento B* **107**, 211.
Zeh, H. D. (1993), *Phys. Lett. A* **172**, 189.

ANALYSES OF CLASSICAL AND THERMODYNAMIC LIMITS OF QUANTUM MECHANICS AND QUANTUM MEASUREMENTS ON THE BASIS OF NONSTANDARD ANALYSIS

Tsunehiro Kobayashi

Institute of Physics, University of Tsukuba
Ibaraki 305, Japan

Classical and thermodynamical limits of quantum mechanics are studied on the basis of nonstandard analysis. Decoherence among classical states are derived in the limit.

Key words: quantum mechanics, classical limit, thermodynamical limit, decoherence, nonstandard analysis.

1. INTRODUCTION

It is known that nonstandard analysis is the theory which provides a suitable framework for the analysis by means of infinitesimals (≈ 0) and infinities (∞). In physical analyses, infinitesimals and infinities are usally defined by the limits. Sometimes we do not understand whether such limits are well-defined or not. A good example is the classical limit of quantum mechanics defined by the limit of $\hbar \to 0$. We know that the wave function having the phase such as $W(x)/\hbar$ with $W(x) = \int^x \sqrt{2m(E - V(x'))}dx'$ loses its meaning when $\hbar = 0$ is taken. A fundamental concept of quantum mechanics such as the wave function has no place in this limit. In order to investigate such limits exactly, nonstandard analysis may be suitable. In this report we shall investigate the classical and thermodynamic limits on the basis of nonstandard analysis.

1.1. Terminology of Nonstandard Analysis.

$^*\mathcal{R}(^*\mathcal{N})$: nonstandard extention of $\mathcal{R}(\mathcal{N})$ (set of real numbers (natural numbers)).

$^*\mathcal{H}$: nonstandard extention of $\mathcal{H}$ (Hilbert space for Hamiltonian H).

M. Ferrero and A. van der Merwe (eds.), Fundamental Problems in Quantum Physics, 169–176.

Eigenpair: For $f \in *\mathcal{H}$ and $\lambda \in {}^*\mathcal{R}, (\{, \lambda)$ is called an eigenpair, if and only if $||Af - \lambda f|| = 0$, where A is a self-adjoint, internal and bounded operator on ${}^*\mathcal{H}$.

Ultra eigenpair: (f, λ) is called an ultra eigenpair, if and only if $||Af - \lambda f|| \approx)$.

Equivalence relation ($\longleftrightarrow$) on $U({}^*\mathcal{H})$: [1] f and g are equivalent on $U({}^*\mathcal{H})$ (set of unitvectors on ${}^*\mathcal{H}$), if and only if $f, g \in U({}^*\mathcal{H})$ and there exists a $\phi \in {}^*\mathcal{R}$ such that $||f - \epsilon^{i\phi} g|| \approx 0$.

Physical space: The physical space on ${}^*\mathcal{H}$ is defined by $S({}^*\mathcal{H}) = U({}^*\mathcal{H})/\longleftrightarrow$. Note that in general $S({}^*\mathcal{H})$ contains both of eigenpairs and ultra eigenpairs.

Expectation values: [1] If $\phi \in S({}^*\mathcal{H})$ is a vector inducing a probability measure with respect to A, the expectation value is given by

$$st(< \phi, A\phi >) = st\left(\int_{\lambda \in {}^*\mathcal{R}} \lambda d < \phi, E_\lambda(A)\phi >\right),$$

where $st(<,>)$ stands for the operation of taking the standard part defined in nonstandard analysis, which corresponds to the operation such as the classical limit $\hbar \to 0$.

1.2. Topology on Topological Space X in Nonstandard Analysis

$\mu(a)$: For $a \in X$ monad of a stands for the set of $x \in {}^*\mathcal{R}$ satisfying $|a - x| \approx 0$. Monad field: Projection of $a \in X \to \mu(a)$.

Theorem: Topology on X is determined by the monad fields.

This theorem means that topology on $S({}^*\mathcal{H})$ depends on the choice of infinitesimal constants. In the classical limits $\hbar$ should be chosen as the infinitesimal to determine the topology, while $1/N$ will be taken as the infinitesimal in the thermodynamical limit. We write the physical spaces with the topologies determined by taking $\hbar \approx 0$ in the classical limit and $1/N \approx 0$ in the thermodynamical limit as $S_\hbar({}^*\mathcal{H})$ and $S_{\text{therm}}({}^*\mathcal{H})$, respectively [2,3,4].

2. ANALYSIS OF CLASSICAL LIMIT

We define the classical limit of quantum mechanics by extended quantum mechanics on $S_\hbar({}^*\mathcal{H})$. The operation for taking the classical limit can be represented by the $st_\hbar$-operation, in which $\hbar$ is taken as the infinitesimal [2,3]. Now we take the Schrödinger equation with the approximate equality as the fundamental equation, i.e.,

$$H\phi \hat{\approx} E\phi$$

for stationary stases, where the approximate equality $\hat{\approx}$ is defined by $\|H\phi - E\phi\| \approx 0$.

2.1. Ultra Eigenfunctions of Extended Schrödinger Equation

Let us investigate classical solutions for stationary states were $\phi(x) = \exp(iW(x)/\hbar)$ can be used. We define the real and imaginary parts of $dW(x)/dx$ by $u(x)$ and $v(x)$, respectively, as $\frac{dW(x)}{dx} = u(x) + iv(x)$. The Schrödinger equation with $H = (-i\hbar\nabla)^2/2m + V(x)$ is reduced to the following two equations for u and v:

$$\hbar\frac{du}{dx} \hat{\approx} 2uv, \tag{A}$$

$$\hbar\frac{dv}{dx} \hat{\approx} 2m\,(E - V(x)) - (u^2 - v^2). \tag{B}$$

From now on we define the orders $O(1)$ and $O(\hbar)$ as follows: $st_\hbar(O(1)) \neq 0$ *and* $\infty, st_\hbar(O(\hbar)) = 0$. From (A) we immediately derive the relation

$$v(x) \hat{\approx} \frac{\hbar}{2u}\frac{du}{dx} = \frac{\hbar}{2}\frac{d\ln u(x)}{dx} \sim O(\hbar).$$

Note that, if u is not a constant, v has the order of $O(\hbar)$ which is independent of the order of u. We can derive solutions of u having the order $O(1)$ from (B) as

$$u_{cl}(x) = \pm\sqrt{2m(E - V(x))}, \quad E - V(x) > 0,$$

which is nothing but the classical momentum of the particle moving in the potential $V(x)$. We can easily estimate the order of $O(\hbar)$ in putting $u(x) \hat{\approx} u_{cl} + \hbar u_1$ from (B) as $u_1(x) = 0$. From these equations the wave function may be represented by

$$\phi_c^E(x) \hat{\approx} \frac{N}{\sqrt{u_{cl}(x)}} \exp\left[i\frac{W_{cl}(x)}{\hbar}\right],$$

for $0 < E - V(x) \in R$, where $W_{cl}(x) = \int^x \sqrt{2m(E - V(x'))}dx'$. Let us discuss the turning points x_0 for satisfying $E - V(x_0) = 0$. In the neighborhood of x_0 where the relations $0 < E - V(x) < O(\hbar^n)$ with $n \geq 2$ are satisfied, the equation for W is reduced to

$$i\hbar\frac{d^2W}{dx^2} \hat{\approx} \left(\frac{dW}{dx}\right)^2.$$

Since this equation has the solution $W(x) \hat{\approx} - i\hbar \ln|x_0 - x|$, we see that

$$\phi_c^E(x) \propto |x_o - x|$$

for $0 < E - V(x) < O(\hbar^n)$ with $n \geq 2$, i.e. $st_\hbar(E - V(x)) = 0$. An important point of our approach is in the fact that we need not solve the Schrödinger equation exactly. We may represent the ultra eigenfunction by

$$\phi_c^E(x) = \frac{N^E}{\sqrt{u_c^E(x)}} \exp\left(iW_{cl}(x)/\hbar\right),$$

where N^E is a constant and $u_C^E(x)$ stands for a function satisfying the conditions

$$u_c^E(x) \hat{\approx} u_{cl}(x) + O(\hbar^2), \quad 0 < E - V(x) \sim O(1),$$

$$u_c^E(x) \propto |x_0 - x|^{-1}, \quad 0 < E - V(x) < O(\hbar^2),$$

such that

$$u_c^E(x) = \frac{u_{cl}(x)^2 + \hbar^2 A(x_o - x)^{-2}}{u_{cl}(x)^2 + \hbar^2 B},$$

with the constants $A, B \in {}^*\mathcal{R}$ being finite and not infinitesimal. The important point is the fact that $\phi_c^E(x = x_0) = 0$. This fact shows that we can connect the eigenfunction given in the region $E - V(x) > 0$ with the function

$$\phi_c^E(x) \approx 0$$

for $E - V(x) < 0$ at $x = x_0$. Let us examine the interesting properties of $\phi_c^E(x)$:

(1) *Distribution function*: The observed distribution is given by

$$st_\hbar(\rho_c) = |N^E|^2/\sqrt{2m(E - V(x))}, \quad E - V(x) > 0,$$

which coincides with that expected from classical mechanics.

(2) *Decoherence among the classical solutions*: Let us consider the superposition of two different classical states with two different energies E and E' satisfying the condition $0 \neq E - E' \in \mathcal{R}$, i.e., $E - E' \sim O(1)$, as $\psi(x) = c_E \phi_c^E(x) + c_{E'} \phi_c^{E'}(x)$. In general the phase factors of off-diagonal terms of density matrix for $\psi(x)$ are described by $\exp[i(1/\hbar)(W_{cl}^E(x) - W_{cl}^{E'}(X))]$. We see that

$$st_\hbar((1/\hbar)(W_{cl}^E(x) - W_{cl}^{E'}(x))) = \infty, \quad 0 \neq E - E' \in, \mathcal{R}.$$

The contributions of the off-diagonal terms disappear in the $st_\hbar$ operation appearing in the expectation values because of the divergent property of the phases. We may conclude that all the pure states represented by the superposition in terms of the classical solutions $\phi_c^E(x)$ are reduced to the mixed states in the $st_\hbar$ operation on $S_\hbar({}^*\mathcal{H})$. The so-called paradox of Schrödinger's cat has no meaning in this theory, because living and dead cats, which have classically observable differences, should be represented by two different classical solutions, and then the decoherence between two states are trivial in this scheme [2,3].

2.2. Classical Solutions for Harmonic Oscillator Potentials

Ultra Eigenfunctions[3]. Let us study the classical solutions for a one-dimensional harmonic oscillator potential $V = \frac{1}{2}m\omega^2x^2$. We can write down the classical solution for $0 < E - V(x) \in {}^*\mathcal{R}$ as

$$\phi_c^E(X)^\pm = \frac{N^E}{\sqrt{u_{cl}^E(x)}} \exp\left[i\frac{W_{cl}^\pm(x)}{\hbar}\right],$$

where

$$u_{cl}^E(x) = \left[(m\omega)^3(x_0^2 - x^2)^{3/2} + \hbar^2 A(x_0^2 - x^2)^{-2}\right] / \left[(m\omega)^2(x_0^2 - x^2) + \hbar^2 B\right]$$

and

$$W_{cl}^\pm(x) = \pm\frac{m\omega}{2}\left[x\sqrt{x_0^2 - x^2} + x_0^2 \arcsin\frac{x}{x_0}\right].$$

Coherent states on the extended Hilbert space[3]. Following the usual technique deriving the coherent states, we can write the coherent states on ${}^*\mathcal{H}$ in therms of the superposition of the eigenfunctions $\phi_n(x)$, which are the extention of the usual eigenfunctions written as Hermite polynomials, such that

$$\psi(x,t) =^* \sum_n c_n\phi_n(x)e^{-iE_nt/\hbar},$$

where c_n are coefficients and ${}^*\sum$ denotes the sum extended in ${}^*\mathcal{N}$. For the choice of $c_n = \sqrt{\frac{\alpha}{\sqrt{\pi}}\frac{(A/\sqrt{2})^2}{n!}}e^{-\frac{1}{2}A^2}$, where A is taken from ${}^*\mathcal{R}$, we derive the coherent state

$$\psi_e(x,t) = \left(\frac{\alpha}{\sqrt{\pi}}\right)^{1/2} \exp\left[-\frac{\alpha^2}{2}\left(x - \frac{\sqrt{2}A}{\alpha}\cos\omega t\right)^2\right] e^{-i[\frac{\omega}{2}t+\sqrt{2}A\alpha\sin\omega t(x-\frac{A}{\sqrt{2}\alpha}\cos\omega}$$

It is well known that the mean energy of the coherent state is given by $< E >= (< n > +1/2)\hbar\omega$, where the mean number $< n >$ is related to A as

$<n>=A^2$. Since $<E>$ must be a classical number of R in the classical limit $\hbar \approx 0$, then we have to take a constraint

$$<n>\hbar \sim O(1).$$

These relations mean that A is a nonstandard number having a infinite magnitude in the right classical limit defined by $\hbar \approx 0$ and $<n>\hbar$ fixed in $\mathcal{R}$. We see that

$$\lim_{(\hbar \to 0, <n>\hbar = E_o/\omega = \text{fixed})} \rho_\psi = \delta\left(x - \sqrt{\frac{2E_0}{m\omega^2}} \cos\omega t\right).$$

It is apparent that $x = \sqrt{2E_0/m\omega^2}\cos\omega t$ is nothing but the classical path. It is shown that the mean values of E, p and x coincide with those of the classical mechanics and their mean squared variances become zero in the classical limit.

We may conclude that the physical space for classical mechanics in the Schrödinger equation, $S_\hbar({}^*\mathcal{H})$, is constructed from the two states corresponding to the stationary states represented by the ultra eigenfunctions and the coherent states represented by the superpositions of the eigenfunctions. We note that the coherent states represent the particle nature of wave functions in the classical limit, while the stationary classical-states do the statistical nature of them.

2.3. Remarks on Classical Solutions

In this report we could not discuss the physical background of the ultra eigenfunctions and many-Hilbert space theory, measurements on the ultra eigenstates, and an interesting relation between detectors in physics and filters in nonstandard analysis [3].

3. ANALYSIS OF THERMODYNAMIC LIMIT

3.1. Thermal Equilibrium in Nonstandard Analysis

Thermal equilibrium is realized in states satisfying *principle of equal a priori probability*. Our question is "*What is the physical space* ($S_{\text{therm}}({}^*\mathcal{H})$) *wherein the principle is realized?*"

Equivalence relation and $S_{\text{therm}}({}^*\mathcal{H})$. On $S_{\text{therm}}({}^*\mathcal{H})$ all states with the same energy (E_M) have the same weight in the density matrix of statistical mechanics. For instance, the density matrix for the canonical ensemble is give by $\rho_{\text{stat}}(N,V) = Q_N^{-1}\sum_n e^{-\beta H}|n><n| = Q_N^{-1}\sum_M e^{-\beta E_M}\sum_l |M,l><M,l|$, where $Q_N = Tr(\sum_n e^{-\beta H}|n><n|)$ and $<M,l|M',l'>=\delta_{MM'}\delta_{ll'}$.

A state in which all the states with the same energy (E_M) have the same weight is described by $|\chi_M>=\frac{1}{\sqrt{W_M}}\sum_{l=1}^{W_M}|M,l>$, where W_M stands for the number of the independent states with the energy E_M. Physical quantities in statistical mechanics can be evaluated as

$$Tr(A\rho_{\mathbf{stat}}) = Tr(A\tilde{\rho}_{\mathbf{stat}}),$$

where

$$\tilde{\rho}_{\mathbf{stat}} \equiv \hat{Q}_N^{-1}\sum_M e^{-\beta H}|\chi_M><\chi_M| \times W_M;$$

$|\chi_M>$ may be understood as the representive state for the subspace with the same energy $E_M(B_M(^*\mathcal{H}))$. Now we can express the equivalence relation ($\longleftrightarrow_{\mathbf{therm}}$) for deriving $S_{\mathbf{therm}}(^*\mathcal{H})$ as ϕ_M^l *and* $\phi_M^{l'}$ *are equivalent* ($\longleftrightarrow_{\mathbf{therm}}$), *if and only if they have the same energy eigenvalue, i.e.,* $\phi_M^l, \phi_M^{l'} \in B_M(^*\mathcal{H})$. The physical states on $S_{\mathbf{therm}}(^*\mathcal{H})$ are represented by ultrapowers based on the above equivalence relation [4]. We see that $S_{\mathbf{therm}}(^*\mathcal{H})$ is represented by the quotient space divided by the equivalence relation as

$$S_{\mathbf{therm}}(^*\mathcal{H}) = S(^*\mathcal{H})/\longleftrightarrow_{\mathbf{therm}}.$$

In order to perform the projection from $S(^*\mathcal{H})$ to $S_{\mathbf{therm}}(^*\mathcal{H})$, we introduce a projection operator

$$\mathcal{P}_M = |\chi_M><\chi_M|,$$

where $\mathcal{P}_M$ has the properties $\mathcal{P}_M\mathcal{P}_M = \mathcal{P}_M$ and $\mathcal{P}_M|M',l>=\delta_{MM'}|\chi_M>/\sqrt{W_M}$. This projection operator selects only one component represented by $|\chi_M>=\sum_{l=1}^{W_M}|M,l>/\sqrt{W_M}$ from the W_M number of the independent components on $B_M(^*\mathcal{H})$. Now we can represent the density matrix on $S_{\mathbf{therm}}(^*\mathcal{H})(\rho_{\mathbf{therm}})$ as

$$\rho_{\mathbf{therm}} = \mathcal{P}\rho\mathcal{P}^\dagger,$$

where the projection operator is given by $\mathcal{P} \equiv \sum_M \mathcal{P}_M$ and ρ stands for a quantum mechanical density matrix for a pure state on $S(^*\mathcal{H})$. Of course, $\rho \neq \rho_{\mathbf{therm}}$ in general. The thermodynamic limit should be derived by taking "$\epsilon \equiv N^{-1} \approx 0$". That is, $N \in^* \mathcal{N} - \mathcal{N}(N = \infty)$. We conclude that, if and only if

$$\text{“}st_\epsilon(\rho_{\mathbf{therm}}) = st_\epsilon(\tilde{\rho}_{\mathbf{stat}})\text{”}$$

is realized, the physical system can be represented by the statistical mechanics, that is to say, the system is in thermal equilibrium.

3.2. Remarks on Thermodynamic Limit

In this report we could not write about internal trace operations for heat bath variables, a model realizing a thermal equilibrium at low temperatures and also the relation between the thermal equilibrium and quantum measurements [4].

4. CONCLUSION

We have two important points in order to specify a physical process on $S({}^*\mathcal{H})$: (1) What kind of physical constants can we treat as infinitesimals in the process? (2) What kind of physical variables are deliberately ignored in the measurement process? Understanding these two points are essential for specifying the physical process on $S({}^*\mathcal{H})$.

REFERENCES

1. M. O. Farrukh, *J. Math. Phys.* **16**, 177 (1975).
2. T. Kobayashi, *The Proceedings of the Symposium on Symmetries on Science VII* (Plenum, New York, 1993), p. 287.
3. T. Kobayashi, preprint of University of Tsukuba, UTHEP-249, 1992.
4. T. Kobayashi, preprint of University of Tsukuba, UTHEP-263, 1993.

A REALISTIC INTERPRETATION OF LATTICE GAUGE THEORIES

Miguel Lorente

Departamento de Física, Facultad de Ciencias
Universidad de Oviedo
E-33007, Oviedo, Spain

Following recent assumptions to unify quantum mechanics and general relativity, the structure of spacetime is suppose to be a consequence of the relations among some fundamental objects, and its concept can be formulated without the reference to the intuition. As physical consequences the continuous laws should be translated in to difference equations and the lattice field theories should be interpreted as a realistic model.

Key words: spacetime, real lattice, process, relations.

1. RECENT ASSUMPTIONS TO UNIFY QUANTUM MECHANICS AND THE STRUCTURE OF SPACETIME

One of the most difficult problems to unify the postulates of QM and general relativity is the different conceptions of spacetime. In QM the spacetime is a container where the fields are distinguished by their position and interactions, in the theory of relativity the gravitational field is identified with the metrical properties of spacetime. Recently several authors have tried to overcome this difficulty by deriving the structure of spacetime from the properties of fundamental processes described by QM.

According to Joseph M. Jauch, the set of propositions of a physical system in QM replaces the phase space [1]. In classical mechanics the underlying spacetime is necessary to impose the initial conditions that determined the solution of equations of motion. In QM the equation of motions are substituted by the set of propositions based on the superposition of the simplest yes-no experiments. The axiomatic form of this structure gives rise to the calculus of propositions, that do not presuppose the space time. The physical state is the result of a series of physical manipulations on the set of propositions.

M. Ferrero and A. van der Merwe (eds.), Fundamental Problems in Quantum Physics, 177–186.

Karl F. von Weizsaecker gives a more explicit connections between quantum theory and the concept of time and space [2]. All the quantum processes can be reduced to binary alternatives (equivalent to yes-no experiments). The interaction among these fundamental entities, which he calls "urs", gives rise to physical system and the structure of spacetime is the set of relations among the binary alternatives.

There are two important postulates: (i) the number of actual alternatives that determine a physical is finite, because they represent real properties; (ii) the number of possible alternatives is infinite due to the indeterministic nature of the quantum processes. As a consequence, the description of facts is given by discrete variables, but the physical laws are given in terms of continuous functions.

Roger Penrose does not pressupose an underlying spacetime for the physical processes [3]. The starting point is the total angular momentum of some fundamental units, the interactions of which produce a discrete network. "My model, says Penrose, works with objects and interactions between objects. An object is thus located either directionally or positionally in terms of its relations with other objects. One does not really need a space to begin with. The notion of space comes out as a convenience at the end."

According to David Finkelstein the world is a network of quantum processes, which he calls "monads"[4]. Every process in nature is a finite assembly of elementary processes, namely, of creation and destruction, and the structure of Spacetime is the set of all elementary processes and their relations.

2. EPISTEMOLOGY OF THESE MODELS

In order to understand better these models it would be useful to consider three levels of human knowledge in the comprehension of the physical world[5]:

Level 1: Physical magnitudes, such as distances, intervals, force, mass, charge, that are given by our sensation and perceptions.
Level 2: Mathematical structures, that are the result of metrical properties given by measurements and numerical relations among them.
Level 3: Fundamental concepts, representing the ontological properties of physical world given by our intelligence in an attempt to know the reality. This level of knowledge is not accepted by some philosophical positions like logical positivismus, conventionalismus, neokantismus.

There must be some connections between the three levels. In QM the theoretical models of microphysics in level 2 are related to observable magnitudes in level 1 by correspondence laws. If we accept level 3 it should be connected to level 2 and to level 1 (through level 2). In fact, the rules governing the constructions of theoretical models in level 2 must be grounded in some fundamental (ontological) properties of the physical world.

We can now raise the following question: in theoretical models of level 2 there are primitive and derived concepts, the last ones are obtained from the first ones by mathematical formulas. Are space and time primitive or derived concepts? If the second answer is given the description of the world in level 2 should not include as primitives the geometrical objects such as lines, planes, surfaces.

3. MODERN THEORIES ON THE STRUCTURE OF SPACE-TIME

In order to answer the last question it is convenient to recall the different interpretations of the concepts of space and time [6]. They are usually divided in three classes.

(a) *Dualistic theories*: Space is a container where the particles and waves are moving. Time is also a separated entity with respect to which the motion takes place. Therefore space and time are absolute and can be thinked of in the absence of particles (Newton).

(b) *Monistic theories*: Spacetime is identified with some properties of matter and can not be concevible without the existence of the later. The field of forces and also the sources are nothing more that geometrical deformations of the Spacetime (Einstein, Kaluza-Klein, Wheeler).

(c) *Relational theories*: Spacetime consists of the set of relations among some fundamental objects: monads (Leibniz), units (Penrose), processes (Weisaecker, Finkelstein), preparticles (Bunge, García Sucre), objects (Hilbert).

In Sec. 1 we have mentioned some of these authors. We expand in some detail Leibniz's and Hilbert's conception. According to Leibniz [7] "time is the order of points (monads) non existing simultaneously and one is the ratio of the other. Space is the order of points that exist simultaneously and are connected by mutual interactions. Space is nothing more that the set of all points and their relations. One point is here if it has relations with some particular points around it. A point changes its position if it changes its relations from some points to different ones. Motion is the change of different positions in time".

In his *Foundation of Geometry*, Hilbert has proposed an axiomatic approach to Euclidean geometry[8], according to which the concept of space is constructed with the help of some logical properties. He distinguishes two types of axiomatization: i) material, by which the concept of space is taken from observation and intuition and ii) formal, in which the concept of space is derived from some formal properties of axioms and inferences without the recourse to the intuition or the observation (his famous expression, "We could say always instead of points, lines and planes, chairs, tables and glasses of beer," confirms his position in favor of the formal axiomatization) The concepts of point, straight line, and plane can be reduced to pure logical relations.

4. A RELATIONAL THEORY OF SPACETIME

Following the assumption of the last section now we give an explicit construction of a formal structure of Spacetime, without the recourse to intuition. We can think of a set of fundamental objects acting among themselves, giving rise to a network of relations. These relations do not pressupose some space. The objects are nowhere if we consider them as elements of the physical world in level 2. In order to be specific we take as a naive network a three-dimensional cubic lattice. Obviously the network can be taken with different structure, such as, triangular, quasiperiodic or random lattices. In order to make connection with the euclidean geometry we take, for simplicity, a infinite set of interacting points in the relation 1 to 4. The set of all relations form a two-dimensional lattice, in which we can define:

A *path* is the connection between two different points, say, A and B, through points that are pairwise neighbours.

The *length* of a path is the numbers of points contained in the path, including the first and the last one.

A *minimal path* is a path with minimal length (in the picture the two paths between A and B are minimal). Between two point there can be different minimal paths.

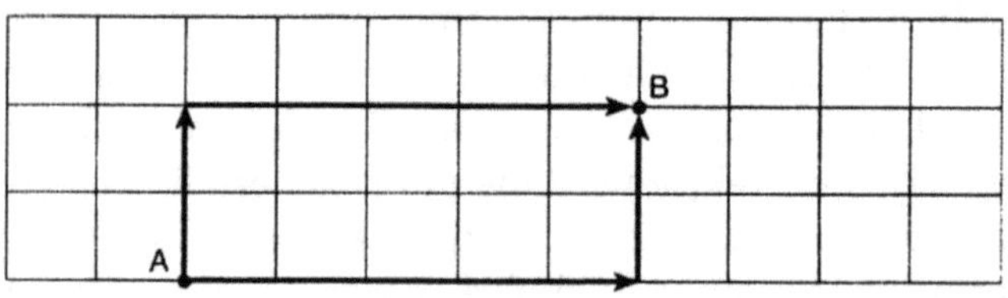

A *principal straight line* is a indefinite set of points in the lattice, such that each of them is contiguous to other two, and the minimal path between two arbitrary points of this line is always unique.

Theorem 1. Through a point of a 2-dimensional square lattice pass only two different principal straight lines (they are called *orthogonal straight lines*).

Theorem 2. Two principal straight lines that are not orthogonal have all the points either in common or separated (in the last case they are called *paralell straight lines*).

From these two theorem we can define Cartesian (discrete) coordinates and an Euclidean space where the postulates of Hilbert can be applied (with the exception of the axioms of continuity). This structure of 2-dimensional space can be easily generalized to 3-dimensional cubic lattice. As we mentioned, those assumptions for the structure of space are given in level 2, but it corresponds to the properties of physical space described in level 1 by our sensations.

In order to introduce the relation that correspond to time we start with only two fundamental objects acting among themselves:

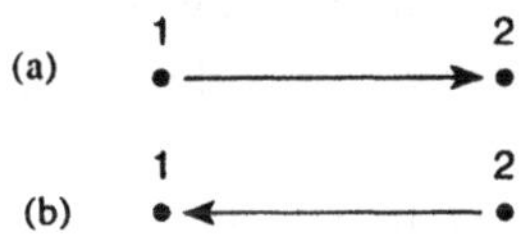

In (a), 1 is acting on 2, and in (b) 2 is acting on 1. But the action of 1 on 2 is supposed to be a necessary condition for the action of 2 on 1, and similarly the action of 2 on 1 is supposed to be a necessary condition for a new action of 1 on 2. Thus we can think of a chain of mutual interactions arranged in a series of necessary conditions. This picture has to be enlarged for the whole lattice. We take a set of interacting objects in the relations 1 to 2.

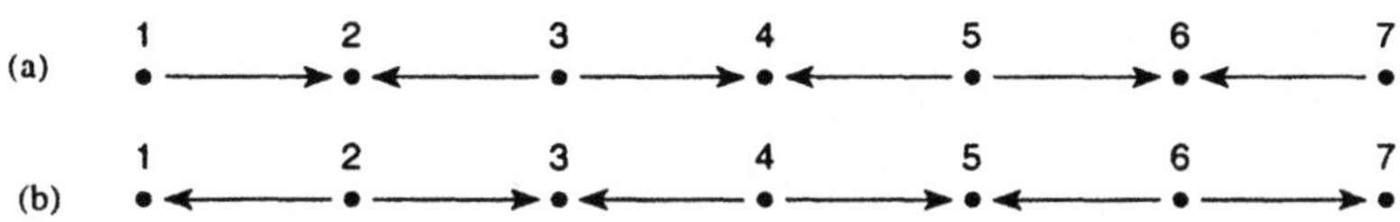

In (a), 1 is acting on 2, 3 is acting on 2 and 4, 5 is acting on 4 and 6, 7 is acting on 6. In (b), 2 is acting or 1 and 3, 4 is acting on 3 and 5, 6 is acting on 5 an 7.

We postulate that the actions of (a) are necessary conditions for the actions of (b) and the actions of (b) are necessary conditions for a further action of type (a) an so on.

Now take a network of objects acting in the relation 1 to 4.

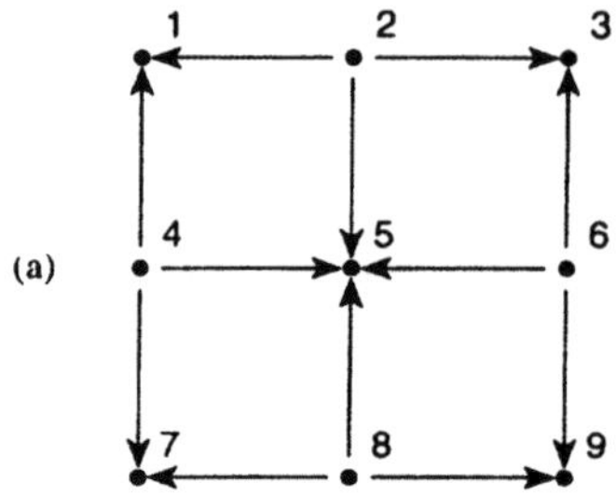

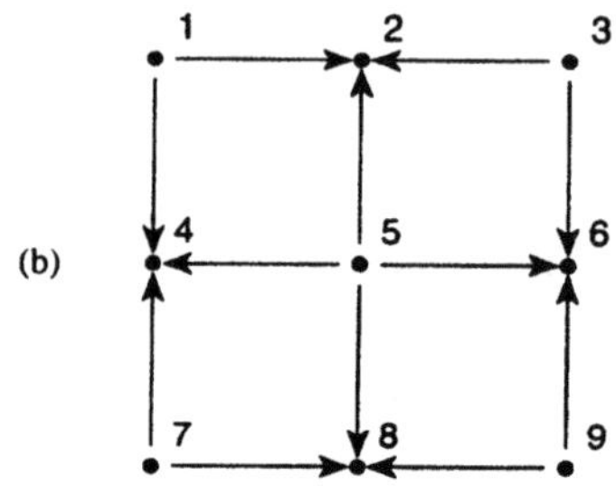

In (a), 2 is acting on 1, 3, 5; 4 is acting on 1, 5, 7; 6 is acting on 3, 5, 9; 8 is acting on 5,7,9. In (b), 1 is acting on 2 and 4; 3 is acting on 2 and 6; 5 is acting on 2, 4, 6, 8; 7 is acting on 4 and 8; 9 is acting on 6 and 8. As before we postulate that the actions of (a) be necessary conditions for the actions of (b) and so on. These logical properties of interactions belong to level 2 and do not pressupose the concept of time, but they can be put in correspondence with the physical properties of time given in level 1.

In level 3 there must be some ontological properties corresponding to the objects and interactions described in level 2.

In our model the most essential character of material entities is not the extension but their capacity to produce effects in other object (external causality). There is a causal relation between cause and effect and the logical necessity that was introduced in the last paragragh for the interpretation of time has its ontological ground in the principle of causality by wich the effect cannot be produced without its cause.

5. PHYSICAL CONSEQUENCES OF THE MODEL

The assumption of relational theory of Spacetime with a particular structure of cubic lattice, implies some physical consequences for the classical as well for the quantum physics:

(i) The Spacetime is discrete, therefore the physical laws are written in the language of finite differences. The solutions have to be described by continuous function of discrete variables [10].

(ii) Lattice gauge theories are not only a mathematical tool but a realistic theory, because they correspond to the underlying discrete structure of Spacetime. Some correspondence law must be given to make connection with the experimental world [11].

(iii) The symmetry of the model is still Poincaré transformation, although one has to select those integral transformations that keep the lattice invariant [12].

(iv) Some experimental test.Although there are infinite number of integral Lorentz transformations, and the continuous Lorentz transformations is a dense set, there are only 24 pure rotations that keep the lattice invariant. Therefore there is a broken SO (3) symmetry that leads to non-isotropy of the world. This means that one could fine in principle some preferred direction either in the microphysical world or in the large scale of the universe. An other physical application of the model could be detected in the discrete mass spectrum as a natural consequence of the elementary time interval an estimation of wich by actual calculations gives about 10^{-36} sec.

In order to prove this we summarized some mathematical results of lattice field theories [13].

We introduce the method of finite differences for the Klein-Gordon scalar field. An explicit scheme for the wave equation consistent with the continuous case (the truncation error is of second orden with respect to space and time variables) can be constructed as follows:

$$\left(\frac{1}{\tau^2}\nabla_n\Delta_n\tilde{\nabla}_j\tilde{\Delta}_j - \frac{1}{\varepsilon^2}\nabla_j\Delta_j\tilde{\nabla}_n\tilde{\Delta}_n + M^2\tilde{\nabla}_n\tilde{\Delta}_n\tilde{\nabla}_j\tilde{\Delta}_j\right)\phi_j^n = 0\,; \qquad (1)$$

here in the field is defined in the grid points of the $(1+1)$-dimensional lattice $\phi_j^n \equiv \phi\,(j\varepsilon, n\tau)$, ε, τ being the space and time fundamental intervals, j, n integer numbers, and Δ_j (∇_j) are the forward (backward) differences with respect to the space index, $\tilde{\Delta}_j$ $\left(\tilde{\nabla}_j\right)$ the forward (backward) averages, and similarly for the time index.

Using the method of separation of variables, it can easily be proved that the following functions of discrete variables are solutions of the wave equation (1):

$$f_j^n\,(k,\omega) = \left(\frac{1+\frac{1}{2}i\varepsilon k}{1-\frac{1}{2}i\varepsilon k}\right)^j\left(\frac{1-\frac{1}{2}i\tau\omega}{1+\frac{1}{2}i\tau\omega}\right)^n, \qquad (2)$$

provided the "dispersion relation" is satisfied:

$$\omega^2 - k^2 = M^2 \qquad (3)$$

M, being the mass of the particle.

In the limit, $j \to \infty$, $n \to \infty$, $j\varepsilon \to x$, $n\tau \to t$, the functions (2) become plane wave solutions

$$f_j^n\,(k,\omega) \to \exp i\,(kx - \omega t)\;. \qquad (4)$$

Imposing boundary conditions on the space indices,

$$f_o^n\,(k,\omega) = f_N^n\,(k,\omega) \qquad (5)$$

we get

$$k_m = \frac{2}{\varepsilon} \tan \frac{\pi m}{N} , \qquad m = 0, 1, \ldots, N-1 , \tag{6}$$

therefore

$$\omega = \pm \left(k_m^2 + M^2\right)^{1/2} . \tag{7}$$

For the positive energy solutions we define

$$\omega_m = + \left(k_m^2 + M^2\right)^{1/2} . \tag{8}$$

Starting from the wave equation (1), we can construct a current vector. Multiplying (1) by $\tilde{\nabla}_n \tilde{\Delta}_n \tilde{\nabla}_j \tilde{\Delta}_j \phi_j^{*n}$ from the left, and multiplying the complex conjugate of the wave equation by $\tilde{\nabla}_n \tilde{\Delta}_n \tilde{\nabla}_j \tilde{\Delta}_j \phi_j^n$ from the right, substracting both results, we obtain the "conservation law"

$$\frac{1}{\varepsilon} \nabla_j j_1 - i \frac{1}{\tau} \nabla_n j_4 = 0 , \tag{9}$$

where

$$\begin{aligned} j_1 \;&\equiv\; i \left[\frac{1}{\varepsilon} \Delta_j \left(\tilde{\nabla}_n \tilde{\Delta}_n \phi_j^{*n} \right) \tilde{\Delta}_j \left(\tilde{\nabla}_n \tilde{\Delta}_n \phi_j^n \right) \right. \\ &\left. - \tilde{\Delta}_j \left(\tilde{\nabla}_n \tilde{\Delta}_n \phi_j^{*n} \right) \frac{1}{\varepsilon} \Delta_j \left(\tilde{\nabla}_n \tilde{\Delta}_n \phi_j^n \right) \right] , \qquad (10) \\ j_4 \;&\equiv\; i\rho \equiv \left[\frac{1}{\tau} \Delta_n \left(\tilde{\nabla}_j \tilde{\Delta}_j \phi_j^{*n} \right) \tilde{\Delta}_n \left(\tilde{\nabla}_j \tilde{\Delta}_j \phi_j^n \right) \right. \\ &\left. - \tilde{\Delta}_n \left(\tilde{\nabla}_j \tilde{\Delta}_j \phi_j^{*n} \right) \frac{1}{\tau} \Delta_n \left(\tilde{\nabla}_j \tilde{\Delta}_j \phi_j^n \right) \right] \qquad (11) \end{aligned}$$

are the spatial and time component, respectively, of the charge vector current on the lattice.

The charge density ρ suggest that we can substitute the scalar field $\phi(x, t)$ by the smeared field $\tilde{\nabla}_j \tilde{\Delta}_j \phi_j^n$ and $\phi^*(x, t)$ by $\tilde{\nabla}_j \tilde{\Delta}_j \phi_j^{*n}$.

A suitable Hamiltonian for the real field ϕ_j^n and its conjugate momentum π_j^n can be defined as follows:

$$\begin{aligned} H_n \;&=\; \varepsilon \sum_{j=0}^{N-1} \frac{1}{2} \left\{ \left(\tilde{\nabla}_j \tilde{\Delta}_j \pi_j^n \right)^2 - \frac{1}{\varepsilon^2} \left(\nabla_j \Delta_j \phi_j^n \right) \left(\tilde{\nabla}_j \tilde{\Delta}_j \phi_j^n \right) \right. \\ &\left. + M^2 \left(\tilde{\nabla}_j \tilde{\Delta}_j \phi_j^n \right)^2 \right\} \equiv \varepsilon \sum_{j=0}^{N-1} \mathcal{H}_j^n . \qquad (12) \end{aligned}$$

As in the continuous case, we can derived the Hamilton equations of motions, varying the Hamiltonian density $\mathcal{H}_j^n$ first with respect to the promediate momentum and secondly with respect to scalar field:

$$\begin{aligned}
\frac{1}{\tau}\Delta_n\left(\tilde{\nabla}_j\tilde{\Delta}_j\phi_j^n\right) &= \frac{\partial\mathcal{H}_j^n}{\partial\left(\tilde{\nabla}_j\tilde{\Delta}_j\pi_j^n\right)} = \tilde{\Delta}_n\left(\tilde{\nabla}_j\tilde{\Delta}_j\pi_j^n\right), \qquad (13)\\
\frac{1}{\tau}\Delta_n\left(\tilde{\nabla}_j\tilde{\Delta}_j\pi_j^n\right) &= -\frac{\partial\mathcal{H}_j^n}{\partial\left(\tilde{\nabla}_j\tilde{\Delta}_j\phi_j^n\right)}\\
&= \tilde{\Delta}_n\left(\frac{1}{\varepsilon^2}\nabla_j\Delta_j\phi_j^n - M^2\tilde{\nabla}_j\tilde{\Delta}_j\phi_j^n\right). \qquad (14)
\end{aligned}$$

Applying the difference operator $\frac{1}{\tau}\nabla_n$ on both sides of (13) and substituting (14) in the result, we recover the wave equation (1).

Using (13) and (14), it can easily be proved that the Hamiltonian (12) is independent of the time index n, namely:

$$\nabla_n H_n = \Delta_n H_n = 0\,. \qquad (15)$$

Since the plane wave solutions $f_j^n(k_m,\omega_m)$ $(m = 0, 1, \ldots, N-1)$ form a complete set of orthogonal functions, we can expand the smeared field and its conjugate momentum as

$$\begin{aligned}
\tilde{\nabla}_j\tilde{\Delta}_j\phi_j^n &= \frac{1}{\sqrt{N\varepsilon}}\sum_{m=-N/2}^{N/2-1}\frac{1}{\sqrt{2\omega_m}}\left(a_m f_j^n(k_m,\omega_m) + a_m^* f_j^{*n}(k_m,\omega_m)\right),\\
\tilde{\nabla}_j\tilde{\Delta}_j\pi_j^n &= \frac{-i}{\sqrt{N\varepsilon}}\sum_{m=-N/2}^{N/2-1}\sqrt{\frac{\omega_m}{2}}\left(a_m f_j^n(k_m,\omega_m) - a_m^* f_j^{*n}(k_m,\omega_m)\right).
\end{aligned}$$

In order to make connection of our scheme with the Einstein-de Broglie relations $E = \hbar\omega,\quad p = \hbar k$, we take, for the period T and wavelength λ of the discrete plane waves functions (2) and (5),

$$T = N\tau, \qquad \lambda = N\varepsilon$$

and, for the phase velocity,

$$v_p = \frac{\lambda}{T} = \frac{\varepsilon}{\tau}\,.$$

We have defined the wave number and the angular frequency of the wave functions as:

$$k_m = \frac{2}{\varepsilon}\tan\frac{\pi m}{N}, \qquad \omega_m = \frac{2}{\tau}\tan\frac{\pi m}{N}, \qquad m = 0, 1, \ldots, N-1\,.$$

Substituting the Einstein-de Broglie relations in the relativistic expresion $E^2 - p^2 = M^2$ (we use natural units $\hbar = c = 1$), we obtain

$$\omega_m^2 - k_m^2 = \omega_m^2 \left(1 - \frac{\tau^2}{\varepsilon^2}\right) = \omega_m^2 \left(1 - \frac{1}{v_p^2}\right) = M^2 .$$

Since the phase velocity and group velocity satisfy $v_p v_g = 1$, we have finally

$$\omega_m^2 = \frac{M^2}{1 - v_g^2} ,$$

hence M has m-dependent discrete spectrum.

REFERENCES

[1] J. Jauch, *Foundations of Quantum Mechanics* (Addison-Wesley, Reading,1968).

[2] C.F. Weizsaecker, "Reconstruction of quantum mechanics," in *Quantum Theory and the Structure of Time and Space*, L. Castell and C. Weizsaecker, eds. (Hanser, Munich, 1986).

[3] R. Penrose, "Angular momentum: an approach to combinatorial analysis" in *Quantum Theory and Beyond*, T. Bastin, ed. (Cambridge University Press, Cambridge, 1971).

[4] D. Finkelstein, "Spacetime code," *Phys. Rev.* **184** (1968) 1261.

[5] M. Lorente, "A causal interpretation of the structure of space and time," *Foundations of Physics*, P. Weingartner and G. Dorn, eds. (Hölder,Pichler & Tempsky, Viena 1986).

[6] M. Lorente, "Modernas Teorías sobre la estructura del espacio-tiempo," *Reunión Matemática en honor de A. Dou* (Universidad Complutense de Madrid, 1989).

[7] M. Jammer, *Concepts of Space* (Cambridge University Press, Cambridge, 1969). According to this author, Leibniz's *Monadology* was inspired by Maimonides, who, in his *Guide for the Perplexed*, chap. 73, describes the theory of a discrete space and time.

[8] D. Hilbert, *Grundlage der Geometrie* (Teubner, Leipzig, 1899). Spanish translation: *Fundamentos de la Geometría* (C.S.I.C., Madrid, 1991).

[9] See Ref. 5.

[10] M. Lorente, *Int. J. Theor. Phys.* **4** (1974) 213; **12** (1976) 927; **25** (1986) 55.

[11] M. Lorente, *J. Group Theory in Phys.***1** (1993) 105.

[12] M. Lorente, "A relativistic invariant scheme for the Klein-Gordon and Dirac fields on the lattice," *XIX Int. Coll. on Group Theor. Meth. in Phys.* (Editorial Ciemat, Madrid, 1992), p. 395–398.

[13] M. Lorente, "Representations of the classical groups on the lattice" in *Symmetries in Science VI*, B. Gruber, ed. (Plenum, New York, 1993), pp. 437-454.

IS THERE A BRIDGE CONNECTING STOCHASTIC AND QUANTUM ELECTRODYNAMICS?

Trevor W. Marshall

Departamento de Física, Universidad de Oviedo
Avda. Calvo Sotelo s/n, 33007 Oviedo, Spain[1]

We review the stochastic (S.E.D.) and quantum (Q.E.D.) electrodynamic formalisms describing the free electromagnetic field. We show that, in spite of their close resemblance, there is a profound difference between a description which considers the zeropoint field to be real and one which considers photons to be real; the former is local while the latter is nonlocal.

We then show how to interpret the Wigner representation of Q.E.D. in such a way that real laboratory states of the electromagnetic field are all represented by positive Wigner densities. This enables us to understand all "single-photon" correlation effects with a purely wavelike description of the field; for the electromagnetic field at least, "wave-particle duality" is an illusion. We retain the quantum-optical theory of detection, which relates the counting rate to a normally ordered product of field operators, but we show that the theory may be recast so that the rate observed is the difference between that for the given light field and that for the vacuum (that is the dark rate).

We illustrate our argument with the experiment, by Grangier, Roger and Aspect (G.R.A.), establishing the phenomenon of anticorrelation in the two channels of a beam splitter. For certain modifications of the G.R.A. experiment we predict divergences between our theory and that used by G.R.A.; these differences arise because our theory predicts interference between the "photons" emitted by different atoms.

Key words: single-photon, anticorrelation, Wigner density

1. INTRODUCTION

Stochastic electrodynamics (S.E.D.), in its original form [1,2,3], had the aim of explaining all the features of the interaction of light with atoms, using a Lorentz, pointlike electron interacting with a classical electromagnetic field with a random zeropoint component. Its fundamental equation

M. Ferrero and A. van der Merwe (eds.), Fundamental Problems in Quantum Physics, 187–198.

of motion, known as the Braffort-Marshall equation, bears a close resemblance to the more recently discovered [4] "quantum Abraham" equation, and also to the Langevin equation [5] of classical Brownian motion. The methods developed for solving the Braffort-Marshall equation follow closely those of the latter theory. We can claim a good understanding, through S.E.D., of the stationary states of the harmonic oscillator; this includes a full description, through the Einstein A- and B-coefficients [6], of the electromagnetically induced transitions. But such progress as has been achieved for nonlinear systems, such as the hydrogen atom, has been of a limited and purely qualitative nature.

One should not discount such progress. S.E.D. is able to explain how the zeropoint field prevents the collapse of the atom through radiative damping of the accelerated electron; this has now been acknowledged in the quantum electrodynamic (Q.E.D.) treatment [4]. S.E.D. also tells us why the atom has a size of the order of the Bohr radius, and can claim credit for having first recognized [7] that the atom is metastable against autoionization; this has also now been confirmed [8] by Q.E.D.

But the outstanding failure of S.E.D. is its inability [9] to explain the line spectra of atoms. This seems to be related to its equally spectacular failure [10] to establish a detailed radiation balance in nonlinear systems; for example, according to S.E.D., the nonlinear oscillator transfers energy systematically [11] from lower to higher frequencies of the radiation field.

In spite of these failures, the real zeropoint electromagnetic field has gained a steadily increasing recognition. Originally proposed by Planck [12] and Nernst [13] in the second decade of this century, in direct opposition to what they perceived as Einstein's incorrect notion of "light quanta" (that is photons), it was reborn when Casimir [14] used it to explain the attraction between two conducting plates. Its rediscovery, as S.E.D., led, in addition to the results mentioned above, to a deeper understanding of van der Waals forces. In the last decade the zeropoint field has shown us, at least qualitatively, how to explain a new phenomenon, known as *enhancement* [15] which is revealed in optical tests of certain Bell-type inequalities [16], thereby indicating a consistently local realist interpretation of such experiments.

It became clear in this latter period that local realism requires us to use a consistently wavelike description of the electromagnetic field. It is precisely when physical phenomena are described in the language of photons that the paradoxes of nonlocality and entanglement arise. (For two opposing ways of expressing this last sentence see Refs. 17 and 18.)

So, can we find a bridge between Q.E.D., a highly successful set of algorithms in which photons and electrons appear and disappear capriciously, and S.E.D., a spectacular failure which, nevertheless, explains several things which Q.E.D. can only calculate? I suggest that we can. Low-energy Q.E.D., in the form of quantum optics, can be presented in such a way that light quanta are completely absent. Such a view vindicates Planck and Nernst, the forerunners of S.E.D. The photon was a bad idea from the beginning!

2. THE WIGNER REPRESENTATION OF THE LIGHT FIELD

The bridge we propose for connecting the Hilbert-space representation of Q.E.D. with the stochastic field of S.E.D. goes via the density matrix of the light field and its Wigner representation.

In S.E.D. the vector potential $\mathbf{A}(\mathbf{x},t)$ is represented as a sum of two mutually conjugate parts

$$\mathbf{A}(\mathbf{x},t)=\sum_{\{\mathbf{k}\}}\sum_{\lambda=1}^{2}\left(\frac{\hbar c}{2L^3\mid\mathbf{k}\mid}\right)^{\frac{1}{2}}\alpha_{\mathbf{k}\lambda}\mathbf{e}_{\mathbf{k}\lambda}e^{i(\mathbf{k}.\mathbf{x}-|\mathbf{k}|ct)} \quad + \quad \text{c.c.,} \tag{1}$$

where $\mathbf{e}_{\mathbf{k}\lambda}$ is a unit polarization vector, $\alpha_{\mathbf{k}\lambda}$ is a complex random variable, L is a normalization volume, and the total number of normal modes is

$$N=\frac{L^3\omega_{\mathbf{max}}^3}{3\pi^2c^3} \quad , \tag{2}$$

$\omega_{\mathbf{max}}$ being a cutoff frequency which, for present purposes, is somewhere in the ultraviolet part of the spectrum.

The light field is specified as a joint probability distribution of the $2N$ real variables $\{\alpha_{\mathbf{k}\lambda}\}$. Our assumption will be that this is simply the Wigner distribution of Q.E.D.[19,20]:

$$W(\{\alpha\})=\pi^{-N}\mathrm{Tr}[\hat{\rho}\hat{M}(\{\alpha\})], \tag{3}$$

where $\hat{\rho}$ is the density matrix of the light field, and $\hat{M}(\{\alpha\})$ is a pseudo-density matrix, also known as the Wigner-Moyal operator

$$\hat{M}(\{\alpha\})=\pi^{-N}\int\prod_{\mathbf{k}\lambda}\exp[\beta_{\mathbf{k}\lambda}(\hat{a}_{\mathbf{k}\lambda}^{+}-\alpha_{\mathbf{k}\lambda}^{*})-\beta_{\mathbf{k}\lambda}^{*}(\hat{a}_{\mathbf{k}\lambda}-\alpha_{\mathbf{k}\lambda})] \quad d^{2N}\beta. \tag{4}$$

Now, of course, there are certain states, notably the "one-photon" state

$$\hat{\rho}_1=\hat{a}_{\mathbf{k}_1\lambda_1}^{+}\mid\text{vac}\rangle\langle\text{vac}\mid\hat{a}_{\mathbf{k}_1\lambda_1}, \tag{5}$$

for which $W(\{\alpha\})$ can take negative values. Does this invalidate our bridge? We think not, and have published elsewhere [21] some arguments supporting this view. Essentially, the quantum state $\hat{\rho}_1$ is a pseudostate which cannot be realized in the laboratory. The building bricks to construct real states are the *dispersion-free states* (Yes, I mean the ones von Neumann said do not exist!) whose density matrices are $\hat{M}(\{\alpha\})$. Then a real laboratory state may be constructed in one of two ways:

$$\hat{\rho}^{\text{LAB}}=\int W(\{\alpha\})\hat{M}(\{\alpha\})d^{2N}\alpha, \tag{6}$$

with

$$W(\{\alpha\})\geq 0, \tag{7}$$

or

$$\hat{\rho}^{\mathrm{LAB}} = \sum_{\{m\},\{n\}} \rho_{\{m\},\{n\}} \mid \{m\}\rangle\langle\{n\} \mid, \tag{8}$$

where $\rho_{\{m\},\{n\}}$ is a positive definite matrix and $\mid \{m\}\rangle$ is a Fock state with $m_{\mathbf{k}\lambda}$ photons in the mode $(\mathbf{k}, \lambda)$. We conjecture that, in all real physical processes, the condition (7) is maintained; for this reason the representation (6) is more natural than (8).

The quantum-optical theory of detection is based on the normally ordered field and intensity autocorrelations. (Note that $\mathbf{E} = c^{-1}\dot{\mathbf{A}}$, but, as explained in [22], the intervention of a light-collecting lens system allows us both to treat $\mathbf{E}$ as a scalar and to ignore its spatial variation.) Thus

$$K_E^{(\mathrm{N})}(t,t') = Tr[\hat{\rho}\hat{E}^-(t)\hat{E}^+(t')], \tag{9}$$

$$K_I^{(\mathrm{N})}(t,t') = Tr[\hat{\rho} : \hat{I}(t)\hat{I}(t') :], \tag{10}$$

where

$$\hat{I}(t) = \hat{E}^-(t)\hat{E}^+(t), \tag{11}$$

and the notation :: signifies that the intervening expression has been normally ordered, that is,

$$: \hat{I}(t)\hat{I}(t') : \quad = \quad \hat{E}^-(t)\hat{E}^-(t')\hat{E}^+(t)\hat{E}^+(t'). \tag{12}$$

In Sec. 4 we shall recast the theory of detection in terms of symmetrically ordered quantities, that is,

$$K_E(t,t') = Tr[\hat{\rho}\mathrm{S}\{\hat{E}^-(t)\hat{E}^+(t')\}], \tag{13}$$

$$K_I(t,t') = Tr[\hat{\rho}\mathrm{S}\{\hat{I}(t)\hat{I}(t')\}], \tag{14}$$

where $\mathrm{S}(\hat{E}^-\hat{E}^+)$ is the average over the two orderings of $\hat{E}^-$ and $\hat{E}^+$, that is

$$\mathrm{S}(\hat{E}^-\hat{E}^+) = \frac{1}{2}(\hat{E}^-\hat{E}^+ + \hat{E}^+\hat{E}^-), \tag{15}$$

and $\mathrm{S}(\hat{I}\hat{I})$ is the average over the 24 orderings of the four $\hat{E}$-operators contained in $\hat{I}\hat{I}$. Symmetrically ordered autocorrelations may be calculated from a general relation, valid for any polynomial, P, of $\hat{a}$ and $\hat{a}^+$:

$$Tr[\hat{\rho}\mathrm{S}P(\hat{a},\hat{a}^+)] = \int W(\alpha)P(\alpha,\alpha^*) \quad d^{2N}\alpha \quad . \tag{16}$$

3. THE "SINGLE-PHOTON" SIGNAL

States which, in some sense, resemble the Fock state (6) have been realized in atomic cascades [23] and in parametric down conversion [24]. We have commented elsewhere [17] on the second of these and will now concentrate on the first.

The apparatus used by Grangier, Roger and Aspect [23] (G.R.A.) to generate a "single-photon" source is shown in Fig. 1. A light signal, from an atomic cascade, S, arrives at a beam splitter (BS) where it is divided into beams b and c; M are ordinary mirrors and D_b, D_c are detectors. For a light signal entering by channel a we measure the probabilities, p_b and p_c, of a count in D_b and D_c respectively, and also the probability, p_{bc}, of a coincidence count. The quantity

$$\alpha = \frac{p_{bc}}{p_b p_c} \tag{17}$$

will be equal to 1 if the detection events are independent. In conventional classical wave theory of light, α is greater than 1, while, in contrast, quantum optics predicts that no coincidences occur if the incoming signal consists of a "single photon". Values of α well below 1 are interpreted as showing the corpuscular behaviour of light. The monitoring procedure is described in the article of GRA, and it consists of recording a detection in D_b and D_c only when a partner photon, from the atomic cascade, is detected by the auxiliary detector D_1. Without monitoring, as pointed out by GRA and as we shall show in more detail below, the beam is chaotic. With monitoring, especially at low source intensities, it approaches the ideal quantum-optical state known as "single-photon".

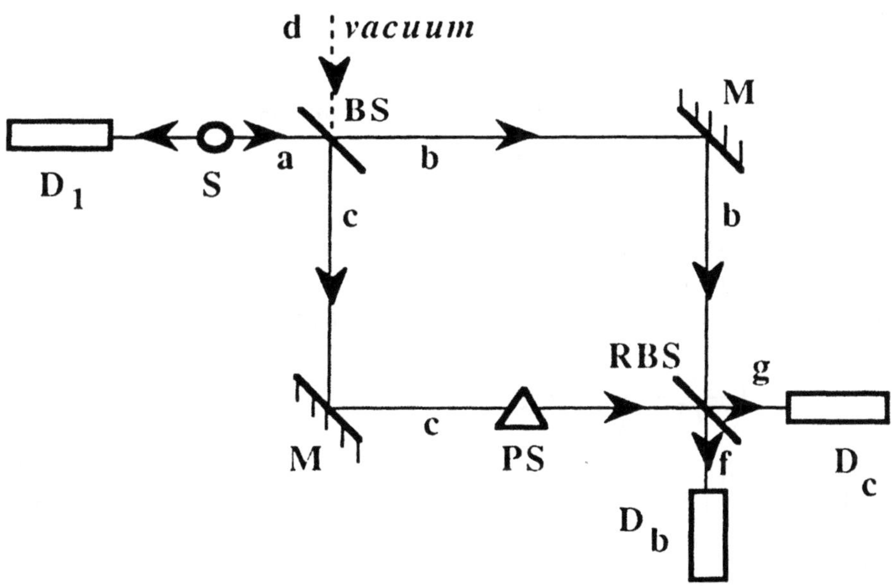

Fig. 1. Arrangement used for displaying wave property (with RBS in place) and particle-like property (with RBS removed) of light. This is a schematic version of the apparatus used in Refs. 23 and 24.

Actually the GRA experiment and its more recent version [24] were designed to show something more than the particle-like behaviour of light; they exhibit something which has been called "wave-particle duality". By placing a second beam splitter (RBS), which recombines the beams b and c to give the ouput channels f and g, either of these latter outputs can exhibit the wave interference between channels b and c. With RBS removed we have the all-or-nothing behaviour ("The photon is either in b or c.") associated with light particles, while with RBS in position we have the interference behaviour ("The photon is in both b and c.") associated with waves. With the purely wave description of the present article, the latter situation presents little difficulty, and we have analyzed it elsewhere [22]. I shall therefore concentrate here on showing that the former, "particle-like" situation is *also* comprehensible within a wave theory, once we take full account of the intervention of the zeropoint field.

Consider first the situation where a single atom makes, at time t_1, the transition from the upper to the intermediate level of the cascade. This gives rise to "the first photon", which is analysed by the detector on the left of Fig.1. The "second photon" is received on the right by a lens system of solid angle Ω; in practice the maximum value, corresponding to a lens semiangle of 45°, is $\Omega/4\pi \simeq 0.15$. This signal is then collimated, by the lens system, to produce a beam with transverse component, $\mathbf{k}^{tr}$, rather small. Then we shall suppose (for details, see [22]) that the collimated light field is represented by the Hilbert-space state

$$\mid t_1\rangle \quad = \quad \hat{A}^-(t_1)\mid \text{vac}\rangle, \tag{18}$$

where

$$\hat{A}^-(t) = \sum_{\mathbf{k}} \quad c_{\mathbf{k}}\hat{a}^+_{\mathbf{k}}e^{-i\omega_{\mathbf{k}}t}. \tag{19}$$

This represents a wave packet of limited extent normalized in the sense that

$$\sum_{\mathbf{k}} \mid c_{\mathbf{k}} \mid^2 = 1. \tag{20}$$

We shall assume that the shape of this packet is given by a Lorentzian (or Wigner-Weisskopf) frequency dependence for the longitudinal component k_ℓ, combined with a uniform density for $\mathbf{k}^{tr}$, that is,

$$c_{\mathbf{k}} = \left(\frac{8\gamma\pi}{TL^2k^2_{\mathbf{max}}}\right)^{\frac{1}{2}}(ck_\ell - \omega_0 - i\gamma)^{-1}(\text{for} \mid \mathbf{k}^{tr} \mid < k_{\mathbf{max}}), \tag{21}$$

where $T = L/c\omega_0$, ω_0 being the frequency of the transition and $\gamma(<< \omega_0)$ its natural linewidth. (The density of $\mathbf{k}^{tr}$ is chosen for computational simplicity only; we do not expect our results to depend critically on this particular choice).

It seems natural to assume that, if n such transitions occur in the time T, the corresponding state vector is

$$\mid t_1, t_2...t_n\rangle = \hat{A}^-(t_1)...\hat{A}^-(t_n)\mid \text{vac}\rangle \tag{22}$$

and that the distribution of n is Poisson. We cannot, however, control these individual transitions, so we average $(t_1, t_2 ... t_n)$ over the time interval $(0, T)$. Our expression for the density matrix, which, as we shall show, corresponds to a chaotic state of the beam, is then

$$\hat{\rho}^{(\mathrm{cha})} = \mathcal{N}(\lambda) \sum_{n=0}^{\infty} \frac{1}{n!} \left(\frac{\lambda}{T}\right)^n \int_0^T dt_1 ... \int_0^T dt_n \mid t_1 ... t_n \rangle \langle t_1 ... t_n \mid, \tag{23}$$

where $\mathcal{N}(\lambda)$ is a normalizing factor and λ is a parameter related to the beam intensity. Now, because of the orthogonality of the normal modes, the integrations result in $\hat{\rho}^{(\mathrm{cha})}$ being diagonal in the number-state representation, that is,

$$\hat{\rho}^{(\mathrm{cha})} = \sum_{\{n_{\mathbf{k}}\}} \mid \{n_{\mathbf{k}}\} \rangle \langle \{n_{\mathbf{k}}\} \mid \prod_{\mathbf{k}} (1 - \lambda \mid c_{\mathbf{k}} \mid^2)(\lambda \mid c_{\mathbf{k}} \mid^2)^{n_{\mathbf{k}}}, \tag{24}$$

where we now put in the explicit value of $\mathcal{N}(\lambda)$. This is indeed a chaotic state, and its Wigner density is [19,20]

$$W^{(\mathrm{cha})}(\alpha) = \prod_{\mathbf{k}} \frac{1}{\pi(\overline{n}_{\mathbf{k}} + \frac{1}{2})} \exp\left(-\frac{\mid \alpha_{\mathbf{k}} \mid^2}{\overline{n}_{\mathbf{k}} + \frac{1}{2}}\right), \tag{25}$$

where

$$\overline{n}_{\mathbf{k}} = \frac{\lambda \mid c_{\mathbf{k}} \mid^2}{1 - \lambda \mid c_{\mathbf{k}} \mid^2}. \tag{26}$$

The mean number of photons in the collimated beam is then

$$\overline{n} = \sum_{\mathbf{k}} \overline{n}_{\mathbf{k}} = \frac{\lambda}{\mu}, \tag{27}$$

where

$$\mu = \left(1 - \frac{2\lambda}{\gamma T}\right)^{\frac{1}{2}}. \tag{28}$$

As we shall see in the next section, this unmonitored signal gives a value of α lying always in the "classical" range ($\alpha > 1$). Let us now consider the case where one of the atomic decays, say t_m which we now rename as t_0, is fixed, so that we average over the remaining $(n-1)$ decays only. If this could be done with one-hundred-percent efficiency, the state would be described by the density matrix

$$\hat{\rho}^{(S)} = \mathcal{N}_S \hat{A}^-(t_0) \hat{\rho}^{(\mathrm{cha})} \hat{A}^+(t_0), \tag{29}$$

where

$$\mathcal{N}_S^{-1} = Tr[\hat{A}^-(t_0) \hat{\rho}^{(\mathrm{cha})} \hat{A}^+(t_0)]. \tag{30}$$

The associated Wigner density may be calculated using (16), the trace operation being calculated in a standard [19,20] manner:

$$Tr[\hat{a}_{\mathbf{r}}^{+}\hat{\rho}\hat{a}_{\mathbf{s}}\hat{\Pi}(\alpha)] = (\alpha_{\mathbf{r}}^{*} - \frac{1}{2}\frac{\partial}{\partial\alpha_{\mathbf{r}}})(\alpha_{\mathbf{s}} - \frac{1}{2}\frac{\partial}{\partial\alpha_{s}^{*}})Tr[\hat{\rho}\hat{\Pi}(\alpha)]. \tag{31}$$

From this it follows that

$$W^{(S)}(\alpha) = \mathcal{N}_S W^{(\mathrm{cha})}(\alpha)\left[4\sum_{\mathbf{jk}}\frac{c_{\mathbf{j}}c_{\mathbf{k}}^{*}\alpha_{\mathbf{j}}^{*}\alpha_{\mathbf{k}}e^{i(\omega_{\mathbf{j}}-\omega_{\mathbf{k}})t_0}}{(1+\lambda\mid c_{\mathbf{j}}\mid^2)(1+\lambda\mid c_{\mathbf{k}}\mid^2)} - \sum_{\mathbf{j}}\frac{\mid c_{\mathbf{j}}\mid^2}{1+\lambda\mid c_{\mathbf{j}}\mid^2}\right], \tag{32}$$

and the normalizing factor, which may be determined from the requirement that the integral of $W(\alpha)$ is 1, is

$$\mathcal{N}_S = \frac{\lambda}{\bar{n}} = \mu. \tag{33}$$

The state $\hat{\rho}^{(S)}$, which may be considered a "single-photon" state (albeit above a chaotic background), is represented by a Wigner density which, for certain α, takes negative values. However, the monitoring process used by GRA produces the state $\hat{\rho}^{(S)}$ only with probability $\xi\Omega/4\pi$, where ξ is the angular correlation factor [25] between the two photons of the cascade ($\xi \simeq 1.2$ if the two lens systems have semiangle of 45°), and otherwise the state remains chaotic. So the monitored state is actually

$$\hat{\rho}^{(M)} = \frac{\xi\Omega}{4\pi}\hat{\rho}^{(S)} + \left(1 - \frac{\xi\Omega}{4\pi}\right)\hat{\rho}^{(\mathrm{cha})}. \tag{34}$$

In accordance with the general conjecture mentioned in the previous section, this state has a positive Wigner density. We have published elsewhere [22] the field and intensity autocorrelations corresponding to this light field, and calculated by putting (16) into (13) and (14).

4. THE DETECTION PROCESS

The standard quantum-optical theory of detection [19,20] gives a coincident counting rate in the channels b and c proportional to the double integral of the normally ordered product

$$K_{bc}^{(\mathrm{N})}(t,t') = \mathrm{Tr}[\hat{\rho} : \hat{I}_b(t)\hat{I}_c(t') :]. \tag{35}$$

We have shown [22] that this is equivalent to using the symmetrically ordered product, but with a subtraction of the zeropoint (that is "dark") background

$$K_{bc}^{(\mathrm{N})}(t,t') = \mathrm{Tr}[\hat{\rho}\mathrm{S}(\hat{I}_b(t) - I_0)(\hat{I}_c(t') - I_0)]. \tag{36}$$

We have also been able to interpret the cross-correlation between I_b and I_c in terms of a completely classical description of the beam splitter. The incoming signal a is mixed (see Fig. 1) with the vacuum field entering through channel d; for more details of this mixing process see Ref. [15]. The GRA parameter α, defined in eq. (17), is obtained in terms of the two parameters

$$\theta = \frac{4\pi\overline{n}w}{\xi\Omega T}, \tag{37}$$

$$v = 2\gamma w, \tag{38}$$

where w is the time window(9ns) used in the monitoring process; following GRA, $\xi\theta$ may be identified as the number of atomic cascades occurring during the window, while v is the window divided by the natural lifetime, τ_s (4.7ns), of the monitored photon. The value we obtain [22] for α is

$$\alpha(\theta, v) = \frac{\alpha^{(\mathrm{cha})}(v)\theta^2 + 2G(v)\theta}{[\theta + F(v)]^2}, \tag{39}$$

where

$$\alpha^{(\mathrm{cha})}(v) = 1 + 2v^{-2}(e^{-v} - 1 + v) = 1.579, \tag{40}$$

$$G(v) = 1 + 2v^{-1}\{1 - (1 + 3v/2)e^{-v}\} = 1.448, \tag{41}$$

$$F(v) = 1 - e^{-v} = 0.853. \tag{42}$$

Note that $\alpha^{(\mathrm{cha})}(v)$ represents the value of α appropriate for chaotic light; this is precisely what the *unmonitored* beam is! Note also that, for suitably high beam intensity, the curve of α against θ approaches this chaotic limit, which, naturally, lies in the "classical" domain ($\alpha > 1$). Thus our theory indicates a continuous transition from the "quantum" ($\alpha < 1$) to the "classical" domain.

It seems pertinent to ask why GRA did not observe this phenomenon. An obvious answer is that they were looking for "quantum" behaviour and were therefore concentrating on low beam intensities. However, this is not the whole story. In taking account only of the time variation, and discarding the spatial variation of the field, for example in eq. (19), we were effectively treating a one-dimensional version of quantum optics. Such a model may only be justified in experimental conditions where the radiating atoms are all located within a rather small source area, and have a rather uniform velocity distribution. Without such a control of the source, the separation between the photons from different atoms may be greater than their lateral coherence length, so that they will not interfere with each other. In that case a simpler analysis is possible leading, instead of eq. (39), to the GRA result

$$\alpha_{GRA}(\theta, v) = \frac{\theta^2 + 2F(v)\theta}{[\theta + F(v]^2}. \tag{43}$$

In Ref.22 we have discussed, in more detail, the conditions in which eqns. (39) and (43) respectively should hold.

5. DISCUSSION

The phenomenon we are predicting in the present article is a direct descendant of the "photon interference" discovered by Brown and Twiss [26] for thermal light and by Pfleegor and Mandel 27] for coherent light. We now predict that even the highly nonclassical "single-photon" states exhibit this property, a very reasonable proposition, but one which we think has not been well understood up to this time. To understand this phenomenon we have found it essential to describe the free electromagetic field in terms of wave packets (see Eq. (19)) rather than plane waves. Thus one realizes that, in a real "photon", the different plane-wave components are phase-related to each other; in every sense of the word, we think that different "photons" interfere with each other.

The above considerations indicate to us that the "particle-like" properties of light are just that; the anticorrelation property is, as we have argued previously [15], fully compatible with a purely wave description. Now I argue that this view would be reinforced if, in conformity with another conjecture of ours [21], the Wigner distribution of the mode amplitudes turns out to be positive in all experimental situations. Such has now been shown to be the case in at least one situation which has been claimed to reproduce the "single-photon" state. Of course, it remains to be seen whether the same is true of more recent claimants [24]. It is a relatively modest extension of the presently reported work to treat the set of pairwise coincident events in an atomic cascade observed by Clauser [28]. These should, after modification of the apparatus along the lines discussed in the previous Section, show a similar transition from "quantum" to "classical" behaviour.

I emphasize that the bridge, between S.E.D. and Q.E.D., which is suggested here does not remove the inadequacies of either theory. S.E.D. is still incapable of giving an adequate description of atomic energy levels; to obtain these levels we have to use nonrelativistic quantum *mechanics*, which is, in some way yet to be explained, derived from Q.E.D. Now nonrelativistic quantum *mechanics* is nonlocal; to say this is no more remarkable than to say that Newtonian gravitation is nonlocal. What *is* remarkable is the uncritical way in which many commentators now assume that the measurement theory of quantum *mechanics* may be extended simply to the measurement of field quantities. This results in a series of contentious and ideologically loaded statements about "violations of locality" in a macroscopic system (like two light detectors separated by several metres). In using the multimode description for light signals we are insisting, from the outset, on the infinity of degrees of freedom of the field, and thus we are going beyond the confines of quantum *mechanics*. This has the consequence of drawing to our attention an important property of detectors (that is, measuring devices), namely their ability to subtract, from the signal intensity, the corresponding intensity of the zeropoint background.

ACKNOWLEDGEMENTS

I wish to thank P. Grangier for drawing my attention to the spatial-coherence question.

[1] Permanent address: Department of Mathematics, University of Manchester, Manchester M13 9PL, UK.

REFERENCES

1. T. H. Boyer, in *Foundations of Radiation Theory and Quantum Electrodynamics*, A. O. Barut, ed. (Plenum, New York, 1980).
2. L. de la Pena, in *Stochastic Processes Applied to Physics and Other Related Fields* B. Gomez et al, eds. (World Scientific, Singapore, 1983).
3. T. H. Boyer, *Sci.Am.*, August 1985.
4. J. Dalibard, J. Dupont-Roc and C. Cohen-Tannoudji, *J. Phys. (Paris)* **43**, 1617 (1982).
5. M. C. Wang and G. E. Uhlenbeck, *Rev. Mod. Phys.* **17**, 113 (1945).
6. H. M. Franca and T. W. Marshall, *Phys. Rev. A* **38**, 3258 (1988).
7. P. Claverie and F. Soto, *J. Math. Phys.* **21**, 1819 (1982).
8. A. Loinger, *Nuovo Cimento A* **98**,679 (1987) and **103**, 1357 (1990).
9. T. W. Marshall and P. Claverie, *J. Math. Phys.* **21**, 1819 (1980).
10. T. W. Marshall and P. Claverie, *Physica A* **104**, 223 (1980).
11. T. H. Boyer, *Phys. Rev. A* **18** , 1228 (1978).
12. M. Planck, *Verh. Deutsch. Phys. Ges.* **13**, 138 (1911).
13. W. Nernst *Verh. Deutsch. Phys. Ges.* **18**, 83 (1916).
14. H. B. G. Casimir, *Proc. K. Ned. Akad. Wet.* **51**, 793 (1948).
15. T. W. Marshall and E. Santos, *Found. Phys.* **18**, 185 (1988).
16. J. F. Clauser and S. Freedman, *Phys. Rev. Lett.* **28**, 938 (1972); A. Aspect, J. Dalibard, and G. Roger, *Phys.Rev. Lett.* **47**, 1804 (1982).
17. T. W. Marshall, E. Santos, and A. Vidiella-Barranco, *3rd Int. Workshop on Squeezed States and Uncertainty Relations*, D. Han et al,eds. NASA Conference Publication Series, 3270, page 581(1994).
18. D. M. Greenberger, M. A. Horne, and A. Zeilinger, *Phys. Today*, August 1993, p. 22.
19. J. Perina, *Quantum Statistics of Linear and Nonlinear Optical Phenomena* (Reidel, Dordrecht, 1984), Chap. 4.
20. W. H. Louisell, *Quantum Statistical Theory of Radiation* (Wiley, New York, 1973), Chap. 3.
21. T. W. Marshall and E. Santos, *Found. Phys. Lett.* **5**, 573 (1992).
22. T. W. Marshall, E. Santos, and A. Vidiella-Barranco, Universidad de Cantabria, preprint, (1994).
23. P. Grangier, G. Roger, and A. Aspect, *Europhys. Lett.* **1**, 173 (1986).
24. Y. Mizobuche and Y. Ohtake, *Phys. Lett. A* **168**, 1 (1992).
25. E. S. Fry *Phys. Rev. A* **8**, 1219 (1973).

26. R. Hanbury Brown and R. Q. Twiss, *Proc. Roy. Soc. A* **242** (1957); **243**, 291 (1957).
27. R. L. Pfleegor and L. Mandel, *Phys. Rev. 159*, 1084 (1967) and *J. Opt. Soc. Am.* **58**, 946 (1968).
28. J. F. Clauser, *Phys. Rev. D* **9**, 853 (1974).

ACTION-ANGLE VARIABLES INHERENT IN QUANTUM DYNAMICS

Jesús Martínez-Linares

Instituto de Estructura de la Materia, CSIC
Serrano 123, Madrid-28006, Spain

A phase-space Poisson-bracket structure induced by the Schrödinger equation is addressed in order to show the existence of a phase space representation of quantum dynamics where action-angle variables can be defined. Quantum dynamics can be embedded into a classical picture of coupled harmonic oscillators in phase space, providing for an internal contextual hidden variable theory. This approach also provides for a natural scenario in order to understand the different manifestations of phase space structures in quantum mechanics as well as to study the ocurrence of chaotic phenomena in the classical limit.

Key words: quantum chaos, quantization maps, invariant toroids.

1. INTRODUCTION

Does chaotic behaviour exist in quantum mechanics? This is an important question that has not yet been understood. Lie linear algebras supporting the formulation of quantum dynamics, the linearity of the Schrödinger equation, and the superposition principle are the basis to attach a quasiperiodic temporal behaviour for quantum states and disregard the ocurrence of chaos in the quantum world [1]. Moreover, a quantum Hamiltonian for a bounded system has a discrete spectrum, opposite to the continuum spectrum of the Liouville operator in classical chaotic systems. On the other hand, if quantum mechanics does not allow for chaotic phenomena there is a need to understand how this richer structure can overcome in the classical limit. The present work is devoted to shed some light into these questions, trying to connect the linear algebraical structure of quantum mechanics with the construction of action-angle variables within a phase space representation of quantum mechanics given by Heslot [2]. The connection with the manifestations of phase space structures in quantum mechanics will then be straightforward.
We restrict ourselves to a bounded system governed by a time independent Hamiltonian. However, the formalism can be extended in a straightforward way to include time dependent quantum Hamiltonians, that will be represented, within

M. Ferrero and A. van der Merwe (eds.), Fundamental Problems in Quantum Physics, 199–203.

certain periodicity conditions, by a linear Floquet system in phase space.
The paper is organized as follows. In Sec. (2) we will describe this linear phase space representation of quantum dynamics. The equation of motion of the associated harmonic oscillators will be given. Section (3) will be devoted to the construction of action-angle variables in this phase space allowing to embed the trajectories into invariant tori.

2. PHASE SPACE STRUCTURE OF QUANTUM DYNAMICS

The cotangent bundle $\Gamma = T^*(M)$, of a differentiable manifold M, equiped with a simplectic form and an associated Poisson bracket holds the Hamiltonian formulation of classical mechanics. A separable Hilbert space, carrying an irreducible representation of the Weyl commutation relation, and Dirac bracket, holds the Hamiltonian formulation of quantum mechanics. Both theories are supported by C^* algebras [3]. The connection between quantum and classical evolution is so close, that canonical transformations supporting classical dynamics can be represented by unitary transformations in the complex representation of phase space [4]. Thus, in order to compare both classical and quantum theories, a line must be drawn between the deterministic evolution of the quantum state and the stochasticity attached to the measurement process. As we are going to show in this section, as longer as the former is concerned, a representation of the evolution of the state in terms of trajectories of an associated system of harmonic oscillators in phase space can be constructed.
Consider a quantum bounded system, *i.e.* a system with a discrete energy spectrum. The corresponding Hilbert space of states can be spanned by an orthonormal base $\{|n>\}_{n=0}^{\infty}$. Let λ_n^o be the complex valued components of a quantum state $|\psi>$ in this base. The evolution of this generic state $|\psi> = \sum \lambda_n^0 |n>$ in the Schrödinger picture is given by the state $|\psi(t)> = \sum_n \lambda_n(t)|n>$, where the coefficients are the solutions of the differential equations

$$i\dot{\lambda}_n(t) = \sum_k H_{nk}\lambda_k(t), \tag{1}$$

with the initial conditions $\lambda_n(0) = \lambda_n^0$ and H is a matrix with elements

$$H_{nk} =< n|H|k > /\hbar. \tag{2}$$

Let us introduce the dimensionless quadrature variables

$$\Lambda(t) = \frac{1}{\sqrt{2}}\left(X(t) + iY(t)\right), \tag{3}$$

where we have defined the vector $\Lambda(t) = \{\lambda_n(t)\}_{n=0}^{\infty}$. In matrix notation, Eq. (1) reads

$$\begin{aligned} \dot{X} &= \overline{W}X + WY, \\ \dot{Y} &= -WX + \overline{W}Y, \end{aligned} \tag{4}$$

where we have used the decomposition $H = W + i\overline{W}$. The Hermitian character of the Hamiltonian allows to write its real and imaginary parts as a symmetric and antisymmetric matrices respectively, thus satisfying

$$\begin{aligned} W &= W^T, \\ \overline{W} &= -\overline{W}^T, \end{aligned} \tag{5}$$

where T stands for the transpose operation.
The set $\{x_k, y_k\}$ plays the role of canonical coordinates for a Hamiltonian given by the quadratic form $\langle\psi|\hat{H}|\psi\rangle$ [2], which can be written in the quadrature representation as

$$\mathcal{H} = \langle\psi|\hat{H}|\psi\rangle = \frac{1}{2}(XWX + YWY) + Y\overline{W}X. \tag{6}$$

Thus, Eqs. (4) can be regarded as the set of Hamiltonian equations

$$\begin{aligned} \dot{x}_n &= \frac{\partial\mathcal{H}}{\partial y_n}, \\ \dot{y}_n &= -\frac{\partial\mathcal{H}}{\partial x_n}. \end{aligned} \tag{7}$$

In spite of the infinite but numerable dimension of the phase space generated, the phase flow is governed by linear dynamics and the evolution of the system can be solved, at least in a formal sense. A formal solution can be easily written for the dynamics of the complex phase space given in Eq. (1) in the form

$$\Lambda(t) = e^{\overline{W}t - iWt}\Lambda(0). \tag{8}$$

The formal solution of Eqs. (4) can be obtained by "breaking" the propagator of this solution. One has then to use the commutator expansion [5]

$$e^{\overline{W}}We^{-\overline{W}} = W + [\overline{W}, W] + \frac{1}{2!}[\overline{W}, [\overline{W}, W]] + ..., \tag{9}$$

in order to identify the real and imaginary part of the Λ vector. A special feature of the model is that the ease of solving the dynamics depends tightly on the commutativity of the real and imaginary parts of the Hamiltonian matrix. In the case that both matrices commute with their commutator, we can apply the Baker-Campbell-Haussdorf theorem to Eq. (8) and obtain the solutions

$$\begin{aligned} X(t) &= e^{\overline{W}}\left\{\cos\left(Wt + \Theta t^2/2\right)X(0) + \sin\left(Wt + \Theta t^2/2\right)Y(0)\right\}, \\ Y(t) &= e^{\overline{W}}\left\{-\sin\left(Wt + \Theta t^2/2\right)X(0) + \cos\left(Wt + \Theta t^2/2\right)Y(0)\right\}, \end{aligned} \tag{10}$$

where we have defined the symmetric matrix

$$\Theta = [W, \overline{W}] = W\overline{W} + (W\overline{W})^T. \tag{11}$$

In order to understand the different roles played by the $\overline{W}$ and W matrices, let us express Eq. (4) as a second order differential equation for the coordinate quadrature vector. Deriving Eq. (4) and solving for the X vector, we find

$$\ddot{X} - (\overline{W}^2 + D)\dot{X} + (W^2 + D\overline{W})X = 0, \qquad D = W\overline{W}W^{-1}. \tag{12}$$

In the case that $\overline{W} = 0$ for the chosen state basis, the equation of motion for the components of the amplitude quadrature vector reduces to the form

$$\ddot{x}_n(t) + \sum_{kl} \frac{\langle n|\hat{H}|l\rangle\langle l|\hat{H}|k\rangle}{\hbar^2} x_k(t) = 0. \tag{13}$$

The system is described by a collection of coupled harmonic oscillators in phase space. In this representation, the diagonal elements of the square of the Hamiltonian matrix give the frequency of each oscillator, meanwhile its off-diagonal matrix elements are mapped to linear couplings among then. As can be seen in Eq. (12), the presence of a nonzero $\overline{W}$ matrix introduces a renormalization for the coupling constants as well as a friction term into the system dynamics. We can associate with each $|n\rangle$ state a phase space harmonic oscillator which displacement around its equilibrium state are given by $x_n(t)$. Initially, the state is defined by the displacement and momentum given by $\Lambda(0)$, and confined by the condition $|\lambda_n| \leq 1$. The evolution, *i.e.*, the flow of probability among the states of the basis, is given by linear interaction among the oscillator plus the friction introduced by the $\overline{W}$ matrix. Conservation of energy and momentum of the whole chain can be related to the normalization and phase arbitrariness conditions of the space of states.

3. CONSTRUCTION OF ACTION-ANGLE VARIABLES

The linear structure of quantum mechanics is involved in the existence of a numerable infinite-dimensional phase space representation where the dynamics is governed by a Hamiltonian quadratic form. Thus, a canonical transformation can be found to transform the canonical coordinates of the Hamiltonian into action-angle variables [6]. A unitary transformation T can be found that diagonalize the Hamiltonian; such normal modes are obtained through the transformation

$$\xi = T^\dagger \Lambda, \tag{14}$$

which can be regarded as a canonical transformation given by the generating function $S = \xi^* T^\dagger \Lambda$. Action variables are given by the modulus of the normal modes, *i.e.*,

$$I_k = |\xi_k(0)|^2 = \sum_{l,n} T_{lk} T^\dagger_{kn} \lambda_n(0) \lambda^*_l(0), \tag{15}$$

which satisfy the normalization condition $\sum I_k = 1$. Expresing Eq. (6) in normal modes with the help of Eq. (14), the Hamiltonian can be written only in terms of the action variables I_k alone as

$$\mathcal{H} = \Lambda^\dagger H \Lambda = \sum_k w_k \, I_k, \tag{16}$$

where $\hbar w_k$ are the eigenvalues of the Hamiltonian. The collection of phase space oscillators associated with the dynamics of a quantum bounded state can provide for an internal hidden-variable theory where the probability of finding the value

$\hbar\omega_k$ in a measurement of the system energy is given by the value of the action variable of the corresponding normal mode oscillating at that energy. Non-separability arises in this theory from the strong collective character of normal modes, as will be shown in a subsequent work.
On the other hand, the existence of action-angle variables allows trajectories in phase space to be confined into invariant tori assuring a quasi-periodical temporal behaviour for the orbits. The degree of periodicity of the orbits is tightly related with the level of degeneracy and incommensurability of the eigen-frequencies. Diverse manifestation of phase space structures in quantum mechanics, such as semi-classical localization, quantum suppression of classical chaos, traps, etc. [7], can be related to the structure of tori inherent in quantum dynamics. The connection between the quantum and classical phase spaces will be achieved in a subsequent work.

4. CONCLUDING REMARKS

We have shown that, due to the linear structure of the Hilbert vector space of states, the dynamics of any quantum nonlinear bounded system can be represented by a collection of harmonic oscillators in phase space subjected to a linear interaction. The equation of motion for the chain of oscillators associated to each operator-eigenstate basis set have been given. Action- angle variables are shown to arise associated to the normal modes of the system. Quantum dynamics can be embedded into a classical picture of coupled harmonic oscillators. As will be shown in a subsequent work, these oscillators will provide for a contextual hidden variable theory of quantum mechanics.
This approach also provides for a natural scenario in order to understand the different manifestations of phase space structures in quantum mechanics as well as to study the ocurrence of chaotic phenomena in the classical limit.

REFERENCES

[1] G. M. Zaslavskii, *Phys. Rep.* **80**, 157 (1981). R. H. G. Helleman, *Fundamental Problems in Statistical Mechanics* V, E. G. Cohen, ed. (North-Holland, Amsterdam, 1980). T. Hogg and B. A. Huberman, *Phys. Rev. Lett.* **A48**, 711 (1982).
[2] A. Heslot, *Phys. Rev. D31*, 1341 (1985).
[3] F. Bayen, M. Flato, C. Fronsdl, A. Lichnerowicz, and D. Sternheimer, *Ann. Phys.* **10**, 11 (1978).
[4] U. Uhlhorn, *Arkiv Fysik* **11**, 87 (1956).
[5] E. Merzbacher, *Quantun Mechanics* (Wiley, New York, 1970), Chap. 8.
[6] H. Goldstein, *Classical Mechanics* (Addison-Wesley, Massachusetts, 1959), Chap. 9.
[7] O. Bohigas, S. Tomsovic, and D. Ullmo, *Phys. Rep.* **223**, 43 (1993).

A PHILOSOPHER STRUGGLES TO UNDERSTAND QUANTUM THEORY: PARTICLE CREATION AND WAVEPACKET REDUCTION

Nicholas Maxwell

Department of Philosophy, Logic and Scientific Method
The London School of Economics
Houghton Street
London WC2A 2AE, United Kingdom

Work on the central problem of the philosophy of science has led the author to attempt to create an intelligible version of quantum theory. The basic idea is that probabilistic transitions occur when new stationary or particle states arise as a result of inelastic collisions.

Key words: philosophy, realism, probabilism, wavepacket collapse, inelastic collision.

I am a philosopher, not a physicist. Nevertheless, for some twenty years I have been struggling with the problem of developing a fully micro-realistic version of quantum theory (QT), free of any reference to measurement in its basic postulates, and thus free of the defects that plague orthodox QT (OQT) [1]. How could a *philosopher* be so foolish as to take on such a project? Before I outline my attempt at developing a trouble-free version of QT, I want first to show how the project arose out of work on the central problem of the philosophy of science - the problem of induction.

The problem of induction is the problem of how it is possible to verify physical laws and theories by means of evidence. Given any physical theory T, however well verified by a body of evidence, there always exist infinitely many rival theories, T_1, T_2,...,$T\infty$, all of which fit the available evidence just as well as T does, but which differ from T, in infinitely many different ways, for

M. Ferrero and A. van der Merwe (eds.), Fundamental Problems in Quantum Physics, 205–214.

phenomena not yet observed [2]. In order to construct such theories, all we need to do is take T and modify it, in an entirely *ad hoc* way, just for some specific kind of phenomenon not yet observed. Granted the ubiquity of this situation, how can there ever be valid grounds for accepting any physical theory T, however tentatively, and rejecting its infinitely many rivals?

In scientific practice, (almost) all such *ad hoc* rivals to T are ignored just because they are all grossly *ad hoc*. In order to be accepted as a part of (conjectural) scientific knowledge, a theory T must satisfy *two* criteria: (1) It must meet with sufficient empirical success, and (2) it must be sufficiently simple, unified, non-*ad hoc*. But this does not in itself solve the problem, for one can ask: What *justification* is there for persistently preferring simple theories to complex theories equally well verified by the phenomena? As long as *evidence* provides the only rational ground for choosing theories in science, it cannot be rational to base choice of theory on grounds of simplicity or unity.

Since David Hume first posed the problem, some two and a half centuries ago [3], philosophers, and some thoughtful physicists and mathematicians, have struggled with the problem without much success [4]. Twenty years ago, Karl Popper resoundingly declared: "I think I have solved a major philosophical problem: the problem of induction" [5]. However, as I [2,6], and others [7], have shown, Popper's putative solution does not succeed.

However, working very much in the spirit of Karl Popper, I now resoundingly declare: I think I have solved the problem of induction! [8]. In order to solve the problem, we need to adopt a new, quasi-Einsteinian conception of science, according to which the whole scientific enterprise is obliged to accept, on pragmatic grounds, as a central component of (conjectural) scientific knowledge, that the universe is comprehensible *in some way or other*. The resulting *aim-oriented empiricist* conception of science (as I have called it) is depicted in diagram 1.

Traditionally, there have been two approaches to the problem: *empiricism*, which holds that, in the end, evidence alone can be permitted to decide what theories are accepted and rejected in science; and *rationalism*, which holds that some truths about the world can be known by reason independently of evidence. Both approaches are hopeless. The decisive point to appreciate is that there is a *third*, largely overlooked possibility. Science adopts, from the outset, the extremely vague, open-ended *a priori conjecture*

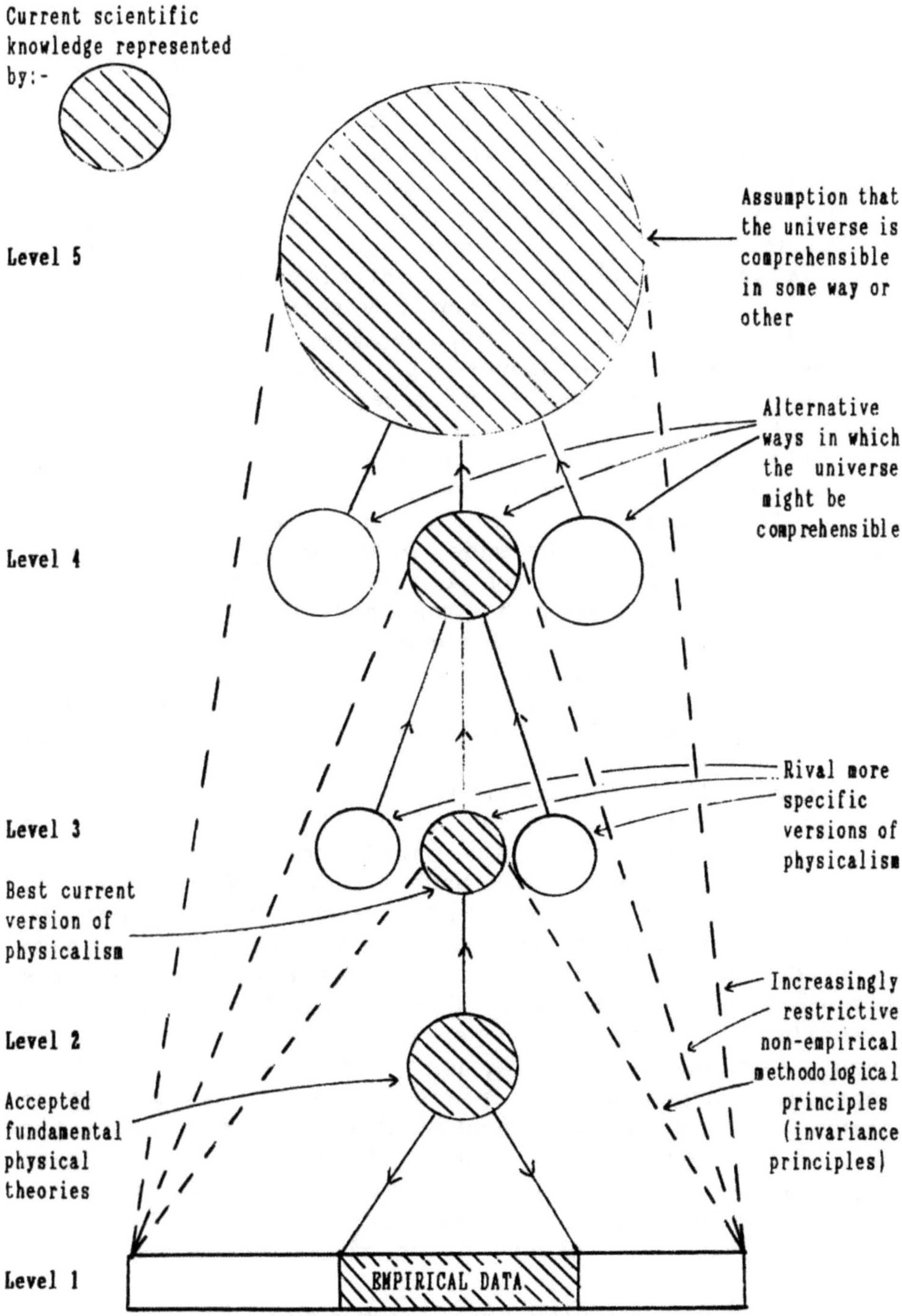

Diagram 1. Aim-oriented empiricism

that the universe is comprehensible in some way or other, it being a basic task of science to try to discover in what precise way the universe is comprehensible as it proceeds. Science is rationally justified in adopting this vague (level 5) conjecture, not because there are good reasons for believing it to be true, but rather because, without the assumption, science becomes impossible. In making the assumption, we have nothing to lose. Even if the assumption is false, and the universe is only partially or approximately comprehensible, nevertheless the best possible way that we can find out in what way this is the case is to assume (some kind of) perfect comprehensibility, and fail in the attempt to discover it. (If the universe is completely incomprehensible science, and even life, become impossible.)

Within the framework of the vague (level 5) assumption of comprehensibility, much more specific rival (level 4 and level 3) assumptions about how the universe is comprehensible need to be made, each having its associated methods, each constituting a rival scientific research programme. That research programme - that level 3 assumption and associated methods - which promotes the greatest growth of (level 1 and 2) empirical knowledge is the one to be preferred. Modern science began with Galileo's adoption of (a) the idea that "the book of Nature is written in the language of mathematics", and (b) associated methods of research.

An elementary requirement for intellectual rigour is that assumptions that are influential, substantial, problematic and implicit are made *explicit* so that they may be criticized, so that alternatives may be developed and assessed, in an attempt to *improve* such assumptions. Aim-oriented empiricist science emerges as a result of putting this elementary requirement for intellectual rigour into scientific practice. The persistent preference that is given in science for simple, unifying theories over *ad hoc* theories, independent of evidence, amounts to making an influential, substantial, problematic assumption about the world, to the effect that it behaves, at least approximately, as if simple or unified. Rigour requires that we make this implicit assumption explicit, criticizable and improvable (the outcome being aim-oriented empiricist science).

Granted aim-oriented empiricism (AOE), the problem of induction disappears. Traditional attempts to solve the problem all failed because they amounted to the attempt to justify the unjustifiable (a seriously *unrigorous* conception of science).

AOE has profound and far-reaching implications for the whole of academic inquiry and education, and even for life [9]. It also has important implications for physics. It implies that physics has *evolving* (level 3) aims and methods within the framework of the fixed (level 4) aim to discover the nature of the unified pattern of physical law that is permanently assumed to run through all phenomena. There is positive feedback between improving *theories* and improving *aims and methods*. As knowledge improves, knowledge about how to improve knowledge improves as well, a point which helps explain the explosive growth of scientific knowledge.

Furthermore, AOE implies that the philosophy of physics, conceived of as the enterprise of proposing and critically assessing possible aims and methods, possible research programmes, for physics, becomes a vital, integral part of physics itself, required for the intellectual rigour of physics. There is, according to AOE, a rational, if fallible and non-mechanical method for the discovery of new fundamental theories in physics, a method first explicitly developed and used by Einstein in creating special and general relativity [10].

Finally, the long-standing failure to put AOE into scientific practice has meant that (level 3) metaphysical ideas, "blueprint" ideas as I have called them, instead of being developed ahead of, and thus aiding the subsequent discovery of, new (level 2) theories, have rather lagged behind the development of good new theories, and as a result have retarded understanding and acceptance of such theories. Thus Newtonian theory did not emerge from the prior development and acceptance of Boscovich's point-atom metaphysical blueprint (according to which the world is made up of point-atoms which have mass and are surrounded by rigid, spherically symmetric fields of force that vary with distance. Rather, the tendency of Newton's Cartesian contemporaries to accept the corpuscular blueprint, according to which the world is made up of rigid corpuscles that interact by contact, served to impede understanding and acceptance of Newtonian theory. And similarly, leaving Faraday on one side, the classical field blueprint was not developed and generally accepted in advance of Maxwell's theory of the electromagnetic field. Rather, persistent attempts were made to interpret Maxwell's theory in terms of the aether, in terms, that is, of ideas that stemmed from Boscovich's Newtonian blueprint. Failure to articulate and scrutinize new possible blueprints for physics, as an integral part of physics itself, has retarded scientific progress.

Having developed and propounded AOE some twenty years ago, as an intellectually rigorous conception of science which avoids the problem of induction, as more traditional conceptions of science do not, it seemed to me that I ought to try to practice what I propounded. I ought to *do* philosophy of physics in the way AOE implied it should be done. That is, I ought to try to develop a new level 3 metaphysical blueprint, and associated research programme, for theoretical physics, if only in an attempt to illustrate AOE in scientific practice. It was in this way that I was seduced into the impossible task of trying to improve QT.

Granted orthodox conceptions of science, the immense empirical success of OQT may seem to render it beyond serious criticism. But granted AOE, with its two, equally important criteria of acceptability for physical theory, it is obvious that OQT, despite its immense empirical success, is an unacceptable theory. This is due to the failure of OQT to specify the nature of quantum systems independent of measurement, due to the failure to solve the wave/particle problem (there being no consistent, level 3 blueprint associated with OQT.)

In fact, as I have argued at length elsewhere [11], failure to solve the wave/particle problem renders OQT imprecise, ambiguous, *ad hoc*, non-explanatory, restricted in scope, and resistant to unification: even those who reject AOE ought to find OQT a seriously defective theory!

In order to be acceptable, QT must specify, precisely and consistently, the nature of quantum systems independent of measurement and classical physics.

It then occurred to me that history might be repeating itself. Just as Newton and some of his contemporaries found his theory of gravitation incomprehensible due to the misguided attempt to interpret it in terms of outdated, inappropriate metaphysics, and just as Maxwell and his successors found his theory of the electromagnetic field incomprehensible due to the misguided attempt to interpret it in terms of outdated, misguided metaphysics, so too with QT. Postulate that the quantum domain is fundamentally probabilistic in character, the basic dynamic laws being irreducibly probabilistic, and it is at once obvious that the *physical entities* of the quantum world must be quite different from anything encountered in *deterministic* classical physics. The attempt to understand the electron or photon in terms of the classical particle, wave or field is utterly misguided from the outset. What is needed is a new (level 3) metaphysical blueprint, an idea for a new kind of inherently *probabilistic* physical entity, which I dubbed, partly following Popper, the *propensiton* [12].

According to propensiton QT (PQT), the ψ-function of QT is to be interpreted as specifying the actual physical state of the individual quantum system; the ψ-function attributes probabilistic properties, or propensities, to quantum systems (such as position probability density), the values of which evolve deterministically, in accordance with the Schrödinger equation, as long as nothing probabilistic occurs. The problem then becomes: What are the quantum conditions for probabilistic transitions to occur? Some physical process is required which occurs all the time in nature, and also whenever probabilistic quantum measurements, that actually detect quantum systems, are made.

My idea here is that probabilistic transitions occur whenever there are inelastic collisions [13]. (A rather different proposal was put forward, independently, by Ghirardi, Rimini and Weber [14].) Consider the interaction (ab) + c → a + b + c (where a, b and c are spinless particles and (ab) is the bound state). In this case, my postulate asserts the following. The superposition of the elastic outcome (A): (ab) + c, and the inelastic outcome (B): a + b + c, that results from the interaction, collapses spontaneously and probabilistically into either (A) or (B), entirely independently of measurement, when the interaction between the particles ends. All quantum measurements that detect particles are just special cases of this process.

The problem with this postulate, as just formulated, is that interactions never do entirely cease. In order to overcome this difficulty, I propose the following.

Let the state of the entire system be $\Psi(t)$, and let the asymptotic states of the two channels (A) and (B) be $\psi_A(t)$ and $\psi_B(t)$ respectively, so that, according to OQT, there are states $\Phi_A(t)$ and $\Phi_B(t)$ such that:

(1) As $t \to +\infty$, $\Phi_A(t) \to \psi_A(t)$ and $\Phi_B(t) \to \psi_B(t)$; and for all t, $\Psi(t) = c_A\Phi_A(t) + c_B\Phi_B(t)$, with $|c_A|^2 + |c_B|^2 = 1$.

Here, the evolutions of $\psi_A(t)$ and $\psi_B(t)$ are governed by the respective channel Hamiltonians, H_A and H_B, respectively. These differ from the Hamiltonian, H, governing the evolution of $\Psi(t)$, $\Phi_A(t)$, and $\Phi_B(t)$, in that forces between particles that are not in bound states are set to zero. We have:

$$(2)\ H = -\{\frac{\hbar^2}{2m_a}\nabla_a^2 + \frac{\hbar^2}{2m_b}\nabla_b^2 + \frac{\hbar^2}{2m_c}\nabla_c^2\} + V_{ab} + V_{ac} + V_{bc}.$$

Here, m_a, m_b, and m_c are the masses of a, b, and c, and V_{ab}, V_{ac}, and V_{bc} are potentials corresponding to the forces between a and b, a and c, and b and c, respectively.

$$(3)\ H_A = -\{\frac{\hbar^2}{2m_a}\nabla_a^2 + \frac{\hbar^2}{2m_b}\nabla_b^2 + \frac{\hbar^2}{2m_c}\nabla_c^2\} + V_{ab}.$$

$$(4)\ H_B = -\{\frac{\hbar^2}{2m_a}\nabla_a^2 + \frac{\hbar^2}{2m_b}\nabla_b^2 + \frac{\hbar^2}{2m_c}\nabla_c^2\}.$$

The condition for probabilistic collapse can now be stated as follows. Let $\Psi^c(t) = c_A\phi_A(t) + c_B\phi_B(t)$. Then:

(5) At the first instant t for which $|\langle\Psi^c(t)|\Psi(t)\rangle| > 1 - \varepsilon$ is satisfied, the state of the system, $\Psi(t)$ jumps probabilistically into either $\phi_A(t)$ with probability $= |c_A|^2$ or into $\phi_B(t)$ with probability $= |c_B|^2$, ε being a universal constant, a positive real number very nearly equal to zero.

It is a straightforward matter to generalize (5) so that it applies to the case of an inelastic interaction with N channel outcomes, with N distinct asymptotic states where $N \geq 2$.

Detection of a micro system as a part of the process of quantum measurement typically involves allowing an unlocalized system to interact inelastically with billions of localized systems, as when a photon interacts with the billions of silver bromide crystals of a photographic plate. In this sort of case, the generalized version of (5) is to be straightforwardly applied. The outcome of the interaction is a superposition of billions of channels; these collapse to just one or other channel when (5) is satisfied. The outcome is the dissociation of one silver bromide molecule somewhere on the photographic plate.

PQT, incorporating postulate (5), but without calling upon measurement or classical physics, is able to reproduce all the empirical success of OQT [15]. PQT does nevertheless yield predictions that differ from those of OQT for experiments not yet performed. In order to detect the difference, it will be necessary to recombine the two channels, (ab) + c and a + b + c, so that the a + b + c channel becomes (ab) + c, and interference effects can be detected. This requires that the experimental set-up is such that no measurement can determine along which route, (ab) + c or a + b + c, the system evolves. Granted that the value of ε is such that, in these circumstances, PQT predicts that the superposition $c_A\phi_A + c_B\phi_B$ collapses into either ϕ_A or ϕ_B (before measurement) then, whereas OQT will predict interference, PQT will predict no interference.

The possibility, in principle, of performing this crucial experiment establishes that OQT and PQT are empirically distinct theories. The experiment is however extremely difficult to do in practice, and has not as yet been performed as far as the present author is aware.

REFERENCES

1. See N. Maxwell, *Am. J. Phys.* **40**, 1431-5 (1972); *Phys. Lett.* **43A**. 29-30 (1973); *Am. J. Phys.* **41**, 1022-5 (1973); *Method. and Science*, **8**, 84-101 (1975); *Found. Phys.* **6**, 275-92 and 661-76 (1976); *Found. Phys.* **12**, 607-31 (1982); *Phil. Sci.* **52**, 23-44 (1985); *Brit. J. Phil. Sci.* **39**, 1-50 (1988); *Phil. Sci.* **55**, 640-5 (1988); A. van der Merwe, F. Selleri, and G. Tarozzi, eds., *Bell's Theorem and the Foundations of Modern Physics* (World-Scientific, Singapore, 1993), pp. 362-70; *Phil. Quart.* **43**, 139-57 (1993); *Phil. Sci.* **60**, 341-348 (1993); *Phys. Lett. A* **187**, 351-5 (1994).
2. See N. Maxwell, *Phil. Sci.* **41**, 123-53 (1974); *Brit. J. Phil. Sci.* **44**, 81-101 (1993).
3. D. Hume, *A Treatise of Human Nature* (Everyman, London, 1959).
4. For a fairly recent discussion of various approaches to the problem see J. Watkins, *Science and Scepticism* (Princeton University Press, Princeton, 1984).
5. K. Popper, *Objective Knowledge* (Routledge, London, 1972), p. 1.
6. N. Maxwell, *Phil. Sci.* **39**, 131-52 (1972).
7. I. Lakatos, in P.A. Schilpp, ed., *The Philosophy of Karl Popper* (Open Court, La Salle, Illinois, 1974), vol. 2, chap. 5; A Grünbaum, *Brit. J. Phil. Sci.* **27**, 105-36 (1976).

8. See N. Maxwell, *Phil. Sci.* **41**, 123-53, 247-95 (1974); *Nature* **265**, 2 (1977); *Scientia* **114**, 629-53 (1979); *From Knowledge to Wisdom* (Blackwell, Oxford, 1984), chap. 9; *Brit. J. Phil. Sci.* **44**, 61-79, 81-101, 275-305 (1993).

9. See N. Maxwell, *From Knowledge to Wisdom* (Blackwell, Oxford, 1984).

10. See N. Maxwell, *Brit. J. Phil. Sci.* **44**, 275-305 (1993).

11. N. Maxwell, *Am. J. Phys.* **40**, 1431-5 (1972); *Found. Phys.* **6**, 275-92, 661-76 (1976); *Found. Phys.* **12**, 607-31 (1982); *Brit. J. Phil. Sci.* **39**, 1-50 (1988); A. van der Merwe, F. Selleri and G. Tarozzi, eds., *Bell's Theorem and the Foundations of Modern Physics* (World-Scientific, Singapore, 1993), pp. 362-70; *Phil. Quart.* **43**, 139-57 (1993).

12. N. Maxwell, *Brit. J. Phil. Sci.* **39**, 1-50 (1988).

13. N. Maxwell, *Found. Phys.* **12**, 607-31 (1982).

14. G. Ghirardi et al. *Phys. Rev. D* **34**, 470-91 (1986).

15. N. Maxwell, *Phys. Lett. A* **187**, 351-5 (1994).

CONSISTENT HISTORIES AND THE INTERPRETATION OF QUANTUM MECHANICS

by Roland Omnès

Laboratoire de Physique Théorique et Hautes Energies
Université de Paris-Sud
Centre-Scientifque d'Orsay
91405 Orsay, France

A general review of the interpretation of quantum mechanics by consistent histories is given, together with its logical aspects.

Key words: consistent histories, logic, microlocal analysis, decoherence, measurement theory, quantum epistemology.

I had first intended to give a talk on some recent results concerning the problem of datum uniqueness in a quantum measurement, when I was told that it would be more useful to review in a general way some significant progress in interpretation that have been made in the last few years, using the idea of consistent histories. This is therefore what I shall try to sketch here, to indicate the main ideas though without examples nor any justification of the results. More details can be found elsewhere [1,2] and I wish only to provoke some interest for looking at these more complete references, as well as the original ones [3-6].

The consistent histories approach can be considered at various levels and from various points of view. The most conservative result of this theory is to cast the somewhat verbal Copenhagen interpretation into an ordinary theory, with a clear-cut and unique basic assumption from which everything is derived, without having to resort to the inspired wisdom of the founders. It is by the way quite remarkable that the results of the new approach recover almost exactly Bohr's practical rules, though they were independent of it (and some of its contributors, including myself, were expecting something different). A more practical result is to obtain precise predictions in the case of experiments for which Bohr's interpretation remained doubtful. This is the case when he only gave hints (delayed-choice experiments), or when the experiments are so clever and tricky that they can lead to endless discussions

M. Ferrero and A. van der Merwe (eds.), Fundamental Problems in Quantum Physics, 215–224.

[7], or for experiments that could not be considered by Bohr, such as a continuous observation of a unique atom. The range of the interpretation even extends to experiments that were initially devised for contradicting the Copenhagen interpretation, such as the quantum behaviour of a macroscopic object such as a SQUID [8-9].

The results of the new interpretation do not always strictly coincide with the older ones and some significant statements by Bohr must be re-evaluated and even corrected. Another remarkable result, more general and ambitious, was to prove that paradoxes are impossible in quantum mechanics, so that one can assert that the theory is logically consistent. This includes of course a removal of the apparent paradox associated with the names of Einstein, Podolsky, and Rosen. Finally, perhaps the deepest consequence of the construction is the drastic change in epistemology it suggests: There is no sense in judging such a general and formal science as quantum mechanics according to the criteria of ordinary intuition, so-called philosophical principles or common sense, as so many people persist in doing (including several speakers in this meeting). It is much more sensible to say that nature obeys some basic laws explaining the ordinary phenomena upon which our intuition and our common sense are based. This new direction has now been followed completely, starting from the basic principles of the theory and finding explicitly when and how classical common sense is valid [2,10].

Four basic ingredients enter in the construction, which I indicate in the order where they are most conveniently used for consistency, not in order of discovery. The first one is the seminal notion of consistent histories [4]. Then comes the derivation of a convenient form of quantum logic to deal with histories [5]. Then there are new results about the derivation of classical physics from quantum principles, involving also their logical aspects, among which determinism is most essential [11]. This was made possible by using the powerful mathematical techniques of microlocal analysis [12]. The last ingredient, and the first one to be discovered, is the famous decoherence effect [3]. A synthesis of these ideas was proposed independently by Gell-Mann and Hartle [5] and by myself [13], though rather differently at first sight. They do not put the same emphasis on these various ideas and they also differ by the present status of their respective proofs, but their similarities are much more striking than their differences. This is basically the same interpretation and, for lack of time, I shall only restrict the present discussion to the approach I proposed.

1. HISTORIES AND LOGIC

One only assumes to start with the most basic principles of quantum theory. They include the Hilbert space formal framework, together with the concept of observables as operators, and the Schrödinger equation. Everything else can be found to follow from this by uniquely defined constructions, which means that these basic abstract axioms do contain in germ their own interpretation.

The first step is to define a *property* (a "predicate" in von Neumann's language): A property is a proposition stating that the value of some observable A is in some range Δ of real numbers at some time t. It is well known that one can associate with it, in a unique way, a projection operator $E(t)$ in Hilbert space. Then come histories: A *history* is simply a sequence of properties occurring at different times. It can be denoted by the corresponding projection operators $\{E_k(t_k)\}(0 \leq t_1 < \ldots < t_n)$. A history looks very much like a motion picture if one compares an individual property with a photograph. It looks exactly like the description of an experiment by a physicist, who says for instance: "A neutron came out from a nuclear reactor at some time, then it crossed a velocity selector, out of which its velocity was restricted into some range; then the neutron collided with a proton and a photon was produced; the photon was detected later on". All the properties occurring in this sentence describe directly the microscopic world, something Bohr would have forbidden, but we shall see why they are nevertheless meaningful.

One can then introduce a probability for these histories. It should be stressed that these probabilities are not yet connected in any way with the existence of any randomness of physical events nor with the results of some measurements. They are just meant presently to provide the cornerstone for a convenient logical framework of quantum mechanics. One can find history probabilities by using a method first proposed by Gleason: The probability of a history should only depend upon the history itself (i.e. its projection operators). It should be non-negative. One must also introduce an exhaustive family of possible histories (the neutron could have missed the target; rather than a reaction on a proton, a scattering could have happened, and so on). The corresponding probabilities should then be normalized (their sum being equal to 1 for an exhaustive set of possible histories) and, most importantly, they should be additive. When this is applied to one-time histories, i. e. properties, additivity means that the probability for the value of an observable A to be in the range $\Delta \cup \Delta'$ is the sum of the probabilities for being in Δ and in Δ', if , if Δ and Δ' are disjoint intervals. Finally, one adds a few simple logical requirements in order to make sure that the sentences expressing the properties and describing a history have their most simple conventional meaning. All together, these conditions completely determine the probabilities. : There should exist a density operator ρ to represent the state of the system at time 0 and the probability of a history is uniquely given by

$$p = Tr\{E_n(t_n)\dots E_1(t_1)\rho E_1(t_1)\dots E_n(t_n)\}. \quad (1)$$

It may be pointed out that this expression is not invariant under time reversal, so that there is a preferred direction of time, the so-called logical direction. This breaking of time symmetry originates in one of the logical assumptions entering in the proof of Eq.(1), namely that if the first property is trivial (i.e., $E_1(t_1)\rho = \rho$), it does not enter into the value of the probability.

Additivity of probabilities has to be used for obtaining Eq.(1), at least as far as the last time t_n is concerned. It turns out however that this additivity is far from being always satisfied for the full histories. One can then distinguish between families of *consistent histories* , which are defined as satisfying additivity. Griffiths has given sufficient *consistency conditions* for this and I have also given necessary and sufficient conditions. One can give a simple example for the case of a family of histories referring only to two different times t_1 and t_2 and only two mutually exclusive properties for each time (one of them stating for instance that the value of A_1 is in a set Δ and the other one doing the same for the complement set $\bar{\Delta}$. There is only one consistency condition in that case, namely

$$Tr\{E_2(t_2), [E_1(t_1), [\rho, \bar{E}_1(t_1)]]\} = 0 \quad (2)$$

where $\bar{E}_1(t_1) = I - E_1(t_1)$. The essential point is of course that these conditions can be explicitly computed.

One can use consistent histories for building up a specific kind of logic allowing to think properly of what happens in a quantum world. Logicians, who had to put order in the much more abstract world of modern mathematics, teach us that a rigorous use of logic must first proceed by defining a field of propositions, then defining upon this field how the propositions can be combined by the logical operations "not, and, or" and finally defining properly what is a logical inference (or implication). An implication means that a proposition a logically implies another proposition b , and this is denoted by $a \Rightarrow b$. Logical equivalence is then a two-sided implication where $a \Rightarrow b$ holds together with $b \Rightarrow a$. Logic leave us free to define all this explicitly at our best but it demands that some twenty or so basic logical axioms, should then be satisfied, most of them going back to Aristotle and Chrysippus.

This kind of formal logical approach is easily performed for quantum mechanics, when relying on consistent families of histories. As a field of propositions, one can choose the histories themselves and the various properties occurring in them. There is a simple short-cut for defining the logical operations: Consider for instance a family of histories where the value of an observable A_1 at time t_1 can be in various intervals, one of them being Δ_1, and the value of another observable A_1 at time t_2 can be in some ranges, one of them being Δ_2. It is convenient to represent the corresponding history by a set, which is the direct product of Δ_1 and Δ_2. One can then use these sets as usual to replace the logical operations "not, or, and" by the corresponding operations of set theory "complement, union, intersection".

It order to define logical inference, one can use the fact that a consistent family of histories allows the existence of genuine probabilities. It is then possible to define a conditional probability $p(b|a) = p(a \text{ and } b)/p(a)$ for proposition b to hold when proposition a holds. This may be for instance the probability for a neutron coming out of a velocity selector with its velocity in some range, under the condition that a photon was inside a detector at a later time. One can then define the implication $a \Rightarrow b$ by the mathematical condition $p(b|a) = 1$. It turns out that this convention, together with the ones already given for the logical operations, satisfy all the rules of ordinary Boolean logic. Quantum mechanics has therefore no need of unconventional exotic logics as were introduced by Birkhoff and von Neumann. Aristotle is good enough.

One gets as a bonus a clear statement of complementarity: It corresponds to the possible existence of different consistent families of histories describing one and the same physical system. The existence of different logical frameworks might easily ring an alarm, since one may run the risk of paradoxes. Such is not the case fortunately since one can prove a non-contradiction theorem removing definitely the ghost of paradoxes: there are none in quantum mechanics! It should also be mentioned at this point that one can slightly relax some too stringent constraints of the construction by allowing a to imply b with a probability of error ϵ when $p(b|a) \geq 1 - \epsilon$. This extension provides the freedom necessary to reconcile the logical quantum way of thinking with the classical one belonging to common sense.

As for this quantum way of thinking, it boils down to a unique and very general axiom of a logical nature: When describing a quantum system, one should be careful that all the properties to be considered belong together to a unique consistent family of histories (this is easily checked by computing the corresponding consistency conditions). When reasoning moreover about these properties, one should make sure that every sentence can be substantiated by valid implications (this is checked by computing some conditional probabilities). A physicist can therefore freely use imagination without caring if Bohr allowed or not what is said. Sense can always be told from nonsense by a policeman writing down a few equations on his notepad during the talk: consistency conditions and conditional probabilities. They allow him to give anybody a fine for saying that a photon went only through one arm of an interferometer before reaching a screen, a penalty for violating consistency conditions and talking nonsense.

2. CLASSICAL PHYSICS

A convenient way of understanding the existence of classical physics is to ask whether we can think in the same conceptual framework as Newton did. One does not however assume with Bohr that this framework should be kept together with quantum concepts and added to them, because we expect only one kind of laws of physics, and not two of them. One needs nevertheless some amount of common sense and determinism, without which no directives for experimenting nor book-keeping of data would be possible. The origin of their validity, even if only an approximate one, must be found somewhere in the intimate connection between logic and physics we have found. One must then discover *when* common sense is valid and to what extent. This programme goes somewhat farther than recovering classical dynamics through Ehrenfest's theorem and it requires much more powerful mathematics, which is why only a sketch of the results will be given.

A classical physicist would say that a pendulum has a position x with an upper error Δx and a momentum p with a similar bound Δp. The point with coordinates (x, p) is therefore assumed to lie within a rectangle in phase space. More generally, an elementary *classical* property states that (x, p) is in some phase space cell C at some time t. We can also say it in our quantum language if we are able to associate with this statement a projection operator in Hilbert space. This is not quite trivial and there are limitations, the cell C must be big enough and its boundary smooth enough, but it can be done. So, even if our native language is the quantum one, we can talk classically if we wish to do so.

What about classical determinism? It says essentially that the previous property is logically equivalent to another one referring to another time t' and to the cell C' into which C is transformed by classical motion. Assume that C and C' are regular enough to be associated with projection operators E and E'. One would then expect determinism to be perfect if $E' = U(t'-t)EU^{-1}(t'-t), U(t)$ being Schrödinger's evolution operator. This equality is of course never exact but it can be almost exact and there is a theorem showing when and how it is so. One can also obtain the corresponding probability of making an error, find the kind of hamiltonians and the kind of cells for which this error is small enough (and most often very small). One also finds with not too much surprise that there are exceptions to this determinism of a quantum origin, since it breaks down typically for chaotic systems with fully established chaos down to the scale of Planck's constant.

One can even translate the basic logical axiom that was given at the end of last section as it applies to classical physics, at least when the correspondence between phase space cells and projection operators is preserved by dynamics: it says that a consistent quantum way of reasoning boils down to common sense in that case. This is why it was said at the beginning that the history approach leads to a drastic change for epistemology: a change of premiss. Physical laws come first and common sense is their consequence, not the other way round.

3. DECOHERENCE

There is only one ingredient left, namely decoherence, for having a complete theory. I assume everybody here knows what it is: a spontaneous and very rapid dynamical diagonalization of a reduced density operator occurring for most macroscopic systems. The reduced density operator is obtained by tracing out the numerous microscopic degrees of freedom from the complete density operator, so as to leave only collective observables. The diagonalization mechanism has the same origin as dissipation, namely a coupling between the collective degrees of freedom and the microscopic ones. There are some interesting questions relating to the definition of collective observables and what are the criteria of "collectiveness"; some other questions are concerned with the preferred basis or bases in which diagonalization takes place, but we have no time for discussing them.

Decoherence is a very important result and Gell-Mann and Hartle are trying to recover the derivation of classical physics that was sketched in the previous section by using only decoherence. I believe that they are most probably right, as far as the physical foundations are concerned, but it will be presumably very difficult to prove it, because the mathematics of decoherence are very crude when compared with the high-tech tools one can buy from the experts in microlocal analysis.

Decoherence can be added to what we already found so as to give new results. It is very often responsible for the validity of useful consistency conditions concerning collective properties of macroscopic objects. It has its own direction of time, the same as dissipation, and one can use the fact that macroscopic consistency can be a consequence of decoherence to show that this direction of time must be the same as the one in logic (as well as in cosmology, as shown by Gell-Mann and Hartle). It seems therefore that we are also learning something interesting about time.

I can only mention very briefly some rather amusing consequences that are obtained when putting together decoherence, microlocal classical theory and algorithmic complexity [2]: Many observables cannot be measured. This is because the instruments that would measure them would have to be much bigger than the universe. This remark concerning a practical impossibility can also be turned into an argument of principle because, if the instrument could be made in some other much larger universe, it would be so massive as to collapse immediately into a black hole. One can thus remove an old objection against decoherence by John Bell [14]: He said that decoherence is a good answer to the problems of measurement theory "for all practical purposes" but not a fundamental one, because quantum interferences persist in the full density operator and some clever and costly measurements could still show their existence. The answer is that these measurements are not compatible with the finiteness of the universe nor with the laws of general relativity. This way of getting rid of a deep objection by resorting to other laws of physics is somewhat reminiscent of Bohr's famous answer to an objection by Einstein against the uncertainty relation between time and energy measurements.

4. MEASUREMENT THEORY

Measurement theory becomes a purely deductive matter in the present theory. A measurement must satisfy two basic conditions. The first one is essentially a definition : Two physical systems Q and M interact. Two observables are dynamically correlated, say A belonging to Q (e. g. a spin component) and B belonging to M (e. g. a pointer position). Their correlation is the following: if the initial state is an eigenstate of A with eigenvalue a_n and an eigenstate of B with eigenvalue 0 (neutral position of the pointer), the final state is an eigenstate of B with an eigenvalue b_n associated in a one-to-one way with a_n. This condition, going back to von Neumann, is supposed to be a direct consequence of the Schrödinger equation for the interacting system $Q+M$. The second necessary condition for a measurement is that M must be a macroscopic system so that B is a collective observable undergoing decoherence (which implies by the way that M also shows dissipation and has an irreversible behaviour).

Measurement theory relies upon two important quantum properties. The first one, according to which $B = b_n$ at the end of the measurement process, is a datum . By using the previous results on classical physics, the *datum* can be also considered as a classical property of the measuring instrument M, something we can observe directly: a phenomenon. The second significant property is the *result* of the measurement, which states that the measured observable A has the value a_n just at the beginning of the measurement. It is perfectly clear that an experiment is always intended to learn something about the result by looking at the datum.

Using the dynamical conditions and relying upon the logical axiom given at the end of Section 1, it is easy to compute a few consistency conditions and conditional probabilities so as to draw various conclusions. Several theorems are obtained, which turn out leave aside what is usually stated as the first axiom. For instance, the result must refer only to the eigenvalues of the measured observable A, because this is a direct consequence of the initial definition of a property in terms of projection operators. There is also a key theorem saying that the datum and the result are logically equivalent, something most often missing in conventional discussions. This logical equivalence refers of course to the basic logical axiom and it turns out that the present theorem is valid in every consistent logic including the statement of the datum (i.e., of an experimental fact) among its field of propositions. The fact that this basic logical equivalence is therefore insensitive to complementarity is can be used to define completely what can be said to be true in quantum mechanics [15] and, as a consequence, to long-standing difficulties associated with the EPR (Einstein, Podolsky and Rosen) experiment.

Another result has to do with the probabilities of various data. They coincide of course with the usual ones. One can say moreover from our standpoint that probabilities cease at that point to be a pure tool of logic to become at last an expression of randomness. The randomness of quantum events is a necessary consequence of quantum logic. Another significant result is concerned with two successive measurements, when an observable A is first measured and a second one A' is measured later on. The theory gives of course explicitly the joint probability for the two data or, equivalently, the conditional probability for a result a'_q under the condition that the result a_n was obtained first. It can be proved that this conditional probability has a simple expression, the one usually written under the assumption of wave function reduction. There is of course no physical reduction effect in the present theory, at least affecting the measured system, but only decoherence in the measuring apparatus.

5. CONCLUSION AND OUTLOOK

At least one problem has not been solved, and could not be in the present approach. It is the so-called objectification or actualization problem, i. e. why and how a unique datum really occurs at the end of a measurement. This remains by far the most important question in quantum mechanics, at least in my opinion. The reader is referred to other publications as to its present status or even perhaps its non-existence as a problem in physics, which would have the most drastic consequences in epistemology [2,10].

REFERENCES

1. R. Omnès, *Rev. Mod. Phys.* **64**, 339 (1992).
2. R. Omnès, *The Interpretation of Quantum Mechanics* (Princeton University Press, Princeton, 1994).
3. R. P. Feynman and F. L. Vernon, *Ann. Phys. (N. Y.)* **24**, 118 (1963); K. Hepp and E. H. Lieb, *Comm. Math. Phys.* **35**, 265 (1974); W. H. Zurek,

Phys. Rev. D **24**, 1516 (1981); **26**, 1862 (1982); A. O. Caldeira and A. J. Leggett, *Physica A* **121**, 587 (1983); E. Joos and H. D. Zeh, *Z. Phys. B* **59**, 223 (1985).

4. R. Griffiths, *J. Stat. Phys.* **36**, 219 (1984).
5. R. Omnès, *J. Stat. Phys.* **53**, 893 (1988).
6. M. Gell-Mann and J. B. Hartle, in *Complexity, Entropy and the Physics of Information* , W. Zurek, ed. (Santa Fe institute Studies in the Science of Complexity, No. 8) (Addison-Wesley, Redwood City, 1991).
7. G. Badurek, H. Rauch and D. Tuppinger, *Phys. Rev. A* **34**, 2600 (1985).
8. A. J. Leggett, *Progr. Theor. Phys. (Suppl.)* **69**, 1 (1980).
9. J. Clarke, A. N. Cleland, M. H. Devoret, D. Estève and J. M. Martinis, *Science* **239**, 992 (1988).
10. R. Omnès, *Philosophie de la science contemporaine* (Gallimard, Paris, 1994).
11. R. Omnès J. *Stat. Phys.* **57**, 357 (1989).
12. L. Hörmander, *The Analysis of Linear Partial Differential Operators, Vol. 3*, (Springer, Berlin, 1985).
13. R. Omnès, *Ann. Phys. (N. Y.)* **201**, 354 (1990).
14. J. S. Bell, *Helv. Phys. Act.* **48**, 93 (1975).
15. R. Omnès, *J. Stat. Phys.* **62**, 841 (1991).

IS QUANTUM MECHANICS A LIMIT CYCLE THEORY?

L. de la Peña and A. M. Cetto

Instituto de Física, UNAM
Apartado postal 20-364, 01000 México, D. F.

Some recent results in the theory of stochastic electrodynamics (SED) are presented. First it is seen that introduction of the SED hypothesis of a real zeropoint field into the thermodynamical description of the equilibrium radiation field in terms of a Maxwell-Boltzmann distribution, leads to Planck's formula and concomitant quantum properties, thus showing that SED is not a classical theory. A new version of SED is then used to study the stationary orbits of the bound particle; under certain explicitly stated conditions it is possible to recover the quantum equations of motion. It is argued that these stationary motions can be interpreted as limit cycles of the mechanical part of the system. The results strongly suggest that quantum mechanics provides an approximate asymptotic description of nature; the kind of approximations involved are briefly discussed.

Key words: Planck's law, zeropoint field, stochastic electrodynamics, Heisenberg equations, limit cycles

1. INTRODUCTION

On this occasion we want to present some recent results of stochastic electrodynamics (SED), the theory built upon the assumption that quantum properties arise as a consequence of the zeropoint radiation field (zpf) . Due to lack of space the presentation will be very sketchy; we invite the interested reader to find the omitted details in Refs. [1,2]. The first part of the paper is devoted to the oldest quantum problem, namely Planck's law, whereas in the second part we direct our attention to the mechanics of the SED system. Although the two parts belong to the same theory they are very different in nature. In dealing with the problem of the statistical properties of the field in equilibrium with matter, we will apply ideas that are well known; the solution we are offering could long ago have been found by any of the proponents of SED and thus counted as another one of the

M. Ferrero and A. van der Merwe (eds.), Fundamental Problems in Quantum Physics, 225–236.

problems that SED has tackled with success –which is characteristic of the linear problems.

As is well known, SED has been able to solve (or at least to offer understandable physical explanations to) a good series of quantum problems, from the simple harmonic oscillator [3], the van der Waals forces [4], the atomic lifetimes and the Lamb shift, both in vacuum and within cavities [5], to those typical effects of quantum optics that have been studied in stochastic optics [6]. However, the passage from the linear problems to the atomic ones revealed to be far more than a simple matter of technicalities, to the extent that it has been expressed more than once that SED simply is not the correct theory of matter—with as much conviction as it has been said that the failure lies not in the essence of SED, but in the approximations and methods used. Some pertinent references and reviews are [7], [8], and [9].

Some of us still believe that SED contains already the essential ingredients to explain the nonrelativistic quantum behaviour of matter, and that the addition of new ones will allow perhaps a more precise and detailed account of the quantum facts, without inducing a deep conceptual change of the basic scheme. In support of this statement we argue as follows. In usual SED it is assumed that the motion of the atomic system should be obtained as a perturbation of the classical solution. However, the zpf and radiation reaction play an essential role, to the extent that in the stationary states, which are reached after a transient, dissipative process, the average rates of energy absorbed from the field and radiated by the particle coincide; the field controls the existence of such states. This situation can hardly be describable by perturbative methods. It is possible to think, as we shall discuss in the second part of the paper, that there exist limit cycles towards which the atomic electrons fall, giving rise to these stationary states of motion. Further, the fact that the random force on the particle comes from a field seems to be crucial. Indeed, in developing its motion the atomic electron continuosly changes the frequencies of the (average) orbital motion due to the nonlinearities of the system. Now, we consider the field as consisting of a set of oscillators, which are physically real and not merely a mathematical artifact; thus, at every moment the orbiting electrons are subject to external excitations to which they resonate. The strong response to such perturbations means fast changes in the orbital motion, except when the system is in one of the stationary states.

In the second part of the paper we sketch an attempt to test these ideas. We show that it is indeed possible to extract a quantum behaviour for the atomic electron subject to the zpf by introducing a couple of demands within the spirit of the above discussion. In this sense, the derivation of Planck's law from the single extra requirement of the existence of the zpf on top of classical elements gives strong support to the conjecture that the inclusion of the zpf is sufficient to generate the quantum properties.

2. THE ZEROPOINT FIELD AND PLANCK'S LAW

Let us consider the radiation field, with the zpf included, in equilibrium with matter at temperature T; our aim is to find the energy distribution for this field. The equilibrium distribution for every mode is assumed to be given by the Maxwell-Boltzmann expression, corresponding to a canonical ensemble (as usual, $\beta = 1/kT$):

$$W(E) = \frac{1}{Z} g(E) e^{-\beta E}. \tag{1}$$

The coefficient $g(E)$ represents an intrinsic probability for the realization of the states characterized by an energy E. In classical physics $g(E) = 1$, and hence equipartition holds among the field oscillators: $\bar{E} = \beta^{-1}$. Also, it follows that $\overline{E^r} = r! \bar{E}^r$. The well-known formula $\bar{E} = \beta^{-1}$ is fundamental for what follows, since it means that $\bar{E}(T = 0) = 0$, *i.e.*, that no zeropoint energy can exist in classical physics. To allow for $\bar{E}(T = 0) \equiv \mathcal{E}_0 \neq 0$, it becomes essential to leave $g(E)$ undefined. (The only other possibility, namely that the coefficient of the exponential in Eq. (1) is modified, would amount to a redefinition of temperature.) Then, for arbitrary $g(E)$:

$$\overline{E^r}' \equiv \frac{\mathrm{d}\overline{E^r}}{\mathrm{d}\beta} = \bar{E}\overline{E^r} - \overline{E^{r+1}}. \tag{2}$$

For $r = 2$ this equation leads to the well known Einstein relation:

$$\bar{E}' = \bar{E}^2 - \overline{E^2}, \tag{3}$$

which in terms of the temperature and the specific heat reads $kT^2 c_V = \sigma_E^2$. It follows that for any system whose specific heat remains finite at $T = 0$ (and this should be the case for any physical system), $\sigma_E^2 \underset{T\to 0}{\to} 0$, *i.e.*,

$$\overline{E^2} = \bar{E}^2 = \mathcal{E}_0^2 \quad \text{at} \quad T = 0. \tag{4}$$

Indeed, an even stronger relation holds, namely: $\overline{E^r} = \bar{E}_0^r$ at $T = 0$ ($r = 0, 1, 2, \ldots$). We skip the demonstration and refer the reader to the literature [1].

To calculate the moments of the field at $T > 0$ we rewrite the recurrence relation Eq.(2) in terms of the energy above the zeropoint, $E_T = E - \mathcal{E}_0$:

$$\overline{E_T^{r+1}} = \overline{E_T}\, \overline{E_T^r} - \overline{E_T^r}'. \tag{5}$$

We look for the most general expression for $\overline{E^r}$ such that (4) holds, along with the correct classical limit: $\overline{E^r_T} = r!\,\overline{E_T}^{\,r}$ at $\beta \to 0$. This requires that $\overline{E^r_T} = r!\,\overline{E_T}^{\,r} + \mathcal{E}_0^r \phi_r$, with ϕ_r a polynomial in $\overline{E_T}$ of degree $r-1$ at most. Then, from (5):

$$\phi_{r+1} = -\frac{1}{\mathcal{E}_0}\phi'_r + \frac{\overline{E_T}}{\mathcal{E}_0}\phi_r + r r!\left(\frac{\overline{E_T}}{\mathcal{E}_0}\right)^{r-1}\phi_2 \tag{6}$$

with $\phi_0 = 0$ and $\phi_1 = 0$. Since the thermal energy vanishes at $T = 0$, it follows that $A_{r0} = 0$ for all r, and we get $\phi_r = \sum_{k=1}^{r-1} A_{rk}\left(\overline{E_T}/\mathcal{E}_0\right)^k$. Thus for ϕ_2 the general solution is

$$\phi_2 = \eta\left(\frac{\overline{E_T}}{\mathcal{E}_0}\right) \qquad (A_{21} = \eta), \tag{7}$$

with η an undetermined coefficient, different from zero. All higher moments follow from the recurrence relation (6).

With the above results the solution of the posed problem is immediate, since by integrating the equation: $\overline{E_T}' = -\overline{E_T}^2 - \eta\mathcal{E}_0\overline{E_T}$ that follows from Eqs. (7) and (3) and demanding that $\overline{E_T} \underset{T\to\infty}{\to} \infty$, one gets

$$\bar{E} = \mathcal{E}_0 + \frac{\eta\mathcal{E}_0}{e^{\eta\mathcal{E}_0\beta} - 1}, \tag{8}$$

which is Planck's law with $\eta = 2$ (recall that $\mathcal{E}_0 = (1/2)\hbar\omega$). Note that the value of η is the only element that could not be determined from thermodynamic considerations. For previous attempts to derive Planck's law within SED see [8] and [10].

Some important consequences of Eq.(8) are immediate; in particular, since the partition function takes the value $Z = \left(1 - e^{-\eta\mathcal{E}_0\beta}\right)^{-1} e^{-\beta\mathcal{E}_0}$, by developing in a power series it follows that:

$$Z = \int g(E)e^{-\beta E}\mathrm{d}E = \sum_{n=0}^{\infty} e^{-((1+\eta n)\mathcal{E}_0\beta} = \sum_{n=0}^{\infty} e^{-E_n\beta}, \tag{9}$$

where

$$E_n = (1+\eta n)\,\mathcal{E}_0, \quad n = 0, 1, 2, 3, \ldots. \tag{10}$$

A comparison of the terms in Eq.(9) gives $g(E) = \sum_{n=0}^{\infty}\delta(E - E_n)$; thus,

$$W(E) = \frac{1}{Z}\sum_{n=0}^{\infty} e^{-E_n\beta}\delta(E - E_n).$$

These results show clearly that the zero-point field changes the nature of the originally classical problem: The *equilibrium* field has acquired quantum properties due to the presence of the zpf. The clue lies in the Einstein formula $\sigma_E^2 = \overline{E_T}^2 + \eta\mathcal{E}_0\overline{E_T}$. It is just from an expression of this sort that Einstein conceived the quantum of radiation. The $\bar{E}^2-$ contribution to the energy fluctuations is predicted by Maxwell theory, when $\eta\mathcal{E}_0 = 0$. The linear term –which Einstein got for the first time by working from Planck's law backwards– he interpreted as coming from independent random events and thus revealing a corpuscular structure (the factor $\eta\mathcal{E}_0$ is precisely the separation between energy levels in the distribution $W(E)$). In the present framework it should be interpreted in wave terms as the extra fluctuations due to the zeropoint field; these additional interferences seem to be the cause of the quantum behaviour of the equilibrium field. Once one has the Planck law, by following Einstein's argument in his noted paper on the AB coefficients [11] one can demonstrate the needle structure of the radiation emitted by an atom during a spontaneous transition, and hence follows the possibility of explaining other typical quantum phenomena as the Compton effect [12] or Debye's law for the specific heat of solids [13], which are considered to require discrete elements for their explanation.

3. ATOMIC SYSTEMS AND THE ZEROPOINT FIELD

Having seen that the introduction of the zpf into an otherwise classical problem gives rise to quantization, let us turn our attention to the atomic problem. To simplify matters as much as possible we start from the Braffort-Marshall equation, which is the Abraham-Lorentz equation for the specific case of SED. It reads:

$$m\ddot{\mathbf{x}} = m\tau\dddot{\mathbf{x}} + \mathbf{F}(\mathbf{x}) + \mathbf{F}_L, \tag{11}$$

where $m\tau\,\dddot{\mathbf{x}} = m(2e^2/3mc^3)\,\dddot{\mathbf{x}}$ is the radiation reaction force and $\mathbf{F}_L$ is the Lorentz force generated by the zpf, which for simplicity is here reduced to its electric component. The vacuum field has spectral energy density $\rho(\omega) = \hbar\omega^3/2\pi^2c^3$, corresponding to an energy of $\mathcal{E}_0 = (1/2)\hbar\omega$ for each mode, as stated previously; in terms of plane waves,

$$\mathbf{E} = i\sum_{n,\sigma}\sqrt{\frac{\pi\hbar\omega_n}{V}}\hat{\epsilon}_{n\sigma}a^0_{n\sigma}\exp i\left(\mathbf{k}_n\cdot\mathbf{x} - \omega_n t\right) + \text{c.c.} \tag{12}$$

For the free field the complex amplitudes $a^0_{n\sigma}$ are taken as independent random variables with a Gaussian distribution around zero and a uniformly distributed phase; we will come back to this important point below.

Now we seek *stationary* solutions to Eq. (11). It is possible to demonstrate very generally that such solutions do no depend on the initial conditions in an essential form, since these are taken care of by the transient part of $x(t)$, which decays in a time of the order of $\tau\Omega^2$ where Ω is a characteristic frequency of the motion. A natural step is to write down the Fourier transform of Eq. (11); however, due to the unknown (nonlinear in general) form of the external force, this may become an entirely formal step of reduced utility. What we do is pay attention to a reduced set of possible solutions, and in this particular case one can go much further. If in a Fourier representation we write $x_\alpha(t) = \sum_\beta \tilde{z}_{\alpha\beta} \exp(i\omega_{\alpha\beta}t)$ and $F_\alpha(x) = \sum_\beta \tilde{\Phi}_{\alpha\beta} \exp(i\omega_{\alpha\beta}t)$, both $\tilde{z}$ and $\tilde{\Phi}$ depend in principle on the $a^0(\omega)$. We have introduced a label α, considering the possibility that there is more than one stationary solution. At the same time, the different Fourier components of a given $x_\alpha(t)$ have been labelled with a second index β; hence the need of two indices for the Fourier amplitudes and frequencies.

As an illustration of the procedure to follow, let us consider the anharmonic oscillator with a non linear force γx^3; this problem is almost the simplest non linear one, but it already contains the ingredients of interest for the present analysis. In the above terminology, the Fourier coefficient of the nonlinear term of frequency $\omega_{\alpha\beta}$ is: $\tilde{\Phi}_\gamma(\omega)\Big|_{\omega_{\alpha\beta}} = \gamma \sum \left(\tilde{z}_{\mu'\nu'}\tilde{z}_{\mu''\nu''}\tilde{z}_{\mu'''\nu'''}\right)_{\omega_{\alpha\beta}}$, where the (triple) sum is realized over the sets of indices such that $\omega_{\alpha\beta} = \omega_{\mu'\nu'} + \omega_{\mu''\nu''} + \omega_{\mu'''\nu'''}$. In terms of Fourier amplitudes $\tilde{x}_{\alpha\beta}$ given by $\tilde{z}_{\alpha\beta} = \tilde{x}_{\alpha\beta}a^0_{\alpha\beta}$, we get from Eq. (11)

$$\left(-\omega_{\alpha\beta}^2 + i\tau\omega_{\alpha\beta}^3 + \omega_0^2\right)\tilde{x}_{\alpha\beta} + \gamma\sum\left(\tilde{x}'\tilde{x}''\tilde{x}'''\right)_{\alpha\beta}\frac{\left(a^{0\prime}a^{0\prime\prime}a^{0\prime\prime\prime}\right)_{\alpha\beta}}{a^0_{\alpha\beta}} = \frac{e}{m}\tilde{E}(\omega_{\alpha\beta}). \tag{13}$$

Here we take a crucial step: we look for solutions $(\tilde{x}_{\alpha\beta}, \omega_{\alpha\beta})$ that are *not stochastic*, *i.e.*, that are insensitive to the fluctuations of the random variables a^0. Of course this can be only an approximation and there is no guarantee right from the beginning that such solutions exist or have any physical meaning or interest; however, the point is that they can exist and are important, as will be seen in what follows. From Eq. (13) we see that they can be constructed by demanding that the field amplitudes satisfy the condition

$$\left(a^{0\prime}a^{0\prime\prime}a^{0\prime\prime\prime}\right)_{\alpha\beta} = a^0_{\alpha\beta}. \tag{14}$$

Then the system responds linearly to the field of characteristic frequencies $\omega_{\alpha\beta}$, and we get a set of nonlinear algebraic equations that determine the response amplitudes $\tilde{x}_{\alpha\beta}$ and characteristic frequencies $\omega_{\alpha\beta}$, namely,

$$-m\omega_{\alpha\beta}^2\tilde{x}_{\alpha\beta} = \tilde{F}_{\alpha\beta} - im\tau\omega_{\alpha\beta}^3\tilde{x}_{\alpha\beta} + e\tilde{E}(\omega_{\alpha\beta}). \tag{15}$$

This result can be cast in a most revealing form in terms of time-dependent amplitudes $\tilde{x}_{\alpha\beta}(t) = \tilde{x}_{\alpha\beta} \exp(i\omega_{\alpha\beta}t)$:

$$m\frac{d^2\tilde{x}_{\alpha\beta}}{dt^2} = \tilde{F}_{\alpha\beta}(\tilde{x}) + m\tau\frac{d^3\tilde{x}_{\alpha\beta}}{dt^3} + e\tilde{E}_{\alpha\beta}. \tag{16}$$

The stochastic amplitudes have disappeared completely; if we leave aside the last two terms of the rhs, which amounts to a neglect of the radiative corrections, we get the Heisenberg equations of motion in matrix form, as will be verified below. Of course, this suggests that the solutions just constructed correspond to the quantum mechanical stationary states (not stationary any more when the radiative corrections are included). This seems to be the general case for binding forces, since a direct generalization of the above procedure, assuming that the force may be represented by a power series, leads to the obvious generalization of the previous results. (For the details, please refer to the literature [2]). Specifically, the generalization of (14) leads to the conditions: i) that each product of amplitudes a^0 must reduce to a single a^0: $a^0_{\alpha\beta} = a^0_{\lambda_1\mu_1}a^0_{\lambda_2\mu_2}\cdots a^0_{\lambda_n\mu_n}$, and ii) that the product of a^0's must be just the required $a^0_{\alpha\beta}$ and not another independent random amplitude. The solution to these requirements is $a^0_{\alpha\beta} = \exp i\varphi_{\alpha\beta}$, with the random phase $\varphi_{\alpha\beta}$ such that

$$\varphi_{\alpha\beta} = \varphi_{\alpha\lambda_1} + \varphi_{\lambda_1\lambda_2} + \varphi_{\lambda_2\lambda_3} + \ldots + \varphi_{\lambda_{n-1}\beta}; \tag{17}$$

in particular, $\varphi_{\alpha\lambda} = -\varphi_{\lambda\alpha}$. Further,

$$a^0_{\alpha\lambda_1}a^0_{\lambda_1\lambda_2}\cdots a^0_{\lambda_{n-1}\beta} = a^0_{\alpha\beta}, \tag{18}$$

$$\omega_{\alpha\beta} = \omega_{\alpha\lambda_1} + \omega_{\lambda_1\lambda_2} + \ldots + \omega_{\lambda_{n-1}\beta}, \tag{19}$$

and $\omega_{\alpha\beta} = -\omega_{\beta\alpha}$. In all these expressions the indices λ_k may take any value from the set $\{\alpha\}$. Further we observe that the general solution to Eqs. (17) and (19) is

$$\varphi_{\alpha\beta} = \varphi_\alpha - \varphi_\beta \quad (\text{modulo} 2\pi), \quad \text{and} \quad \omega_{\alpha\beta} = \Omega_\alpha - \Omega_\beta, \tag{20}$$

with φ_α and φ_β random phases uniformly distributed in the interval $(0, 2\pi)$, and hence,

$$a^0_{\alpha\beta} = \exp i(\varphi_\alpha - \varphi_\beta). \tag{21}$$

The above results mean that the relevant modes of the random field that support the stationary states of a given quantum system must have statistical properties and correlations that strongly depend on the specific problem; thus, for each stationary system there exists an equilibrium field. This seems to indicate that just as the mechanical system is affected by the field during the evolution towards an equilibrium situation, also the mechanical system affects the field and impresses upon it its characteristic features.

Introducing a matrix for p with elements $\tilde{p}_{\alpha\beta} = im\omega_{\alpha\beta}\tilde{x}_{\alpha\beta}$ one gets from Eq. (16), in the radiationless approximation:

$$\tilde{p} = m\frac{d\tilde{x}}{dt}, \quad \frac{d\tilde{p}}{dt} = \tilde{F}(\tilde{x}). \tag{22}$$

To these equations one must add another one that fixes the scale of the solutions, which depends on the scale of the fluctuations. It is not difficult to get it from general considerations [2]; here we simply quote the final result:

$$-\sum_{\beta}\omega_{\alpha\beta}\left|\tilde{x}_{\alpha\beta}\right|^2 = \frac{\hbar}{2m}, \quad \text{or} \quad [\tilde{x};\tilde{p}]_\alpha = i\hbar. \tag{23}$$

An important consequence of the above equations is $\tilde{H}_{\alpha\beta} = \hbar\Omega_\alpha\delta_{\alpha\beta}$, which indicates that the (Hamiltonian) matrix H is diagonal, with non random eigenvalues $H_\alpha = \hbar\Omega_\alpha a_{\alpha\alpha} = \hbar\Omega_\alpha = \mathcal{E}_\alpha$. Hence, from Eq. (20), we have $\hbar\omega_{\alpha\beta} = \mathcal{E}_\alpha - \mathcal{E}_\beta$, which means that the characteristic frequencies are just the transition frequencies of quantum mechanics, related to the $\mathcal{E}_\alpha$ as usual by Bohr's formula.

Before discussing the meaning of the theory let us illustrate how the passage to the Hilbert space formulation can be realized. Consider a set of matrices $\hat{a}^{\alpha\beta}$:

$$\left(\hat{a}^{\alpha\beta}\right)_{\mu\nu} = a_{\alpha\beta}\delta_{\alpha\mu}\delta_{\beta\nu}, \tag{24}$$

with $a_{\alpha\beta} = \exp[i(\varphi_\alpha - \varphi_\beta) + i\omega_{\alpha\beta}t]$. A product of such matrices gives another matrix of the set:

$$\left(\hat{a}^{\alpha\beta}\hat{a}^{\gamma\delta}\right)_{\mu\nu} = a_{\alpha\delta}\delta_{\alpha\mu}\delta_{\beta\gamma}\delta_{\delta\nu} = \delta_{\beta\gamma}\left(\hat{a}^{\alpha\delta}\right)_{\mu\nu}. \tag{25}$$

Thus, for a dynamical variable, say x, we have

$$\hat{x} = \sum_{\alpha,\lambda}\tilde{x}_{\alpha\lambda}\hat{a}^{\alpha\lambda} \equiv \sum_{\alpha}\hat{x}_\alpha; \quad \hat{x}_\alpha = \sum_{\lambda}\tilde{x}_{\alpha\lambda}\hat{a}^{\alpha\lambda}, \tag{26}$$

where the matrix elements of $\hat{x}$ are just $\hat{x}_{\alpha\lambda} = \tilde{x}_{\alpha\lambda}a_{\alpha\lambda}$; then the square of x is represented by $\left(\hat{x}^2\right)_{\mu\nu} = \sum_\lambda \tilde{x}_{\mu\lambda}\tilde{x}_{\lambda\nu}a_{\mu\lambda}a_{\lambda\nu} = \left(\tilde{x}^2\right)_{\mu\nu}a_{\mu\nu}$, and so on, which once more reveals the matrix structure of the theory. Note further that $\hat{a}^{\alpha\beta}$ can be written as the product of two vectors $|\alpha\rangle$, $\langle\beta|$, where $\langle\alpha| = (|\alpha\rangle)^\dagger$ and

$$|\alpha\rangle = \begin{pmatrix} 0 \\ \vdots \\ a_\alpha \\ \vdots \end{pmatrix}, \tag{27}$$

with $a_\alpha = \exp i(\varphi_\alpha + \Omega_\alpha t)$, $a_\alpha a_\beta^* = a_{\alpha\beta}$. From Eqs. (24) and (27),

$$\hat{a}^{\alpha\beta} = |\alpha\rangle\langle\beta|\,, \quad \sum_\alpha |\alpha\rangle\langle\alpha| = \sum_\alpha \hat{a}_{\alpha\alpha} = \hat{1}, \quad \langle\alpha\mid\beta\rangle = \delta_{\alpha\beta},$$

and thus we have constructed the corresponding Hilbert space; note that the vectors $|\alpha\rangle$ involve not the $\omega_{\alpha\beta}$, but the $\mathcal{E}_\alpha$. An arbitrary observable f is now represented by

$$\hat{f} = \sum_\alpha \hat{f}_\alpha = \sum_{\alpha,\beta} \tilde{f}_{\alpha\beta}\hat{a}^{\alpha\beta} = \sum_{\alpha,\beta} \tilde{f}_{\alpha\beta}\,|\alpha\rangle\langle\beta|\,, \tag{28}$$

as usual, with matrix elements $\tilde{f}_{\alpha\beta} = \langle\alpha|\,\hat{f}\,|\beta\rangle$.

4. DISCUSSION OF THE RESULTS

We have shown that the quantum mechanical stationary states are recovered as approximate, time-asymptotic solutions for the mechanical part of the SED system. This seems to corroborate the idea that the quantum behaviour of matter can be understood as a result of its permanent interaction with the zpf. Thus, both field and matter acquire their respective quantum properties as a consequence of the reality of this field. One of the most important points is that the theory leads in a natural way to the discovery of certain orbital motions that are stable against fluctuations and are essentially independent of the initial conditions, a stability that is reached through the interplay of the stochastic forces, counterbalanced by the radiation reaction. Thus, as was stated in the Introduction, it seems reasonable to consider that the stationary states of the quantum systems represent limit cycles of the dynamical system described by SED.

The description in terms of stationary states is of course a result of the approximations made, the most important ones being the demand of stationarity itself, the narrow-band approximation implicit in expressions such as (15) and the total neglect of the fluctuations in solving these equations. In a more complete description one would obtain also a very slow diffusion of such orbits, so slow that times of order $(\tau\Omega^2)^{-1}$ are required to get a significative change in the system; this diffusion of the orbits was identified already within the old version of SED (E. Santos in [3]). Since the times involved in the diffusion are very long compared with the characteristic periods of the orbital motions –for atoms they may be 10^8 times larger or even more– a description in terms of stationary states makes sense for many purposes. The 'continuous' (as seen at the larger time scale) evolution of the motions is reinserted in the quantum description by means of the formalism of spontaneous transitions between states, a very useful tool to describe the basic effects of the passage from one limit cycle to another; indeed, taking $1\Omega \sim E/\hbar$ it is easily seen that $\tau\Omega^2$ is of the order of the natural widh of atomic levels. Of course, these complex problems require much further study.

The above theory can be criticized on the grounds that it mimics what Heisenberg did when passing from classical to quantum mechanics to construct matrix mechanics; however, there are several important differences. The old transition was from classical to quantum physics and entirely formal, the only justification being that it worked; here, the transition is from SED and leads to quantum mechanics as an approximation. Moreover, as was just shown, our procedure clarifies the meaning of previous formal steps and reveals several aspects of the physics of atomic systems that had remained hidden, among them the properties acquired by the field that sustains the stationary states and the linear response of the system to the field once it has reached such a state. Also, the knowledge that the approximations work seems to be entirely new in SED.

From a fundamental perspective, however, we are still in trouble. SED is a physical theory free of conceptual difficulties (at least neglecting the infinite energy of the zpf); in particular, it is genuinely statistical and local, whereas the quantum description is neither, to say the least. In quantum theory we have negative probabilities, there is no consistent and uniquely defined phase-space distribution, we face non local descriptions (recall *e.g.* the quantum potential introduced by Bohm in his attempts to construct a causal description), etc. How is it possible that a basic theory that is free of such fundamental difficulties leads to a theory that is full of them? Of course, our answer is that the use of approximations and the loss of the internal connections between both descriptions due to the lack of a smooth and detailed transition from SED to QM are to be blamed. Unfortunately, both problems remain unsolved for the time being and we have not much more to offer; however, about the first point some comments seem in order.

One finds similar situations, even in the domain of classical physics, where from physically correct theories one gets approximate descriptions which, though useful to perform calculations and valid under limited conditions, nevertheless suffer from essential diseases. A clear example is afforded by the radiation reaction in the Abraham-Lorentz (or Lorentz-Dirac) equation of motion. It is a well-known fact that this equation is non-causal, that it predicts strange behaviours such as preacceleration, runaway solutions and the like. Nothing of this pertains to healthy physics, but the equation is derived from the most important branches of classical physics, Newtonian mechanics and Maxwellian electrodynamics. All these problems arise of course from the approximate character of the equations: the ghosts enter into physics with the approximations. Let us indulge ourselves for a moment in a *gedankenexperiment* in the history of physics. Assume that Maxwell got instead of the full description of electrodynamics, only the Abraham-Lorentz equation of motion. By studying its predictions many might feel tempted –at least those who take the model for the real thing– to conclude that nature is noncausal, that nonlocality (preacceleration) is a fact of nature and the like. Of course, we laugh at such a proposal, because we know the source of the error.

Is it not possible that something similar is happening with quantum theory? According to the perspective offered by SED, the final theory must be genuinely statistical, local and causal. As seen from this side, quantum theory is an approximate description –even if exceedingly good and comprehensive– and we may be confident that the conceptual difficulties of principle associated to it might be overcome with the development of a more exhaustive and refined description. Even if still incomplete, the results that we offer here seem to indicate that SED may be a valuable tool to try a programme of this kind.

This work was supported in part by Conacyt through grant No. 0682-E9208.

REFERENCES

1. L. de la Peña and A. M. Cetto, "Planck's law as a consequence of the zeropoint radiation field," preprint IFUNAM FT93-023.
2. L. de la Peña and A. M. Cetto, *Rev. Mex. Fís.* **39** (1993); **37**, S26 (1991); *Found. Phys.* **24** (1994), "Quantum phenomena and the zeropoint radiation field II," preprint IFUNAM FT93,012. A. M. Cetto and L. de la Peña, *Found. Phys. Lett.* **4**, 73 (1991).
3. T. W. Marshall, *Proc. Roy. Soc.* **276A**, 475 (1963); E. Santos, *Nuovo Cim.* **19B**, 57 (1974); L. de la Peña and A. M. Cetto, *J. Math. Phys.* **18**, 1612 (1977).
4. T. H. Boyer, *Ann. Phys. (N.Y.)* **56**, 474 (1970).
5. A. M. Cetto and L. de la Peña, *Phys. Rev. A* **37**, 1952, 1960 (1988); H. França, T. W. Marshall, and E. Santos, *Phys. Rev. A* **45**, 6436 (1992).
6. T. W. Marshall and E. Santos, in *Problems in Quantum Physics; Gdansk '87*, L. Kostro et al., eds. (World-Scientific, Singapore, 1988); *Phys. Rev. A* **39**, 6271 (1989).
7. T. H. Boyer, in *Foundations of Radiation Theory and Quantum Electrodynamics*, A. O. Barut, ed. (Plenum, New York, 1980).
8. L. de la Peña in *Stochastic Processes Applied to Physics and Other Related Fields*, B. Gómez et al., eds. (World-Scientific, Singapore, 1983).
9. T. W. Marshall and P. Claverie, *J. Math. Phys.* **21**, 1819 (1980); P. Claverie and F. Soto, *J. Math. Phys.* **23**, 753 (1982).

10. T. H. Boyer, *Phys. Rev.* **182**, 1374 (1969); **186**, 1304 (1969); O. Theimer, *Phys. Rev. D* **4**, 1597 (1971).
11. A. Einstein, *Phys. Z.* **18**, 121 (1917).
12. A. V. Barranco and H. M. França, *Found. Phys. Lett.* **5**, 25 (1992).
13. R. Blanco, H. M. França, and E. Santos, *Phys. Rev. A* **43**, 693 (1991).

REALIZATION AND CHARACTERIZATION OF QUANTUM NONDEMOLITION MEASUREMENTS IN OPTICS

J. Ph. Poizat, J. F. Roch and P. Grangier

Institut d'Optique
B.P. 147
F91403 Orsay Cedex, France

The efficiency of an optical quantum non-demolition (QND) measurement can be characterized using three criteria, which describe respectively the quality of the quantum measurement, the non-destruction of the signal, and the conditional variance of the output signal beam, given the output meter beam (quantum-state-preparation criterion). Quantitative limits can be defined with respect to these criteria, delimiting "classical" and "quantum" domains of operation. We describe the implementation of two experiments which fulfill these criteria, using either three-level atoms inside a doubly-resonant optical cavity, or semiconductors light emitters and receivers.

Key words: quantum optics, quantum measurements.

1. INTRODUCTION

A quantum measurement usually perturbs the quantity which is measured, adding "back-action noise" to the system under study. However, "quantum non-demolition" (QND) measurements can be performed, leaving the observed quantity unperturbed, while adding the back-action noise into another complementary observable. Quantum non-demolition measurements were first proposed to monitor the motion of mechanical oscillators [1–3], but it was then realized that they were easier to implement in the optical domain [4–6,8], where quantum noise sensitivity is easily reached. In this case, the measurements are made on a mode of the electromagnetic field considered as a quantum harmonic oscillator, and the coupling responsible for the measurement is usually created using either χ^2 or χ^3 optical non-linearities [9–11]. In this paper, we will focus on the QND measurements of "travelling wave" light beams, rather than fields stored in high Q cavities [12]. Moreover, we will assume that the quantum

M. Ferrero and A. van der Merwe (eds.), Fundamental Problems in Quantum Physics, 237–256.

fluctuations of the beams are very small compared to the mean intensities, so that a linear treatment of these quantum fluctuations is possible [13]. The aim of a QND-travelling-wave measurement scheme is to detect the whole information contained in a quadrature component of the input field, without having degraded this information on the outgoing signal. However, such a device does not exist at present time, and it appears necessary to be able to characterize the "non-ideality" of practical QND schemes. In Sec. 2, we will briefly recall the criteria [14–16] which have been proposed for this characterization, and we will define quantitative limits with respect to these criteria, delimiting "classical" and "quantum" domains of operation. Finally, two experiments, meeting all criteria for QND measurements, will be described in the last sections of the paper. The first one (sec. 3) is based on the coupling of two modes of the field via a crossed-Kerr effect (χ^3 non-linearity) in a three-level atomic system. The second one (Sec. 4) utilizes semiconductor light receiver and emitter to detect and then recreate the intensity fluctuations of the input signal.

2. QND CRITERIA

A. Presentation

Quantum mechanics allows in principle perfectly nondemolition measurements of an observable, but the actual implementation of such measurements has to put up with the non-ideality of any real experimental situation. It is thus useful to define quantitative criteria capable of evaluating the efficiency of an experimental set-up as a QND measurement device. Such criteria have been proposed for the first time in 89 by Holland et al. [15]. In this paper [15] the authors identify three properties, that can be quantitatively determined by correlation coefficients, characterizing the effectiveness of a QND measurement. But some of these quantities are only calculable theoretically, and can not be directly measured in an experiment. Consequently, Grangier et al. [16] have proposed in 92 an experimentally accessible evaluation of these properties using equivalent input noise.

As we said before, the goal of a QND measurement in optics is to measure the variations δX_s^{in} of a quadrature component of a signal beam (labelled s), without changing them, by coupling them, to a quantity δY_m^{out}, called the meter (labelled m). These variations can be either quantum fluctuations or a classical modulation.

The effectiveness of a QND measurement can be characterized by three properties [15]:

(1) The meter beam should pick up some information about the signal beam with a good precision so that a measurement is actually performed (efficiency of the measurement).

(2) The signal should not be excessively degraded by the measurement (non-demolition property).

(3) The meter output should be quantum-correlated with the signal output so that the readout of the meter does give some information about the outgoing signal (output quantum correlation).

We will present here two different sets of such criteria. Actually, the evaluation of the efficiency of a QND system designed to measure δX_s^{in} depends, in the most general case, on the whole quantum state of the input beam, i.e., not only on δX_s^{in}, but also on the complementary quadrature δY_s^{in} and their correlations $\langle \delta X_s^{in} \delta Y_s^{in} \rangle$. This appears as a fundamental problem, that we shall not discuss here, but that is presented in great details in [17]. Here, we shall avoid this problem by imposing a not too restrictive (and experimentally fulfilled) condition on the input state.

As said previously, in the limit of small fluctuations compared to the mean values, the problem can be linearized, and amplitude fluctuations are proportional to intensity fluctuations.

B. Equivalent Input Noises

The idea for an experimental characterization of a QND measurement device was introduced by Grangier et al. [16] using the concept of equivalent input noise, which is well known in electronics. The key-point is to consider the QND device as an input signal beam amplifier with two outputs, the signal output on one hand, and the meter output on the other hand, and to look at the noise properties of these amplifiers. As linearized descriptions of the fluctuations will always be used in this paper, we shall deal with linear amplifiers.

In order to evaluate the quality of the measurement, we shall be concerned with the meter amplifier, i.e., the one having the measurement channel as an output, and the signal beam as an input. We can write:

$$\delta Y_m^{out} = g_m \delta X_s^{in} + B_m^{ad}, \tag{1}$$

where δX_s^{in} and δY_m^{out} are respectively the input signal and the meter read out fluctuations, g_m is the gain of the *meter* amplifier, and B_m^{ad} the added noise. This added noise can itself be split into two terms:

$$B_m^{ad} = f_m \delta Y_s^{in} + B_m, \tag{2}$$

where the first term comes from the signal complementary quadrature, which can eventually be correlated with δX_s^{in}. Indeed, in the case of an input beam squeezed along a direction θ different from these two quadratures (*i.e.*, $\theta \neq 0, \pi/2$), the correlations $\langle \delta X_s^{in} \delta Y_s^{in} \rangle_{sym}$ are non-zero. But the value of f_m depends on measurement system of interest, and is of course zero for an ideal system. The second term B_m accounts for noise which is not correlated with the input signal. Equation (1) can be rescaled to the level of the input signal δX_s^{in},

$$\frac{\delta Y_m^{out}}{g_m} = \delta X_s^{in} + \frac{B_m^{ad}}{g_m}. \tag{3}$$

The *meter* equivalent input noise is then defined by:

$$N_m^{eq} = \frac{\langle \delta Y_m^{out^2} \rangle}{| g_m^2 |} - \langle \delta X_s^{in^2} \rangle. \tag{4}$$

It should be noted that this quantity is different from the variance of B_m^{aj}/g_m when B_m^{ad} is correlated with δX_s^{in}, that is when $f_m \langle \delta X_s^{in} \delta Y_s^{in} \rangle_{sym} \neq 0$. For a perfect measurement system, one has $N_m^{eq} = 0$.

The degradation of the signal is evaluated by considering the noise properties of the *signal* amplifier, *i.e.* the one having the signal beam as an output. We have:

$$\frac{\delta X_s^{out}}{g_s} = \delta X_s^{in} + \frac{B_s^{ad}}{g_s}, \tag{5}$$

where g_s is the gain of the *signal* amplifier, and:

$$B_s^{ad} = f_s \delta Y_s^{in} + B_s, \tag{6}$$

like Eq. (2). The equivalent input noise then reads:

$$N_s^{eq} = \frac{\langle \delta X_s^{out^2} \rangle}{| g_s^2 |} - \langle \delta X_s^{in^2} \rangle. \tag{7}$$

For a perfectly nondemolition system, one has $N_s^{eq} = 0$.

It is to be noticed that when $f_{s,m} \langle \delta X_s^{in} \delta Y_s^{in} \rangle_{sym} \neq 0$, equivalent input noises can become negative. A detailled analysis of this problem is presented in [17].

The third criterion tells us how well the meter read out is correlated with the signal output. A convenient quantity for characterizing it, is the variance in the signal output, given a measured value of the meter (conditional variance). We first define the following correlation coefficient,

$$C^2(\delta X_s^{out}\delta Y_m^{out}) = \frac{|\langle \delta X_s^{out}\delta Y_m^{out}\rangle_{sym}|^2}{\langle \delta X_s^{out^2}\rangle\langle \delta Y_m^{out^2}\rangle}. \tag{8}$$

The conditional variance can then be written (for a linearized system) as :

$$V(\delta X_s^{out} \mid \delta Y_m^{out}) = \langle \delta X_s^{out^2}\rangle(1 - C^2(\delta X_s^{out}\delta Y_m^{out})). \tag{9}$$

To simplify the notation, the conditional variance will be written $V_{s|m}$. This criterion gives in fact the state preparation ability of the system. It gives the best squeezing one can obtain on the outgoing signal beam by using the information on the meter to correct the quantum noise of the signal. For a perfect state preparation device $V_{s|m} = 0$, whereas $V_{s|m} = 1$ (no quantum effect) corresponds to the shot noise level, and $V_{s|m} < 1$ indicates that the two output beams are quantum-correlated.

The conditional variance can be measured experimentally by looking at the correlations between the meter read out and the signal output. The two equivalent input noises can also be measured experimentally, just like they are for amplifiers in electronics, that is by applying a small classical modulation to the input signal, and by looking at how its signal-to- noise ratio is being processed towards the different outputs. This is why we have introduced, for convenience, the transfer coefficients that we shall now describe.

C. Transfer Coefficients

These $T_{m,s}$ coefficients characterize the transfer of the signal-to-quantum-noise ratio (SNR or R in the equations) of the input signal beam towards the meter output for T_m, and the signal output for T_s. One defines the SNR as the ratio of the power of a small classical modulation at a given frequency[1] by the quantum noise power at the same frequency for a given quadrature,

$$R(\omega) = \left| \frac{\langle X(\omega)\rangle^2}{\langle \Delta X(\omega)^2\rangle} \right|, \tag{10}$$

where X is the modulated quadrature and where all quantities are defined in the frequency domain [16]. The transfer coefficient are then defined by:

$$T_m = \frac{R_m^{out}}{R_s^{in}}, \quad T_s = \frac{R_s^{out}}{R_s^{in}}. \tag{11}$$

These coefficients are actually inverse from the noise figures F usually used in electronics. They are positive numbers normally smaller than unity. They are related to the equivalent input noise in a very simple manner:

$$T_m = \frac{\langle \delta X_s^{in^2} \rangle}{\langle \delta X_s^{in^2} \rangle + N_m^{eq}}, \quad T_s = \frac{\langle \delta X_s^{in^2} \rangle}{\langle \delta X_s^{in^2} \rangle + N_s^{eq}}. \tag{12}$$

when the input signal is at the shot noise level, *i.e.*$\langle \delta X_s^{in^2} \rangle = 1$, this comes down to:

$$T_m = \frac{1}{1 + N_m^{eq}}, \quad T_s = \frac{1}{1 + N_s^{eq}}. \tag{13}$$

These equations show that there is a one to one correspondance between the transfer coefficients and the equivalent input noises. It can be notes that $N_{s,m}^{eq} < 0$ is equivalent to $T_{s,m} > 1$ [17].

The interest of these coefficients is that they are directly accessible in an experiment. Signal-to-noise ratios can indeed be very easily visualized on a spectrum analyser, and it is then straight forward to compare the various SNR and to work out the transfer coefficients [8].

D. Heisenberg Inequalities and Classical Limits

We shall show, in this section, what are the limits set by Heisenberg inequalities on a measurement apparatus, and define a classical limit beyond which a system can be labelled as QND. To do so we shall derive expressions between equivalent input noises of various transformations [7].

In the most general case, the input-output relations for a measurement device read (after an eventual rotation of the output signal quadratures, to eliminate the dependence on the complementary input quadrature,

$$\begin{aligned} \delta X_s^{out} &= g_s \delta X_s^{in} + \mathcal{F}, \\ \delta Y_s^{out} &= f_s \delta Y_s^{in} + \mathcal{H}, \\ \delta Y_m^{out} &= g_m \delta X_s^{in} + f_m \delta Y_s^{in} + \mathcal{G}. \end{aligned} \tag{14}$$

We have not written the expression of δX_m^{out} because we will not need it. The operators $\mathcal{F}, \mathcal{H}$ and $\mathcal{G}$ have zero mean value. Let us assume (see subsection 2.B) that $\langle \delta X_s^{in} \delta Y_s^{in} \rangle_{sym} = 0$. The added noises $\mathcal{F}, \mathcal{H}$ and $\mathcal{G}$, which could depend on the complementary quadrature (*cf.* Eq. 2), are then not correlated with the input signal.

In a completely quantum calculation, it is necessary that the output operators satisfy the same commutation relation as the input operators, that is:

$$[\delta X_i^{out}, \delta Y_j^{out}] = 2i\delta_{ij}, \tag{15}$$

where i and j are s or m. One then, gets from (14),

$$\begin{aligned} [\delta X_s^{out}, \delta Y_m^{out}] &= 0 = 2ig_s f_m + [\mathcal{F}, \mathcal{G}], \\ [\delta Y_s^{out}, \delta Y_m^{out}] &= 0 = -2if_s g_m + [\mathcal{H}, \mathcal{G}]. \end{aligned} \tag{16}$$

which lead to the dispersion relations:

$$\langle \mathcal{F}^2 \rangle \langle \mathcal{G}^2 \rangle \geq \mid g_s f_m \mid^2, \quad \langle \mathcal{H}^2 \rangle \langle \mathcal{G}^2 \rangle \geq \mid f_s g_m \mid^2 . \tag{17}$$

Besides, as we have assumed that $\langle \delta X_s^{in} \delta Y_s^{in} \rangle_{sym} = 0$ and that the added $\mathcal{F}, \mathcal{H}$ and $\mathcal{G}$ are not correlated with the input signal, the equivalent input noises for the $\delta X_s^{in} \rightarrow \delta X_s^{out}$ and $\delta Y_s^{in} \rightarrow \delta Y_s^{out}$ transformations are given by :

$$N^{eq}(\delta X_s^{in} \rightarrow \delta X_s^{out}) = \frac{\langle \mathcal{F}^2 \rangle}{\mid g_m^2 \mid}, \quad N^{eq}(\delta Y_s^{in} \rightarrow \delta Y_s^{out}) = \frac{\langle \mathcal{H}^2 \rangle}{\mid f_s^2 \mid}. \tag{18}$$

We also have:

$$\begin{aligned} N^{eq}(\delta X_s^{in} \rightarrow \delta Y_m^{out}) &= \frac{1}{\mid g_m^2 \mid}(\mid f_m^2 \mid \langle \delta Y_s^{in^2} \rangle + \langle \mathcal{G}^2 \rangle), \\ N^{eq}(\delta Y_s^{in} \rightarrow \delta Y_m^{out}) &= \frac{1}{\mid f_m^2 \mid}(\mid g_m^2 \mid \langle \delta X_s^{in^2} \rangle + \langle \mathcal{G}^2 \rangle). \end{aligned} \tag{19}$$

On multiplying the two equations (18) with each other and using inequality (17), one obtains:

$$N^{eq}(\delta X_s^{in} \rightarrow \delta X_s^{out}) N^{eq}(\delta Y_s^{in} \rightarrow \delta Y_m^{out}) \geq \frac{\langle \mathcal{F}^2 \rangle \langle \mathcal{G}^2 \rangle}{\mid g_s^2 \mid\mid f_m^2 \mid} \geq 1. \tag{20}$$

It appears that a good transmission on δX_s^{in}, *i.e.* a small $N^{eq}(\delta X_s^{in} \rightarrow \delta X_s^{out})$, is unavoidably to the detriment of the measurement of the complementary quadrature δY_s^{in}, because $N^{eq}(\delta Y_s^{in} \rightarrow \delta Y_m^{out}) \geq 1/N^{eq}(\delta X_s^{in} \rightarrow \delta X_s^{out})$ becomes then very large. This corresponds to a situations where one could measure at the same time δX_s^{in} by directly detecting the output signal and its complementary quadrature δY_s^{in} on the meter output. It is therefore rather logical to have this type of inequality. Similarly, a "symetrical" inequality can be derived:

$$N^{eq}(\delta Y_s^{in} \to \delta Y_s^{out}) N^{eq}(\delta X_s^{in} \to \delta Y_m^{out}) \geq 1. \tag{21}$$

It says that a good measurement of δX_s^{in} is at the detriment of the transmission of its complementary quadrature δY_s^{in} and the interpretation is the same as for (20).

For a phase-insensitive measurement device, which we shall take as our definition of a classical device, one has:

$$N^{eq}(\delta X_s^{in} \to \delta X_s^{out}) = N^{eq}(\delta Y_s^{in} \to \delta Y_s^{out}), \tag{22}$$

which means that the quality of the transmission is the same for the two complementary quadratures. It is then possible to replace $N^{eq}(\delta Y_s^{in} \to \delta Y_s^{out})$ by $N^{eq}(\delta X_s^{in} \to \delta X_s^{out})$ in (21), in order to obtain an inequality that has to be satisfied by any phase-insensitive apparatus:

$$N^{eq}(\delta X_s^{in} \to \delta X_s^{out}) = N^{eq}(\delta X_s^{in} \to \delta Y_m^{out}) \geq 1, \tag{23}$$

where $N^{eq}(\delta X_s^{in} \to \delta X_s^{out})$ corresponds to the signal equivalent input noise N_s^{eq}, and $N^{eq}(\delta X_s^{in} \to \delta Y_m^{out})$ to N_m^{eq}. This limits shall be taken as the definition of the classical standard limit for a measurement device. It applies, in particular for any combination of beamsplitters and phase-insensitive amplifiers, like laser amplifiers for example.

In the case of a coherent input signal, inequality (23) can be written very easily with transfer coefficients using their expression with the equivalent input noises (*cf.* Eq. (13)). One obtains:

$$T_s T_m \leq 1, \tag{24}$$

for any phase-insensitive device. For a ideal QND device, one would have $T_s = T_m = 1$,*i.e.*, $T_s + T_m = 2$ [7,18].

A measurement device deserves therefore the lable QND when the two conditions $T_s + T_m > 1$, and $V_{s|m} < 1$ are fulfilled at the same time.

3. QND EXPERIMENTS WITH THREE-LEVEL SYSTEMS

The basic idea of this experiment is to use the optical Kerr effect to couple the intensity fluctuations of the signal beam to the phase fluctuations of the meter beam ("crossed-phase modulation effect") [11,19–21]. The information is then read out using some kind of phase-sensitive (interferometric) detection scheme of the meter beam. The first attempt to uswe the optical Kerr effect for QND measurements has been made using non-resonant $\chi^{(3)}$ in optical fibers [4,22,23]. Other experiments were done using resonant $\chi^{(3)}$ in solid-state systems [15,24]. Despite the simplicity of the scheme, it has not been possible to obtain noiseless $\chi^{(3)}$ media : either guided acoustic wave Brillouin scattering (GAWBS) in optical fibers, or spontaneous emission noise in resonant systems, contribute to the added noise B^{ad}. In nearly resonant three-level systems, it is possible to obtain an almost pure crossed-Kerr coupling, without self-phase modulation, by using two-photon non-linearities [20,21,25,26]. For a proper choice of the detunings between the two laser beams and the two transitions, the two-photon dispersion can be made dominant, and the ideal transformation given above can be recovered in principle [26,27]. We will now describe and experiment [8], using a cascade three-level system in a sodium atomic beam [6,7]. This atomic system has the advantage that the resonant enhancement of the non-linear effect is much larger than in solid-state non-linear systems, and the phase damping of the transitions much smaller.

The experimental set-up used for this experiment Fig. 1 is described in detail in [8]. We use the *ghost transition* configuration of [27] (see also [28]). The meter beam is tuned around the lower atomic transition and has a very small power ($5\mu W$) compared to the signal beam (about 300 μW) which is tuned around the upper transition. In this configuration, the absorption of the signal beam is very small, and two-photon bistability effects [29] are avoided due to the weakness of the signal beam. Therefore, this configuration guaranties a theoretically nearly perfect non-demolition interaction, with quite a large coupling between the amplitude of the signal and the phase of the meter beam [27]. The sign of the detunings are chosen in order to be in the moste favorable conditions taking into account the hyperfine structure of the sodium atom. Typical Rabi frequencies normalized to the transverse linewidth of the intermediate level are 8 for the meter beam, and 80 for the signal beam, while the detunings are respectively -300 and $+100$ in the same units (or 1.5 GHz and +0.5 GHz in frequency units).

The cavity and the lasers are adjusted so that both fields are simultaneously resonant in the cavity. A classical amplitude modulation at 10 MHz is applied by an electro optical modulator on the signal beam and the measurement is read out on the phase of the 589 nm laser beam. The levels of the modulation transferred onto the output signal and meter beams are then compared with the signal input modulation level using a spectrum analyser Fig. 2. The level of the signal modulation reflected off the off-resonant cavity is

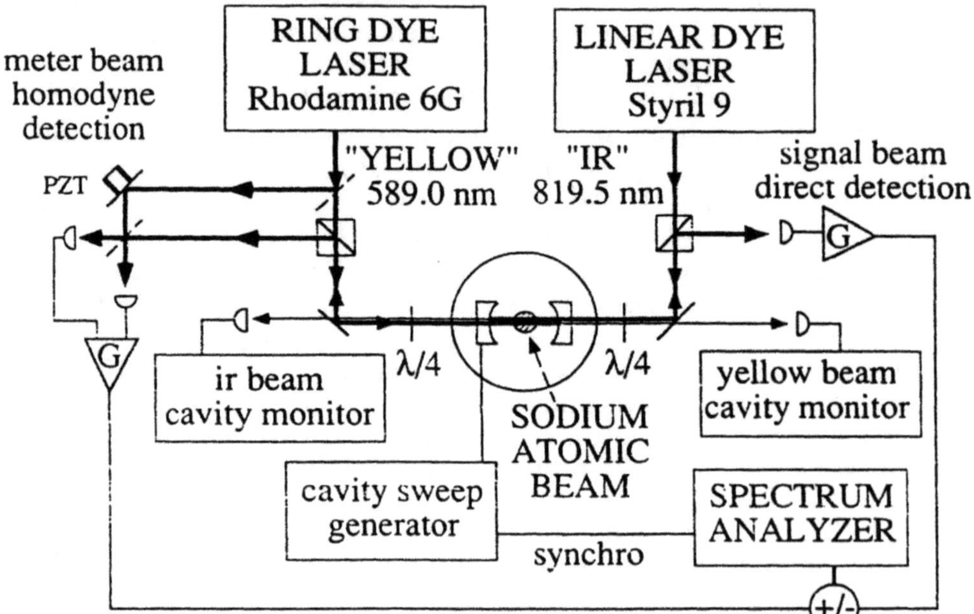

Fig. 1. Experimental set-up. Two lasers at 589.0 nm ("signal" beam) and 819.5 nm ("meter" beam) are tuned close to the resonances of the 3s1/2-3p3/2-3d5/2 cascade in a sodium atomic beam. The quarter-wave plates are used to polarize circularly the two beams in the interaction region, and also to separate the input and output beams.

taken as the reference input modulation level.

Typical results are presented in Fig. 2. The level difference between the peak and the background noise corresponding to the signal beam (recorded off-resonance) gives the input SNR_{sig}^{in} (trace (a)). The read-out of the meter beam is achieved via the homodyne detection, whose phase is continually scanned, giving the fringes in trace (b) of Fig. 2. The transferred SNR_{met}^{out} is then measured as the level difference between the peak and background noise of the phase quadrature, which corresponds to the superior envelope of the fringes. From Fig. 2, the raw transfer from the signal onto the meter is of -3dB which gives after corrections due to amplifiers noise a measurement transfer coefficient $T_{met} = 0.4$. On the other hand, no significant degradation of the signal could be observed. Therefore the signal degradation can be estimated according to overall losses due to the imperfection of optical elements along the propagation of the beam. This gives a raw SNR transfer for the signal $T_{sig} = 0.9$. As a consequence, we obtain a sum of the signal and measurement SNR transfers $T_{sig} + T_{met} = 1.3$, which can be seen as a 1.1 dB improvement of the overall information transfer of any classical op-

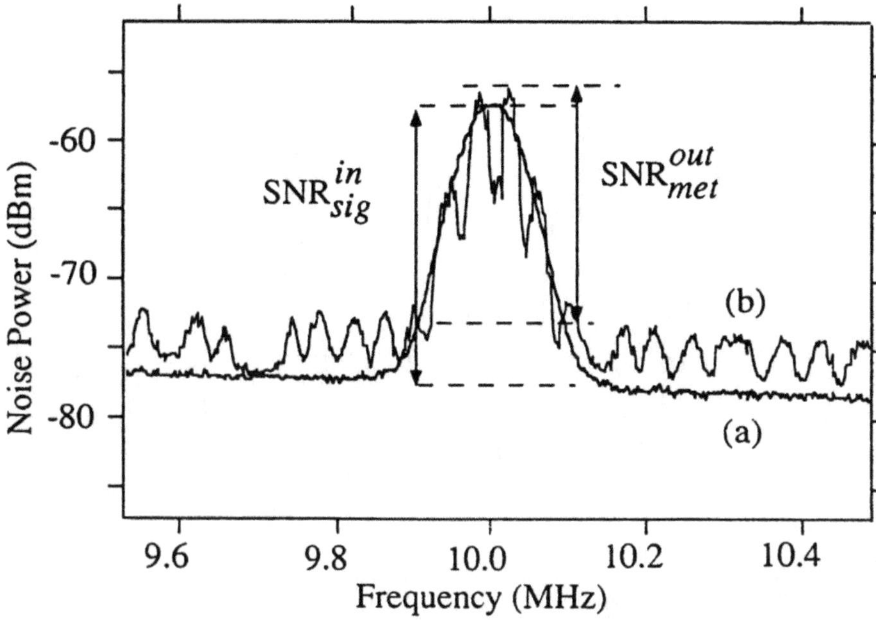

Fig. 2. Transfer of the Signal-to-Noise Ratios from the signal input (smooth curve) to the meter output (wigglering curve, due to the scan of the homodyne detection phase). The meter output SNR is obtained as the superior enveloppe of this curve.

tical tap (a beam splitter for example). A perfect quantum optical tap would have $T_{sig} = T_{met} = 1$ which would give a perfect 3 dB improvement, and could also be regarded as a quantum duplicator.

The experimental results concerning the quantum state preparation ability for this device can be characterized by the conditional variance $V_{s|m}$ introduced in Sec. 3. This quantity is measured by the noise reduction of the signal when corrected by the adequately attenuated photocurrent coming from the meter beam [7]. As it can be seen in Fig. 3, the noise drops by 0.6 dB, leading to a conditional variance $V_{s|m} = 0.85$.

4. QUANTUM OPTICAL REPEATERS USING SEMICONDUCTOR EMITTERS AND RECEIVERS

It has been known for several years that the conversion from light to electrical current and from electrical current to light can be realized below the optical standard quantum limit, using high efficiency semiconductor electro-optical devices. Indeed, the ability of photodiodes to detect sub-shot-noise light by turning it into sub-

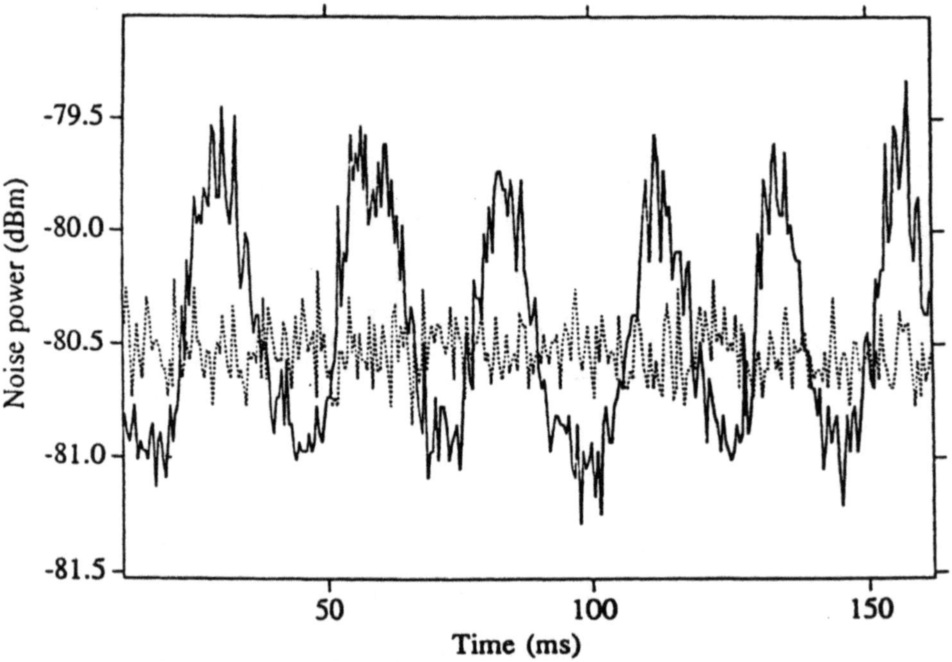

Fig. 3. Combined noise levels with 12 dB attenuation of the meter photocurrent. The fringes are due to the phase of the homodyne detection, and yield summed and differenced photocurrents (resp. upper and lower envelope of the curve), which go respectively 0.9 dB above and 0.6 dB below the shot-noise-level of the signal alone (flat middle curve).

shot-noise current is a fundamental property of quantum optics [30], and has been experimentally demonstrated by the observation of squeezing [31]. The quantum properties of the inverse transformation are due to the work of Yamamoto and coworkers on semiconductor lasers [32] (see also [33] and [34]). More recently, the appearance of high efficiency light emitting diodes (LED) made possible to use these simple devices for generating non-classical light [35–37]. A crucial idea in these experiments is that the noise of an electrical current at RF frequencies is usually dominated by thermal (Johnson-Nyquist) noise [38], whose power can be made such smaller than the shot-noise level corresponding to the same average current intensity [34,35,39]. In this regime, the electrical current behaves classical, and can therefore be measured, duplicated, or amplified without the quantum constraints attached to a light field. If one is concerned by the non-destructive measurement of the intensity of a light beam, one may imagine to convert the light via a photodiode into an electrical current, to measure this current, eventually to amplify it, and to convert it back to a light field using a semiconductor light emitter [40–42]. This would be a perfect quantum- nondemolition (QND) device [3] if both the conversion rates were unity.

A check of these ideas can be carried out using the experimental set-up [40] schematically depicted in Fig. 4. The measurement apparatus itself consists in a large area p-i-n silicon photodiode PD_1 of quantum efficiency $\eta = 0.90 \pm 0.05$, followed by a low noise electronic amplifier, and a light efficiency emitting diode LED_2. The current to current conversion rate of the LED-PD system is $\varepsilon = 0.16$ at room temperature and $\varepsilon = 0.28$ at 100 K. The detectors are placed as close as possible to the LEDs to maximize light collection, and detector + LED are inserted in a liquid nitrogen cryostat. The bandwith of operation is from 150 kHz to 350 kHz, the lower limit being due to electronic cut-off, and the upper one to the finite response time of the LEDs. From these numbers, it appears that if the LED is series connected with the photodiode, without any amplifier, the transfer coefficients for a shot-noise-limited input will be respectively $T_s = \eta\varepsilon$ and $T_m = \eta$, leading to a rather small effect ($T_s + T_m = 1.15 \pm 0.06$ at 100 K) mainly due to the low conversion rate of the LED. In order to overcome this limitation, the current coming from the photodiode is amplified prior to being sent to the LED, which makes the conversion back to light less sensitive to vacuum fluctuations [44], and therefore greatly improves the non-degradation property of the system without changing the others.

For a shot-noise-limited input, and assuming a negligible amplifier noise, the transfer coefficients read:

$$T_m = \eta, \tag{25}$$

$$T_s = \eta \left(1 + \frac{1 - \varepsilon}{\varepsilon g_{eff}}\right)^{-1}, \tag{26}$$

where $g_{eff} = g^2/G$, which g being the AC gain of the amplifier and G the DC gain. The distinction between AC and DC gain, allowed in our experimental set-up, is an extra degree of freedom given by electronics compared to optics. Note that for practical purposes in information transmission, one has to choose $G = g$, which leads to $g_{eff} = g$. In the high gain limit ($g_{eff} \to +\infty$), the sum of these two coefficient goes to $TQs + T_m = 2\eta$, and do not depend on $LED_2's$ conversion rate ε anymore. On the other hand, the conditional variance is limited by the conversion rate of LED_2, and is given by :

$$V_{s|m} = 1 - \varepsilon. \tag{27}$$

In our experiment, the input light beam is the light coming from LED_1, and can therefore be either squeezed (switch $S\omega$ in po-

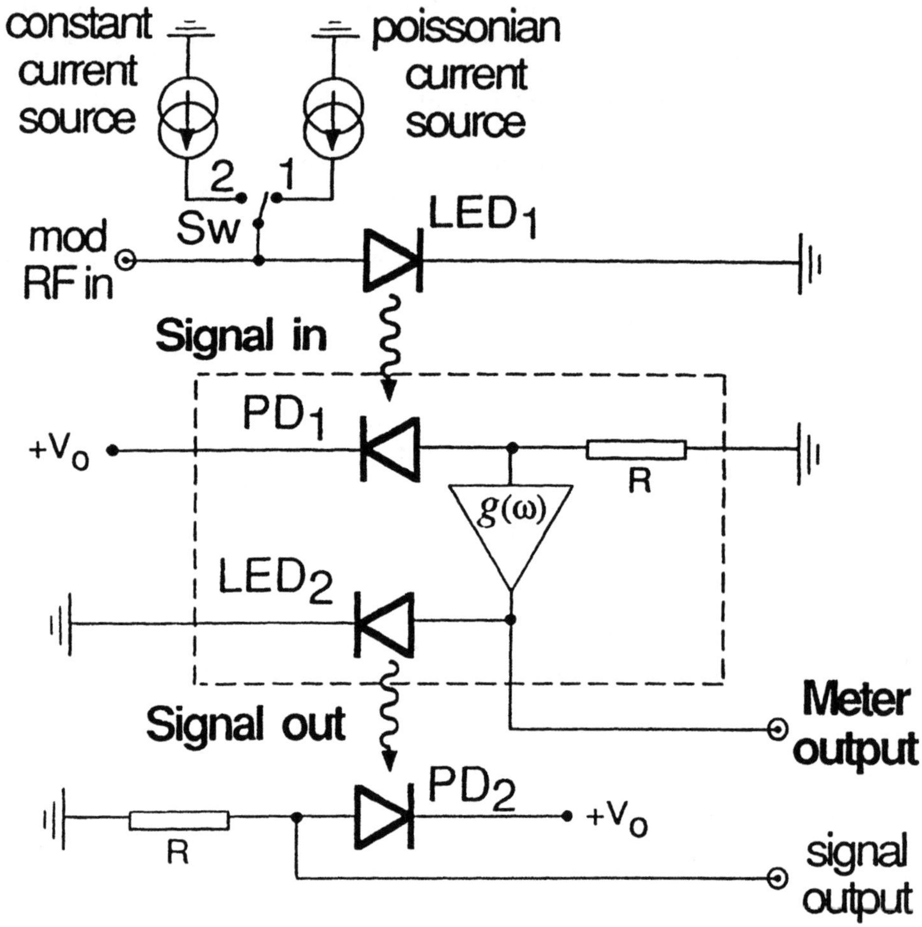

Fig. 4. Schematic experimental set-up. The measurement device is enclosed in the dashed box. $LED_{1,2}$ and $PD_{1,2}$ are cooled at 100K. The poissonian current source consists in three parallel connected photodiodes, illuminated by white incandescent lamps. When switch Sw is in position 1, LED_1 is driven by a shotnoise limited current ; when in position 2, LED_1 is driven by a‘ constant (filtered) current, and shines therefore squeezed light on PD_1. LED_2 can be also driven by the poissonian current source, in order to get a shotnoise reference of PD_2. Note that, for clarity, all the electrical filtering is not represented in the figure.

sition 2), or shot-noise limited ($S\omega$ in position 1). The evaluation of the transfer coefficient is performed by modulating the input signal light (at 250 kHz) by addition of a small RF modulation to LED_1 poissonian driving current. The input signal-to-noise-ratio R_s^{in} is de-

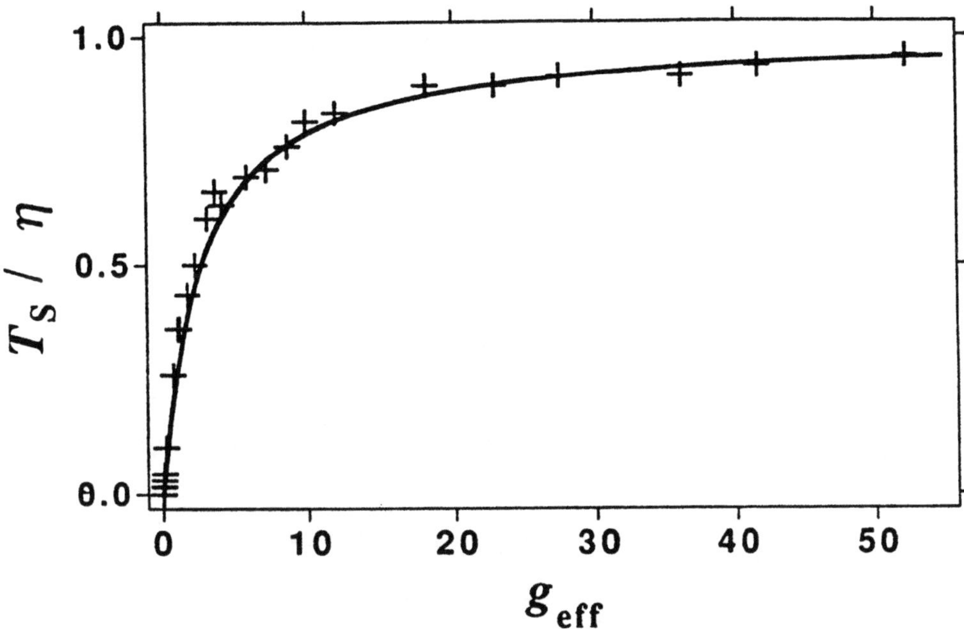

Fig. 5. Evolution of the signal transfer coefficient T_s versus the intermediate electronic gain g_{eff}. The experimental points are compared with the theoretical prediction given in the text. The DC currents in the experiment are 25 mA for the LEDs and 7mA for the PDs.

duced from the direct measurement of R_m^{out} on a spectrum analyser, taking into account the quantum efficiency of PD_1, and the SNR of the signal output R_s^{out} is visualised on the output of PD_2. For high gain ($g_{eff} = 100$), no distinction between R_m^{out} and R_s^{out} can be made within our experimental precision (about 3 $T_s + T_m = 1.8 \pm 0.1$. We show in Fig. 5 the experiemntal evolution of T_s/η with the gain g_{eff}, and compare it with the theoretical expression given by Eq. (24), demonstrating thereby a good theory-experiment agreement.

For the observation of the conditional variance (cf. Fig. 6), we have to correct the signal output by substracting the meter output, using a 0 - 180° RF power combiner, in order to get the signal fluctuations down to below the shotnoise. This requires a careful adjustment of the relative RF phase, since the fluctuations to be substracted are both about 20 dB above the shot noise. We obtain thus a noise reduction down to $1.2 \pm 0.1 dB$ below the shot-noise of the signal beam. This yields $V_{s|} = 0.76 \pm 0.02$, while the theoretical value is $V_{s|m} = 0.72$.

Owing to the possibility to have an amplitude squeezed signal input light, when LED_1 is driven by a constant current, we are

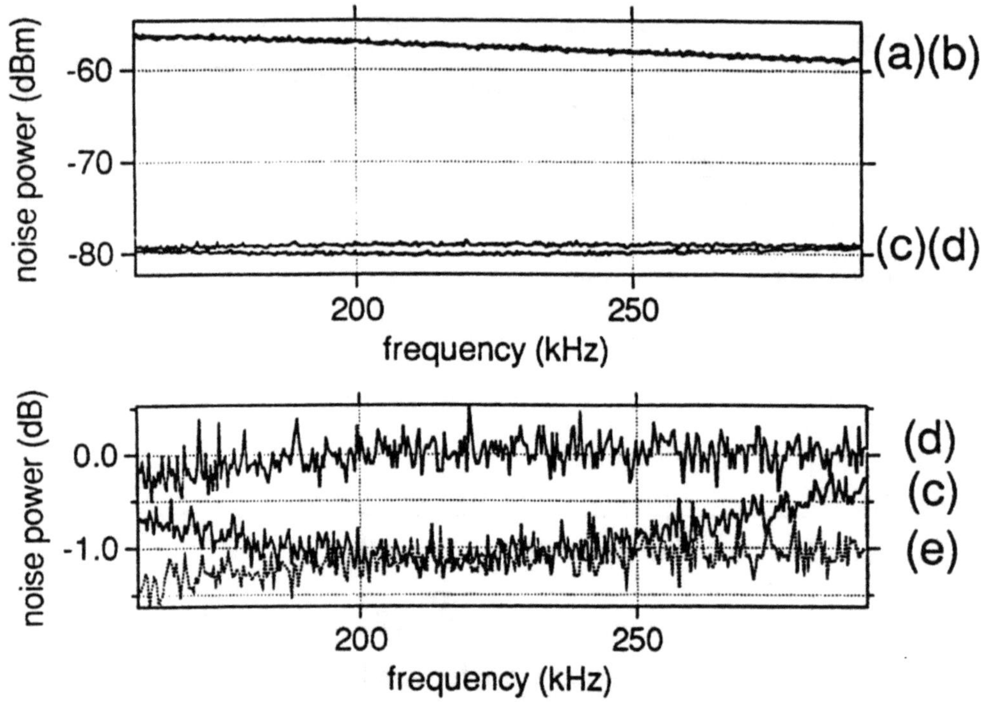

Fig. 6. The upper diagram shows the amplified signal output (trace (a)) and the meter output (trace (b)), for an effective gain $g_{eff} = 100$. The noise spectra are recorded with a resolution bandwith of 10 kHz and a 25 Hz video filter. Trace (c) is the combined noise level, which goes $1.2 \pm 0.1 dB$ below the shotnoise level on PD_2 (trace (d)). The lower diagram is a close up, showing the noise power spectrum normalized to the shotnoise. Trace (e) is the squeezing on PD_2 obtained by driving LED_2 with a constant current, which is also the best achievable noise reduction. The frequency dependence of trace (c) when compared to the other ones is due to electronic phase-shifts.

also able to perform QND measurement of a squeezed light. In the high gain regime, the signal output beam is far above shot noise, but a squeezed output beam can be recovered by decreasing the gain g_{eff} to unity, where our system still operates in the quantum domain, with $T_s + T_m = 1.15$. The input variance W_s^{in} can be inferred from the measured degree of squeezing on the meter channel $W_m^{out} = 0.80$, as it is usually done for a non-ideal photodetector. The directly measured signal output squeezing is then $W_s^{out} = 0.93$, while the experimental value of the conditional variance is $V_{s|m} = 0.79 \pm 0.02(-1.0 \pm 0.1 dB)$. All these results clearly show that our scheme does operate in the "QND" regime, while having a squeezed signal beam, both at the input and the output.

It is worth pointing out that having a signal gain close to one is an extra requirement with respect to the QND criteria, which appears when one is concerned by preserving the squeezing of the input signal. On the other hand, the first QND criteria would rather lead to strong amplification of the intermediate electrical current, because this improves the value of the SNR transfer, by making the system less sensitive to downstream losses. We propose to use the term "quantum repeater" for a device which fulfills the QND criteria with large signal gain, and therefore is not a QND scheme in the original sense [1,3]. Finally, the value of the conditional variance does not depend on the gain, but only on the quantum efficiency of the reemitting LED and detection system. Another point that makes this PD-LED scheme different from a usual QND scheme is that the phase information is completely lost, what ever the precision of the measurement is. The color of the light reemitted can even be different from the one of the input beam ! In other words, the system is not left into a minimum uncertainty state. The perspective of emergence of very high efficiency single- mode LED [45], or alternatively the technological realization of low threshold laser diodes, makes this type of system very promising for a realistic implementation in ultra-low noise optical telecommunication networks.

5. CONCLUSION

In Fig. 7, the results obtained is some optical QND experiments realized to present date are plotted as a function of the values of the conditional variance and of $T_s + T_m$. Some other experiments [23,24,47], which are not in the quantum domain, are not represented here.

The best QND result in the usual sense (signal gain close to unity) is the three-level experiment described above (point f), while the LED experiment just described (point g) appears to be the best "quantum repeater" experiment (large signal gain). However, a fair comparison should also take into account the frequency bandwidth over which the QND effects are achieved. On that respect, pulsed experiments (e.g., point e) have been so far the most successful.

ACKNOWLEDGEMENTS

This work was completed in the framework of the EC Contract ESPRIT III BRA 6934, with partial support from the Centre National d'Etude des Télécommunications (France Telecom). We acknowledge efficient technical assistance of M. Allain, F. Contet, J.C.

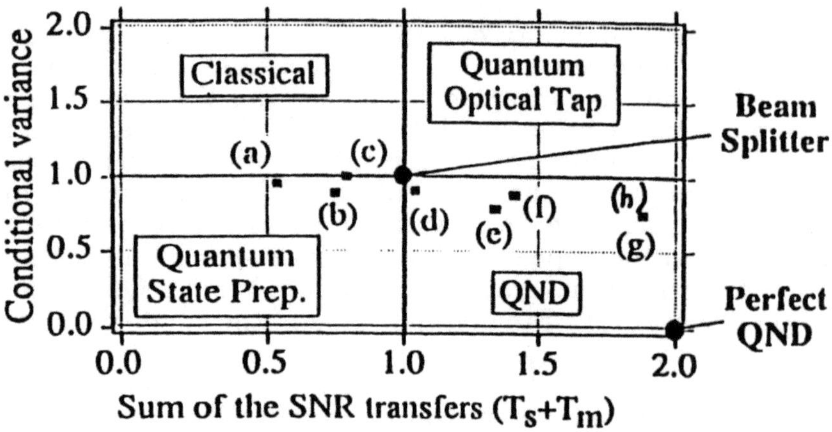

Fig. 7. Experimental results plotted in the plane $x = T_s + T_m, y = V_{v|m}$. The point (x=1, y = 1) is simply a beamsplitter (lossless classical coupler, see Fig. 1). Points (a), (b), (c), (d), (e) and (h) correspond, respectively, to the experiments described in [5], [6], [7], [46] and [41]. Note that [41] uses the same type os set-up as the one described in Sec. 4 [40]. Points (f) and (g) are the results described in the present paper (resp. [8] and [40]), which clearly fulfill all criteria for QND measurements.

Rodier, G. Roger and A. Villing. Institut d'Optique Théorique et Appliquée is URA 14 of Centre National de la Recherche Scientifique.

REFERENCES

1. C. M. Caves, K. S. Thorne, R. W. P. Drever, V. D. Sandberg, and M. Zimmermann, *Rev. Mod. Phys.* **52**, 341 (1980).
2. P. Meystre and M. O. Scully, eds., *Quantum Optics, Experimental Gravitation and Measurement Theory*, (Plenum, New York, 1983).
3. V. B. Braginsky, Y. L. Vorontsov, and K. S. Thorne, *Science*, **209**, 547 (1980). V. B. Braginsky, F.Y. Khalili, *Quantum Measurement*, K.P. Thorne, ed (Cambridge University Press, Cambridge, 1992).
4. M. D. Levenson, R. M. Shelby, M. Reid, and D. F. Walls, *Phys. Rev. Lett.* **57**, 2473 (1986).
5. A. La Porta, R. E. Slusher, and B. Yurke, *Phys. Rev. Lett.* **62** 28 (1989).
6. P. Grangier, J. F. Roch, and G. Roger, *Phys. Rev. Lett.* **66**, 1418 (1991).

7. J. F. Roch, G. Roger, P. Grangier, J. M. Courty, and S. Reynaud, *Appl. Phys. B* **55**, 291 (1992).
8. J. Ph. Poizat and P. Grangier, *Phys. Rev. Lett.* **70**, 271 (1993).
9. B. Yurke, *J. Opt. Soc. Am.* B**2**, 732 (1985).
10. J. H. Shapiro, *Opt. Lett.* **5**, 351 (1980).
11. G. J. Milburn and D. F. Walls, *Phys. Lett. A* **28**, 2065 (1983).
12. M. Brune, S. Haroche, V. Lefèvre, J. M. Raimond, and N. Zagury, *Phys. Rev. Lett.* **65**, 976, (1990).
13. J. M. Courty, P. Grangier, L.Hilico, S. Reynaud, *Opt. Lett.* **65**, 976, (1990).
14. N. Imoto and S. Saito, *Phys. Rev. A* **39**, 675, (1989).
15. M. J. Holland, M. J. Collett, D. F. Walls, M. D. Levenson, *Phys. Rev. A* **42**, 2995, (1990).
16. P. Grangier, J. M. Courty, and S. Reynaud, *Opt. Comm.* **89**, 99 (1992).
17. J. Ph. Poizat, J. F. Roch, and P. Grangier, *Ann. Phys. (Paris)* **19**, 265 (1994).
18. Y. Yamamoto, *Trans. IEICE E* **73**, 1598 (1990).
19. N. Imoto, H. A. Haus, and Y. Yamamoto, *Phys. Rev. A* **32**, 2287 (1985).
20. P. Grangier, J. F. Roch, and S. Reynaud, *Optics Comm.* **72**, 387 (1989).
21. P. Grangier and J. F. Roch, *Quant. Optics* **1**, 17 (1989).
22. N. Imoto, S. Watkins, and Y. Sasaki, *Opt. Comm.* **61**, 159 (1987).
23. H. A. Bachor, M. D. Levenson, D. F. Walls, S. H. Perlmutter, and R. M. Shelby *Phys. Rev. A* **38**, 180 (1988).
24. M. D. Levenson, M. J. Holland, D. F. Walls, P. J. Manson, P. T. H. Fisk, and H. A. Bachor, *Phys. Rev. A* **44**, 2023 (1991).
25. C. A. Blockley, and D. F. Walls, *Optics Comm.* **79**, 241 (1990).
26. J. Ph. Poizat, M. J. Collett, and D. F. Walls, *Phys. Rev. A* **45**, 5171 (1992).
27. K. M. Gheri, P. Grangier, J. Ph. Poizat, and D. F. Walls, *Phys. Rev. A.* **46**, 4276 (1992).
28. H.-A. Bachor, P. Rottengatter, and C. M. Savage, *Appl. Phys. B* **55**, 258 (1992).
29. P. Grangier, J. F. Roch, G. Roger, L. A. Lugiato, E. M. Pessina, G. Scandroglio, and P. Galatola, *Phys. Rev. A.* **46**, 2735 (1992).
30. R. J. Glauber, in Ecole d'Eté des Houches 1964, C. De Witt, A. Blandin, and C. Cohen-Tannoudji, eds., (Gordon & Breach, New-York, 1965).
31. Special issues on squeezed light, *J. Opt. Soc. Am. B* **4**, 1450 (1987), *J. Mod. Optics* **34**, 709 (1987), *Appl. Phys. B* 55, 189-303 (1992).
32. S. Machida and Y. Yamamoto, *Optics Comm.* **57**, 290 (1986). S. Machida and Y. Yamamoto and Y. Itaya, *Phys. Rev. Lett.* **58**,

1000 (1987).
33. M. C. Teich and B. E. A. Saleh, *J. Opt. Soc. Am B* **2**, 275 (1985).
34. P. R. Tapster, J. G. Rarity, and J. S. Satchell, *Europhys. Lett.* **4**, 293 (1987).
35. P. J. Edwards, *Int. J. Optoelectronics* **6**, 23 (1991). P. J. Edwards and G. H. Pollard, *Phys. Rev. Lett.* **69**, 1757 (1992).
36. H. A. Bachor, P. Rottengatter and C.M. Savage, *Appl. Phys. B* **55**, 258, (1992).
37. E. Goobar, A. Karlsson, G. Björk, and P-J. Rigole, *Phys. Rev. Lett.* **70**, 437 (1993).
38. J. B. Johnson, *Phys. Rev.* **32**, 97 (1928), H. Nyquist, *Phys. Rev.* **32**, 110 (1928).
39. H. B. Callen and T.A. Welton, *Phys. Rev. A* **83**, 34 (1951).
40. J.-F. Roch, J.-Ph. Poizat, and P. Grangier, *Phys. Rev. Lett.* **71**, 2006, (1993).
41. E. Goobar, A. Karlsson, and G. Björk, *Phys. Rev. Lett.* **71**, 2002 (1993).
42. F. Capasso and M. C. Teich, *Phys. Rev. Lett.* **57**, 1417 (1986).
43. W. H. Richardson, S. Machida, and Y. Yamamoto, *Phys. Rev. Lett.* **66**, 2867 (1991).
44. H. P. Yuen, *Phys. Rev. Lett.* **56**, 2176 (1986).
45. E. Yablonovitch, T. J. Gmitter, R. D. Meade, A. M. Rappe, K. D. Brommer, and J. D. Joanopoulos, *Phys. Rev. Lett.* **67**, 3380 (1991), and references therein.
46. J. A. Levenson, I. Abram, T. Rivera, P. Fayolle, J. C. Garreau, and P. Grangier, *Phys. Rev. Lett.* **70**, 267 (1993).
47. S. R. Friberg, S. Machida, and Y. Yamamoto, *Phys. Rev. Lett.* **69**, 3165 (1992).

NOTES

1. The amplitude of this modulation is small compared to $|\langle X\rangle|$. The frequency ω of this modulation (typically a few MHz) is very much slower than the optical frequency.

FUZZY SETS AND INFINITE-VALUED ŁUKASIEWICZ LOGIC IN FOUNDATIONS OF QUANTUM MECHANICS

Jarosław Pykacz

Institute of Mathematics, University of Gdańsk
ul. Wita Stwosza 57, 80-952 Gdańsk, Poland

The logic of quantum mechanical systems is recognized as a family of propositions which belong to the domain of infinite-valued logic studied by Łukasiewicz in the beginning of XX century. A traditional quantum logic, i.e., an orthocomplemented σ-orthocomplete orthomodular partially ordered set is recognized as a Lindenbaum-Tarski algebra of this logic. A "concrete" model for this algebra in the form of a family of fuzzy subsets of a set of states of a physical system is given.

Key words: quantum logic, many-valued logic, fuzzy sets

1. INTRODUCTION

The purpose of the present paper is to reexamine once more the fundamental problem: What kind of logic is quantum mechanics based on? The study of this problem was originated by Birkhoff and von Neumann already in 1936 in their historic paper [1] and later developed in numerous papers and textbooks (cf. Pavičić's "Bibliography on quantum logics and related structures" [2], containing 1854 references). However, in the vast majority of these papers the mathematical object called "quantum logic" is not a *logic* in the original sense of this word but rather its algebraic counterpart - the so called Lindenbaum-Tarski algebra of a not-yet-existing genuine *quantum logic* (i.e., propositional calculus adequate to describe properties of quantum systems). The situation encountered in classical mechanics is such that we are able to express properties of a classical physical system as propositions which belong to the domain of traditional two-valued logic with well defined traditional connectives ("not," "and," "or," "if...then"). The Lindenbaum-Tarski algebra of this family of

M. Ferrero and A. van der Merwe (eds.), Fundamental Problems in Quantum Physics, 257–265.

propositions occurs to be a Boolean algebra and a family of suitably chosen subsets of a phase space of a system is a "concrete" model of this algebra. In quantum mechanics both the "algebraic" part of this construction (orthocomplemented σ-orthocomplete orthomodular partially ordered set) and a "concrete" model (lattice of closed subspaces of a suitable Hilbert space) are well-established, but the proper propositional calculus in which we could express and study properties of quantum systems is still lacking. The purpose of the present paper is to indicate that a version of infinite-valued Łukasiewicz logic could be the proper candidate.

Fuzzy sets are as closely related to many-valued logics as traditional ("crisp") sets to two-valued logic. We proceed in our construction from an abstract orthomodular poset to families of propositions about quantum systems passing via fuzzy sets. As a by-product we obtain "concrete" models for the families of quantum propositions which consist of fuzzy subsets of sets of states of physical systems. Since such models are (fuzzy-)set-theoretic, they are of the same type as Boolean algebras of subsets of phase spaces of classical systems, contrary to up-to-now used models consisting of families of closed subspaces of Hilbert spaces.

2. QUANTUM LOGICS - THE TRADITIONAL APPROACH

There are many slightly different definitions of a quantum logic encountered in the literature [2]. For our purpose we have chosen the following general one:

Definition 1. A *quantum logic* (or simply a *logic*) is an orthocomplemented σ-orthocomplete orthomodular poset, i.e., a partially ordered set L which contains the smallest element **0** and the greatest element **1**, in which the orthocomple-mentation map $':L\to L$ satisfying the conditions (i) - (iii) exists:

(i) $(a')' = a$ (idempotency);

(ii) if $a\leq b$, then $b'\leq a'$ (order reversing);

(iii) the greatest lower bound (*meet*) $a\wedge a'$ and the least upper bound (*join*) $a\vee a'$ with respect to the given partial order exist in L for any $a\in L$ and they satisfy the excluded middle law: $a\vee a' = \mathbf{1}$, and the law of contradiction: $a\wedge a'=\mathbf{0}$;

the σ-orthocompletness condition:

(iv) if $a_i\leq a_j'$ for $i\neq j$, then the join $\vee a_i$ exists in L;

and orthomodular identity:

(v) if $a \leq b$, then $b = a \vee (a' \wedge b) = a \vee (a \vee b')'$
hold.

Elements of a logic represent elementary statements about physical system. *Probability measures* on a logic L, i.e., mappings $s: L \rightarrow [0,1]$ such that

(i) $s(1) = 1$;

(ii) $s(\vee_i a_i) = \Sigma_i s(a_i)$ for any sequence $\{a_i\}$ of *pairwise orthogonal elements*, i.e., elements such that $a_i \leq a_j'$ for $i \neq j$,

represent states of a physical systems and, therefore, are often themselves called *states* on a logic L. A set of states S on a logic L is called *ordering* iff $s(a) \leq s(b)$ for all $s \in S$ implies $a \leq b$.

The following theorem by Mączyński [3,4] yields a possibility of expressing the object defined by Definition 1 (provided that it admits an ordering set of probability measures) in the language of fuzzy set theory:

Theorem 1. (i) If L is a logic with an ordering set of probability measures S, then each $a \in L$ induces a function $\underline{a}: S \rightarrow [0,1]$, where $\underline{a}(s) = s(a)$ for all $s \in S$. The set of all such functions $\underline{L} = \{\underline{a} : a \in L\}$ satisfies the following condition:

Orthogonality Postulate: If $\underline{a}_1, \underline{a}_2, \ldots$ is a sequence of functions such that $\underline{a}_i + \underline{a}_j \leq 1$ for $i \neq j$, then there exists $\underline{b} \in \underline{L}$ such that $\underline{b} + \underline{a}_1 + \underline{a}_2 + \ldots = 1$.

$\underline{L}$ equipped with the natural partial order: $\underline{a} \leq \underline{b}$ iff $\underline{a}(s) \leq \underline{b}(s)$ for all $s \in S$ and complementation $\underline{a}' = 1 - \underline{a}$ is isomorphic to L.

(ii) Conversely, if $\underline{L}$ is a set of functions from X into [0,1] for which the Orthogonality Postulate holds, then it is a logic with respect to the natural partial order and complementation. Every point $x \in X$ induces a probability measure m_x on $\underline{L}$ where $m_x(\underline{a}) = \underline{a}(x)$ or all $\underline{a} \in \underline{L}$ and the set of all such measures is ordering.

Mączyński [3,4] showed that the Orthogonality Postulate is equivalent to the conjunction of the following three conditions:

I. $\underline{0} \in \underline{L}$ ($\underline{0}(x) = 0$ for all $x \in X$),

II. $\underline{a} \in \underline{L}$ implies $1 - \underline{a} \in \underline{L}$,

III. for any sequence $\{\underline{a}_i\} \subseteq \underline{L}$ such that $\underline{a}_i + \underline{a}_j \leq 1$ for all $i \neq j$, $\Sigma_i \underline{a}_i \in \underline{L}$.

These conditions make translation of an abstract quantum logic into the language of fuzzy set theory particularly easy.

3. FUZZY QUANTUM LOGICS

Let $U \neq \emptyset$ be a fixed (crisp) set called *universe*. According to Zadeh [5] *fuzzy set* A in U is defined by its *membership function* $\mu_A: U \to [0,1]$ in such a way that, for any $x \in U$, the number $\mu_A(x) \in [0,1]$ represents the *degree of membership* of x to the fuzzy set A. Many authors identify fuzzy sets with their membership functions and write A(x) instead of $\mu_A(x)$. This convention is adopted throughout the rest of the paper.

The most natural way to define various relations and operations on fuzzy sets is to define them pointwisely on membership functions of these sets. The following ones were defined already by Zadeh in his historic paper [5]:

$$A = B \text{ iff } A(x) = B(x) \text{ for all } x \in U, \tag{1}$$
$$A \subseteq B \text{ iff } A(x) \leq B(x) \text{ for all } x \in U, \tag{2}$$
$$B = A' \text{ iff } B(x) = 1 - A(x) \text{ for all } x \in U, \tag{3}$$
$$C = A \cup B \text{ iff } C(x) = \max[A(x), B(x)] \text{ for all } x \in U, \tag{4}$$
$$C = A \cap B \text{ iff } C(x) = \min[A(x), B(x)] \text{ for all } x \in U. \tag{5}$$

One can easily notice that all above-mentioned relations and operations coincide with traditional, set-theoretic ones when membership functions of fuzzy sets are replaced by {0,1}-valued characteristic functions of traditional crisp sets. This shows that the theory of fuzzy sets is a generalization of the traditional set theory and that crisp sets form a proper subclass of a class of fuzzy sets.

The fuzzy set union (4) and intersection (5) introduced by Zadeh are not the only possible operations of this type that can be defined on fuzzy sets. Actually, there are infinitely many such fuzzy set generalizations of traditional set-theoretic union and intersection. We found the following ones defined by Giles [6] the most suitable for describing properties of physical systems.

$$C = A \oplus B \text{ iff } C(x) = \min[A(x) + B(x), 1] \text{ for all } x \in U, \tag{6}$$
$$C = A \otimes B \text{ iff } C(x) = \max[A(x) + B(x) - 1, 0] \text{ for all } x \in U. \tag{7}$$

Giles operations in general are, contrary to Zadeh ones, not distributive, which actually should be the case in the description of quantum systems. Moreover, when combined with the standard fuzzy complementation (3), they satisfy not only de Morgan's laws

$$(A \oplus B)' = A' \otimes B', \quad (A \otimes B)' = A' \oplus B' \tag{8}$$

(which are satisfied by Zadeh operations as well), but also the excluded middle law:

$$A \oplus A' = U, \tag{9}$$

the law of contradiction:

$$A \otimes A' = \varnothing, \tag{10}$$

and the formal counterpart of the orthomodular identity:

$$\text{if } A \subseteq B, \text{ then } B = A \oplus (A' \otimes B). \tag{11}$$

Since obviously $A''=A$ and $A \subseteq B$ is equivalent to $B' \subseteq A'$, we see that the standard fuzzy complementation combined with Giles operations satisfies all formal conditions required for an orthocomplementation map. Moreover, the condition $A \subseteq B' \Leftrightarrow A+B \leq 1$, i.e., the formal counterpart of orthogonality of elements in a quantum logic, is equivalent to the condition

$$A \otimes B = \varnothing, \tag{12}$$

which was called *weak disjointness* of fuzzy sets A and B by Giles in [6]. Therefore, we expect that suitably chosen families of fuzzy sets endowed with the standard fuzzy complementation (3) and Giles operations (6) and (7) could provide "concrete" models of quantum logics different from customarily studied families of closed subspaces of Hilbert spaces.

Indeed, it is easy to notice [7,8,9] that functions $\underline{a}: S \rightarrow [0,1]$ that appear in Mączyński Theorem or in the conditions I - III equivalent to the Orthogonality Postulate can be treated as membership functions of fuzzy subsets of the universe S. Therefore, from the second part of Mączyński Theorem we infer that any family $\Lambda(U)$ of fuzzy subsets of an arbitrary universe U which satisfies conditions

- Ia. $\Lambda(U)$ contains the empty set $\varnothing$,
- IIa. $A \in \Lambda(U)$ implies $A' \in \Lambda(U)$,
- IIIa. for any sequence $\{A_i\} \in \Lambda(U)$ of pairwise weakly disjoint fuzzy sets their algebraic sum $\Sigma_i A_i$ (i.e. the fuzzy set B such that $B(x) = \Sigma_i A_i(x)$ for all $x \in U$) belongs to $\Lambda(U)$:

$$\Sigma_i A_i \in \Lambda(U), \tag{13}$$

is a quantum logic in the traditional sense with respect to the fuzzy set inclusion (2) as partial order and the standard fuzzy set complementation (3) as orthocomplementation. Any such family of fuzzy sets was called a *fuzzy quantum logic* in [10]. Later on it was noticed [11] that the condition (13) can be replaced by the conjunction of conditions:

$$\Sigma_i A_i \leq 1 \text{ and } \oplus_i A_i \in \Lambda(U), \tag{13'}$$

and that the Giles union of pairwise weakly disjoint sets coincides with the join of these sets with respect to the fuzzy set inclusion (2) as the partial order [12]. In such a way the notion of a fuzzy quantum logic was almost homogenously expressed in terms of the standard fuzzy set complementation (3) and Giles operations (6), (7). The last step was done in [13] where it was shown that one of the conditions (13') saying that $\Sigma_i A_i \leq 1$ can be omitted if we add to the definition of a fuzzy quantum logic any of the following equivalent conditions:

IV.1 for any $A \in \Lambda(U)$, if $A \subseteq A'$, then $A = \varnothing$;
IV.2 for any $A \in \Lambda(U)$, if $A' \subseteq A$, then $A = U$;
IV.3 for any $A \in \Lambda(U)$, if $A \otimes A = \varnothing$, then $A = \varnothing$.

Since, because of the condition IIa the condition Ia can be replaced by the requirement that $U \in \Lambda(U)$, the "homogenous" definition of a fuzzy quantum logic based on the above-mentioned conditions can be formulated as follows [13]:

Definition 2. A *fuzzy quantum logic* on a universe U is a family $\Lambda(U)$ of fuzzy subsets of U such that:

Ib. $\Lambda(U)$ contains the universe U,
IIb. $\Lambda(U)$ is closed under the standard fuzzy set complementation,
IIIb. $\Lambda(U)$ is closed under Giles unions of pairwise weakly disjoint sets,
IVb. $\Lambda(U)$ does not contain any nonempty set weakly disjoint with itself.

It was proved in [13] that any family of fuzzy sets defined by Definition 2 is a quantum logic in the sense of Definition 1 with respect to the standard fuzzy set inclusion as partial order and the standard fuzzy set complementation as orthocomplementation. Orthogonality of elements of $\Lambda(U)$ is equivalent to their weak disjointness and for an arbitrary sequence of pairwise weakly disjoint elements of $\Lambda(U)$ their join exists in $\Lambda(U)$ and coincides with their Giles union.

4. QUANTUM LOGICS AS PARTIAL INFINITE-VALUED ŁUKASIEWICZ LOGICS

As it was mentioned in the Introduction, fuzzy sets are as closely related to many-valued logic as crisp sets to two-valued logic [6]. Actually, if we apply traditional two-valued logic to evaluate the truth-value $\tau(p)$ of a proposition $p = "x \in A"$ the set A has to be crisp, while, if we apply to this proposition many-valued logic, the set A can be fuzzy with $A(x) = \tau("x \in A")$. It was pointed out by Giles [6] that operations on fuzzy sets arise as immediate consequence of building fuzzy set theory on the basis of Łukasiewicz infinite-valued logic and that various logical disjunctions and conjunctions give rise to various fuzzy set unions and intersections. Two pairs of fuzzy set connectives mentioned in this paper: Zadeh (4), (5), and Giles (6), (7) occur to have the common multiple-valued logical root. Indeed, let us follow Łukasiewicz [14] and choose negation ~ and implication $\rightarrow$ as basic connectives with the following rules of counting their truth-values:

$$\tau(\sim p) = 1\text{-}t(p), \tag{14}$$
$$\tau(p \rightarrow q) = \min[1-\tau(p)+\tau(q),1]. \tag{15}$$

We can obtain now either disjunction and conjunction, studied already by Łukasiewicz [14], which generate Zadeh union and intersection:

$$\tau(p \cup q) = \tau[(p \rightarrow q) \rightarrow q] = \max[\tau(p),\tau(q)], \tag{16}$$
$$\tau(p \cap q) = \tau[\sim(\sim p \cup \sim q)] = \min[\tau(p),\tau(q)], \tag{17}$$

or other disjunction and conjunction, studied by Frink [15] in 1938, which generate Giles opereations on fuzzy sets:

$$\tau(p \oplus q) = \tau(\sim p \rightarrow q) = \min[\tau(p)+\tau(q),1], \tag{18}$$
$$\tau(p \otimes q) = \tau[\sim(\sim p \oplus \sim q)] = \max[\tau(p)+\tau(q)-1,0]. \tag{19}$$

Frink called connectives (18), (19) *arithmetic* to distinguish them from original Łukasiewicz connectives (15), (16) which he called in turn *logical*. Nevertheless, it can be mentioned that in one of the earliest Łukasiewicz papers on many-valued logic [16] we can find formulas which lead straightforward to arithmetic operations (18) and (19).

We denote by F, as usual, the proposition which is false in any valuation, i.e. such that $\tau(F) = 0$ for any τ, and by T the proposition which is true in any valuation: $\tau(T) = 1$ for any τ. Since two fuzzy sets A and B which are weakly disjoint, i.e. are such that $A \oplus B = \varnothing$, are defined, respectively, by propositions $p = "x \in A"$ and $q = "x \in B"$ such that

$$\tau(p \otimes q) = 0, \tag{20}$$

we call *weakly exclusive* any two propositions whose arithmetic conjunction (19) is the false proposition.

In the quantum logic approach to foundations of physical theories it is assumed that for a given element $a \in L$ and a probability measure "s" on L the number $s(a) \in [0,1]$ expresses the probability that an elementary statement about a physical system represented by "a" is true when the system is in a state represented by "s". The same "probabilistic" interpretation was considered by Łukasiewicz as a proper interpretation of truth-values encountered in many-valued logic (see [16], and also the paper "A numerical interpretation of the theory of propositions," both reprinted in [14]). This fact, as well as connections between fuzzy set operations (3) - (7) and many-valued logical connectives (14) - (19), suggest that an abstract quantum logic defined by Definition 1 could be the Lindenbaum-Tarski algebra of a suitable family of propositions belonging to the infinite-valued Łukasiewicz logic $Ł_\infty$. Indeed, since Definition 2 was formulated entirely in terms of fuzzy set operations generated by Łukasiewicz negation (14) and arithmetic disjunction (18) and conjunction (19), and since the empty set $\varnothing$ and the universe U correspond, respectively, to the always false proposition F and the always true proposition T, the desired family of infinite-valued logical propositions $Ł_q \subset Ł_\infty$ is at hand:

$Ł_q$ is a partial Łukasiewicz logic, i.e., a family of propositions belonging to $Ł_\infty$ with partially defined connectives, such that:

- Ic. $Ł_q$ contains the always true proposition T,
- IIc. $Ł_q$ is closed under the Łukasiewicz negation (14),
- IIIc. $Ł_q$ is closed under arithmetic disjunctions (18) of pairwise weakly exclusive propositions,
- IVc. $Ł_q$ does not contain, except the always false proposition F, any proposition weakly exclusive with itself.

REFERENCES

[1] G. Birkhoff and J. von Neumann, "The logic of quantum mechanics," *Ann. of Math.* **37**, 823-843 (1936).

[2] M. Pavičić, "Bibliography on quantum logics and related structures," *Int. J. Theor. Phys.* **31**, 373-461 (1992).

[3] M. J. Mączyński, "The orthogonality postulate in axiomatic quantum mechanics," *Int. J. Theor. Phys.* **8**, 353-360 (1973).

[4] M. J. Mączyński, "Functional properties of quantum logics," *Int. J. Theor. Phys.* **11**, 149-156 (1974).

[5] L. A. Zadeh, "Fuzzy sets," *Information and Control* **8**, 338-353 (1965).

[6] R. Giles, "Łukasiewicz logic and fuzzy set theory," *Int. J. Man-Machine Studies* **8**, 313-327 (1976).

[7] J. Pykacz, "Quantum logics as families of fuzzy subsets of the set of physical states," in: *Preprints of the Second IFSA Congress, Tokyo 1987*, Vol. 2, pp. 437-440.

[8] J. Pykacz, "Fuzzy set ideas in quantum logics," *Int. J. Theor. Phys.* **31**, 1767-1783 (1992).

[9] J. Pykacz, "Fuzzy quantum logic I," *Int. J. Theor. Phys.* **33**, 1691-1708 (1993).

[10] J. Pykacz, "Quantum logics and soft fuzzy probability spaces," *BUSEFAL* **32**, 150-157 (1987).

[11] J. Pykacz, "Probability measures in the fuzzy set approach to quantum logics," *BUSEFAL* **37**, 81-85 (1988).

[12] J. Pykacz, "Fuzzy quantum logics and the problem of connectives," *BUSEFAL* **43**, 49-53 (1990).

[13] J. Pykacz, "Fuzzy quantum logics and infinite-valued Łukasiewicz logic," *Int. J. Theor. Phys.* **33**, 1419-1432 (1994).

[14] J. Łukasiewicz, *Selected Works* (North-Holland, Amsterdam, 1970).

[15] O. Frink, "New algebras of logic," *Amer. Math. Monthly* **45**, 210-219 (1938).

[16] J. Łukasiewicz, *Die logischen Grundlagen der Warscheinlichkeitsrechnung* (Akad. d. Wiss., Kraków, 1913); reprinted as "Logical foundations of probability theory," in [14], pp. 16-63.

A MODEL OF TOPOLOGICAL QUANTIZATION OF THE ELECTROMAGNETIC FIELD

Antonio F. Rañada

Departamento de Física Teórica
Universidad Complutense
28040 Madrid, Spain

A recently proposed topological model of the electromagnetic field is described, which is based in the existence of electromagnetic knots, standard solutions of the Maxwell equations, characterized by the linking numbers n and m of the electric and magnetic vectors, the magnetic and electric helicities having the values $h_{mag} = 2n\hbar$, $h_{el} = 2m\hbar$. The model coincides locally with the classical Maxwell standard theory, but it is globally nonequivalent because its topological properties imply what is called a hidden nonlinearity: the fields span only a nonlinear subset of the solutions of a linear equation. Two consequences of the topological structure are important: the classical expression for the difference between the numbers of right handed and left handed photons is equal to $n + m$, having thus a topological interpretation, and the electric charge is quantized, the fundamental value being close to 14/3 times de electron charge.

Key words: topological quantization, electromagnetic fields, knots.

1. INTRODUCTION

Many physicists are convinced now that topology will play an imporant role in future quantum physics (Atiyah 1990). Far from being a new idea, this is older than quantum physics, in a certain sense. As has been recently rediscovered, Kelvin proposed in 1868 a model in which the atoms were knots or links in the vorticity lines of the aether (Rañada 1992a). Maxwell himself was very much in favour of this approach which he considered basically well oriented. However neither the phenomenology of atoms and molecules nor the topology were develped enough to follow this very deep insight and the idea was forgotten, although present interest in knot theory can be traced back to Tait attempts to classify knots (1911) be-

M. Ferrero and A. van der Merwe (eds.), Fundamental Problems in Quantum Physics, 267–277.

cause of his interest in this model, which stimulated the development of the Alexander polynomials in 1923.

The model which I am discussing, proposed in 1989 (Rañada 1989), is based on a similar idea, since it takes, as an important propriety of the electromagntic field, the linking number of the magnetic and electric lines, which provides with a topological quantization (Rañada 1990a, 1990b, 1991, 1992a, 1992b, 1993).

2. DESCRIPTION OF THE MODEL

In the standard approach to electromagnetism, the basic element is either the vector A_μ or the pair $\mathbf{E}, \mathbf{B}$, the magnetic lines being a secondary concept. In this model it is the other way round, the primary idea is the force line, while the vector fields are secondary concepts. This would please Faraday, for whom the magnetic and electric lines were very real, because he believed that something very special, although still to be understood, was happening along them. This gave rise to many attempts to explain the electromagnetic phenomena in terms of the streamlines and the vorticity lines of the aether (Whittaker 1910). According to these views, the lines of force were something real and tangible, since they were associated to the aether particles.

In order to follow Faraday's original view, the simplest thing to do is to represent the lines of force as the intersections of families of surfaces or, equivalently as the level curves of complex scalar fields, one for the magnetic lines $\phi(\mathbf{r}, t)$, another for the electric ones $\theta(\mathbf{r}, t)$. The magnetic field has then the form $\mathbf{B} = f\nabla\phi^* \times \nabla\phi$, where f is some function which, because $\nabla\cdot\mathbf{B} = 0$, must depend on $(t,\mathbf{r})$ only through ϕ and ϕ^*, that is

$$\mathbf{B} = f(\phi, \phi^*)\nabla\phi^* \times \nabla\phi, \tag{1}$$

which can also be written as $B_k = -\epsilon_{ijk}F_{ij}/2$, where F_{ij} is the antisymmetric tensor

$$F_{ij} = -f(\phi, \phi^*)(\partial_i\phi^*\partial_j\phi - \partial_j\phi^*\partial_i\phi). \tag{2}$$

Covariance implies then that the electromagnetic tensor must have the form

$$F_{\mu\nu} = -f(\phi, \phi^*)(\partial_\mu\phi^*\partial_\nu\phi - \partial_\nu\phi^*\partial_\mu\phi), \tag{3}$$

the electric field being

$$\mathbf{E} = -f(\phi, \phi^*)(\partial_0\phi^*\nabla\phi - \partial_0\phi\nabla\phi^*). \tag{4}$$

As is seen, $\mathbf{E}\cdot\mathbf{B} = 0$, the electric and the vector field being orthogonal, which means that in this case the Faraday 2-form is degenerate and the field is of radiation type. In this talk, only degenerate radiation fields are considered. However, this is not an important restriction, since any electromagnetic field can be expressed always as the sum of two such degenerate radiation fields.

In this topological model, the electromagnetic field is always of the form (3). Moreover ϕ can be interpreted as a map $S^3 \times R \mapsto S^2$, after identifying, via stereographic projection, $R^3 \cup \infty$ with S^3 and the complete complex plane C with S^2, what can be done because it is assumed that the limit of ϕ when $r \to \infty$ does not depend on the direction. The area 2-form in the sphere S^2, pulled back to the $R^3 \times R$ (the spacetime) is then (taking the complex number ϕ as a coordinate in S^2, depending on $\mathbf{r}, t$)

$$\mathcal{W}_\phi = \frac{1}{2\pi i} \frac{d\phi^* \wedge d\phi}{(1+\phi^*\phi)^2} \tag{5}$$

and is identified with the Faraday 2-form $\mathcal{F} = \frac{1}{2} F_{\mu\nu} dx^\mu \wedge dx^\nu$, up to a normalization constant factor $\sqrt{a}$, so that

$$\mathcal{F} = \sqrt{a}\, \mathcal{W}_\phi, \tag{6}$$

and, consequently,

$$F_{\mu\nu} = \frac{\sqrt{a}}{2\pi i} \frac{\partial_\mu \phi^* \partial_\nu \phi - \partial_\nu \phi^* \partial_\mu \phi}{(1+\phi^*\phi)^2}. \tag{7}$$

Note that the action constant a must be introduced in order that $F_{\mu\nu}$ will have the right dimensions. Note also that it serves as normalizing factor (in natural units, a is a pure number, the electric and magnetic fields being inverse square lengths). As is seen, $F_{\mu\nu}$ is of the form (3) with $f = -\sqrt{a}/((2\pi i)(1+\phi^*\phi)^2)$. Because $\mathcal{F}$ is closed in S^3, the second cohomology group of which is trivial, there exists a 1-form $\mathcal{A} = A_\mu dx^\mu$, such that $\mathcal{F} = d\mathcal{A}$, so that the fourvector A_μ is the electromagnetic field, $F_{\mu\nu} = \partial_\mu A_\nu - \partial_\nu A_\mu$ and $\mathbf{B} = \nabla \times \mathbf{A}$ (Bott and Tu 1982; Nash and Sen 1982; Rañada 1989, 1992a).

As long as no charges are present, we can play the same game with the electric field $\mathbf{E}$ and a scalar field θ. In that case, if the pull back of the area 2-form in S^2 by θ is

$$\mathcal{W}_\theta = \frac{1}{2\pi i} \frac{d\theta^* \wedge d\theta}{(1+\theta^*\theta)^2}, \tag{8}$$

and the dual to the Faraday form is taken to be

$${}^*\mathcal{F} = \sqrt{a}\, \mathcal{W}_\theta, \tag{9}$$

the dual to the Faraday tensor being then

$$ {}^*F_{\mu\nu} = \frac{\sqrt{\not{a}}}{2\pi i} \frac{\partial_\mu \theta^* \partial_\nu \theta - \partial_\nu \theta^* \partial_\mu \theta}{(1 + \theta^* \theta)^2}, \tag{10} $$

and the following duality condition must be fulfilled

$$ {}^*F^{\mu\nu} = \frac{1}{2} \epsilon^{\mu\nu\alpha\beta} F_{\alpha\beta}. \tag{11} $$

The conditions for the existence of the pair ϕ, θ do not pose any problem (Rañada 1992a). What is important is that, if this pair exists, $F_{\mu\nu}$ and ${}^*F_{\mu\nu}$ obey automatically the Maxwell equations.

A point must be stressed. This model makes use only of what were called *admissible fields*, which are standard electromagnetic fields (that is solutions of the Maxwell equations), which can be written as (7), (10), the scalar ϕ, θ being interpretable as two continuous maps $S^3 \mapsto S^2$. It is clear that not all the standard fields are admissible, which might seem disastrous since many solutions of Maxwell equations are then discarded. However, it turns out that the set of the admissible fields coincide locally with the solutions of the Maxwell equations in the sense that, given any standard field, there is an admissible field which coincides with it in a region around any point, except perhaps in a zero measure set. This means that this model and the standard Maxwell theory are locally equivalent. The details of this important property can be found in Rañada (1992a).

A striking consequence follows: as all the maps $S^3 \mapsto S^2$, ϕ and θ can be classified in homotopies labelled by the value of the so called Hopf indexes n and m which are equal to the linking number of the level curves of ϕ and θ, respectively. (Hopf 1931; Moffatt 1969, 1981; Kuznetsov and Mikhailov 1980; Bott and Tu 1982; Rañada 1989, 1990b, 1992a, 1992b). These topological numbers are equal to

$$ \int_{R^3} \mathbf{A} \cdot \mathbf{B}\, d^3r = n\not{a}, \quad \int_{R^3} \mathbf{C} \cdot \mathbf{E}\, d^3r = m\not{a}, \tag{12} $$

where $\mathbf{C}$ is a vector potential for $\mathbf{E}$ and the left hand sides are called *the magnetic* and *the electric helicities* and n, m are equal to the linking number of any pair of magnetic and electric lines, respectively. Note that (12) are two quantum conditions of topological origin. A solution characterized by the topological numbers n, m can be appropriately called *electromagnetic knot of class* (m, m), since it can be said that it is knotted to itself because of the linking of the force lines. This clarifies the meaning of the admissible fields: they are the electromagnetic knots.

The basis of these identifications is the following: complex fields in R^3, with only one value at infinity, can be interpreted as maps $S^3 \mapsto S^2$

at any time and are therefore classified in homotopies labelled by the Hopf index n, which appears physically as the corresponding helicity, their level curves being the integral lines of a vector field having the properties of a magnetic field (the same can be said of the electric field in vacuum).

Following this argument we fall upon an unexpected bonus: the quantization of the electric charge. This quantization is usually expressed by saying that the charge q of any particle is always an integer multiple of the electron charge e, that is $q = \tilde{n}e$, $\tilde{n}$ being an integer. However, it is more enlightening, although equivalent, to state that the electric flux across any closed surface Σ is always an integer times $\sqrt{4\pi/137.04} = 0.3028$, since this suggest that the number of elementary charges $\tilde{n}$ could be given the meaning of the degree of a map from Σ on some closed surface. This is indeed the case in this model *because of the double role of the Faraday 2-form and its dual which are both flux forms and area forms.* If the electromagnetic field is expressed through (7), (10), the Faraday 2-form and its dual are two pull-backs of the area form in S^2. Furthermore, as ${}^*\mathcal{F}$ is the electric flux form, the flux of $\mathbf{E}$ through any closed surface Σ is the product of $\sqrt{a}$ times the degree of the map $\Sigma \mapsto S^2$ induced by θ which is always an integer number $\tilde{n}$. However it can be different from zero only if θ has some point singularities at the position of the charges inside Σ. Let us assume that it has just one such point singularity at $\mathbf{r}_0$ (and another one at ∞) and that Σ is a sphere centered at $\mathbf{r}_0$. Let $\tilde{n}$ be the degree of the induced map $\theta : \Sigma \times R \mapsto S^2$. The consequence is important: since, because of the Gauss theorem, the electric flux is equal to the total charge inside Σ, it follows that any charge has the value $q = \tilde{n}\sqrt{a}$, so that, in this model, there is a fundamental charge

$$q_0 = \sqrt{a}. \tag{13}$$

To obtain the field equations for the scalars ϕ, θ, the most natural election for the lagrangian density is a numerical coefficient times the difference of the norms of the Faraday form anf its dual

$$L = -\frac{1}{8}\left(F_{\mu\nu}F^{\mu\nu} - {}^*F_{\mu\nu}\,{}^*F^{\mu\nu}\right), \tag{14}$$

since this is equal to the standard lagrangian density $-F_{\mu\nu}F^{\mu\nu}/4$. It must be submitted to the duality condition

$${}^*F^{\mu\nu} - \frac{1}{2}\epsilon^{\mu\nu\alpha\beta}F_{\alpha\beta} = 0. \tag{15}$$

The method of the Lagrange multipliers must be used accordingly and it turns out that the duality condition does not contribute to the Euler-Lagrange equations, what means that it is conserved naturaly under the time evolution. In other words, if it is verified at time $t = 0$ it is verified always for $t > 0$. As was shown in I, the Euler-Lagrange equations are equivalent to

$$\partial_\alpha F^{\alpha\beta} = 0, \quad \partial_\alpha {}^*F^{\alpha\beta} = 0, \quad \beta = 0,1,2,3, \tag{16}$$

which is the second Maxwell pair for both tensors, the first pair being identically satisfied because of the definitions (7) and (10). The important point not to be missed is that, although the equations (16) are nonlinear from the point of view of the scalars ϕ, θ, the tensor $F_{\mu\nu}$ obeys the linear Maxwell equations. If we accept as a working hypothesis that the scalars are the fundamental fields, the theory is nonlinear, in spite of the fact that it looks linear. This is what was called *hidden nonlinearity* (Rañada 1992a). A noteworthy consequence is that not all the solutions of Maxwell equations are admissible. This property can also be stated by saying that although the electromagnetic fields of this model obey the linear Maxwell equations, they do not form a vector space, but just a nonlinear subspace of a vector space.

3. THE MEANING OF THE HELICITIES

Let us consider now the question of the physical meaning of the helicities of the knots or, more precisely, of how are them related to observable quantities. The vector potential can be written in the form

$$\mathbf{A}(\mathbf{r},t) = \frac{\hbar^{1/2}}{(2\pi)^{3/2}} \int \frac{d^3k}{\sqrt{2\omega}} \left((\epsilon_R a_R + \epsilon_L a_L) e^{-ik\cdot x} + (\epsilon_L a_R^* + \epsilon_R a_L^*) e^{ik\cdot x} \right), \tag{17}$$

where $\epsilon_R = \epsilon_R(\mathbf{k})$ and $\epsilon_L = \epsilon_L(\mathbf{k})$ are right handed and left handed polarization vectors and $a_R = a_R(\mathbf{k})$ and $a_L = a_L(\mathbf{k})$ are Fourier transform functions which in the quantum version of the theory are taken to be the annihilation operators.

From this expression, we can calculate the magnetic field which turns out to be

$$\mathbf{B}(\mathbf{r},t) = \frac{\hbar^{1/2}}{(2\pi)^{3/2}} \int \frac{d^3k}{\sqrt{2\omega}} \omega \left((\epsilon_R a_R - \epsilon_L a_L) e^{-ik\cdot x} + (\epsilon_L a_R^* - \epsilon_R a_L^*) e^{ik\cdot x} \right). \tag{18}$$

To compute the helicities we must substitute in (12). A bit of simple algebra leads to (the details are given in Rañada 1993)

$$h_{mag} = U + V, \quad h_{el} = U - V, \tag{19}$$

where

$$U = \hbar \int d^3k \, (a_R^*(\mathbf{k}) a_R(\mathbf{k}) - a_L^*(\mathbf{k}) a_L(\mathbf{k})), \tag{20}$$

$$V = -\hbar\Re\int d^3k\, e^{-2i\omega t}\,(a_R(\mathbf{k})a_R(-\mathbf{k}) - a_L(\mathbf{k})a_L(-\mathbf{k}))\,, \qquad (21)$$

$\Re$ meanig real part. Note that, if a_R, a_L are taken to be the annihilation operators for right and left handed photons (and their complex conjugate to be the creation operators), then

$$U = \hbar(N_R - N_L), \qquad (22)$$

where N_R and N_L are the numbers of photons with right and left polarization, respectively. As $h_{mag} = n\not{a}$, $h_{el} = m\not{a}$, where n, m are integer numbers and $\not{a}$ is the action constant of the model, it turns out that

$$N_R - N_L = \frac{\not{a}}{2\hbar}(n + m). \qquad (23)$$

In order that (i) $N_R - N_L$ will be always an integer and (ii) it can take any integer value, it is necessary and sufficient that $\not{a} = 2\hbar$. These conditions give a precise criterion to fix the value of $\not{a}$, normalizing thus the model. Consequently, the model predicts then the topological quantization of $N_R - N_L$ as

$$N_R - N_L = \int d^3k\,(a_R^*(\mathbf{k})a_R(\mathbf{k}) - a_L^*(\mathbf{k})a_L(\mathbf{k})) = n + m, \qquad (24)$$

and also the quantization of the quantity

$$\Re\int d^3k\, e^{-2i\omega t}\,(a_R(\mathbf{k})a_R(-\mathbf{k}) - a_L(\mathbf{k})a_L(-\mathbf{k})) = -n + m. \qquad (25)$$

It should be stressed that this applies only to degenerate radiation fields (which verify $\mathbf{E}\cdot\mathbf{B} = 0$). The general field can always be expressed as the sum of two degenerate radiation fields, the topological quantum rules (24), (25) applying then to each one of those two components. Note that this topological quantization of $N_R - N_L$ has a classical meaning in terms of c-number fields, which does not depend on the use of q-number fields.

In answer to the question posed at the beginning of this section, we can say that the sum of the magnetic and the electric helicities is equal to $N_R - N_L$ as given by the integral in (24).

4. THE FUNDAMENTAL CHARGE

The topological structure of this model implies that the electromagnetic fields can only be coupled to electric charges which are integer multiples of the fundamental value

$$q_0 = \sqrt{\not{a}}. \qquad (26)$$

By taking, following the results of the previous section, $\not{d} = 2\hbar$, this has the value, in natural units,

$$q_0 = \sqrt{2} = 4.67e \tag{27}$$

close to 14/3 times the electron charge $e = \sqrt{4\pi/137.04}$. It seems encouraging that q_0 turn out to be of the same order of magnitude as e. Moreover it is greater. To assess this result, one must not forget the need to renormalize the classical value of q_0 because of the quantum effects. This suggests that q_0 could be interpreted as the bare electron charge, which is known to be partially screened by the charge density due to the polarization of the vacuum by the virtual pairs, continuously being created around any free charge.

5. THE ELECTROMAGNETIC KNOTS AND THE HOMOTOPY CLASSES

The set of the knots can be classified in homotopy classes $C_{n,m}$, each one labelled by two integer numbers, n and m, which have two meanings: (1) they give the magnetic and electric helicities as $h_{mag} = 2n\hbar$, $h_{el} = 2m\hbar$ and (2) they are the linking numbers of any pair of magnetic or electric lines, respectively. Moreover the difference between the numbers of right-handed and left-handed photons, as given by (23), is equal to $n + m$. It seems fit to call electromagnetic knots of order (n, m) to the solutions in class $C_{n,m}$.

A representative of class $C_{1,1}$ was already studied in detail (Rañada 1990b, 1992a) by means of the so called Hopf map. By using the same method one can obtain the expression in terms of **B**, **E** of families of knots of order (n, m), member of the class $C_{n,m}$ (Rañada 1993, Rañada and Trueba 1994).

6. SUMMARY AND CONCLUSIONS

In this topological model of electromagnetism, the degenerate radiation fields are electromagnetic knots, defined as solutions of the Maxwell equations in which the magnetic (electric) lines are linked n (m) times. It has some very interesting properties. Including some of those already proved in previous work (Rañada 1989, 1992a, 1993), it turns out that:

1. The model coincides with Maxwell standard theory from the local point of view. However, it is globally nonequivalent to it because of the existence of topological constants of the motion which imply a quantization of topological origin. Moreover an action constant $\not{d}$ must be introduced necessarily in the model because of dimensional reasons. In order that the topological quantization agree with the known quantization, the dimensional

action constant of the model must be $\phi\!\!\!/ = 2\hbar$.

2. The topological quantization is expressed by the discretization of the values of the magnetic and electric helicities h_{mag} and h_{el} as

$$h_{mag} = \int_{R^3} \mathbf{A} \cdot \mathbf{B}\, d^3r = 2m\hbar, \quad h_{el} = \int_{R^3} \mathbf{C} \cdot \mathbf{E}\, d^3r = 2n\hbar, \tag{28}$$

where **A** and **C** are vector potentials for **B** and **E** and n and m are integers equal to the linking numbers of the magnetic and electric lines, respectively.

3. Although the electromagnetic tensor obeys the linear Maxwell equations, there is a hidden nonlinearity. This means that the fields do not form a vector space, but just a nonlinear subset of a vector space. This is so because although the Maxwell equations can be obtained as the linearization of an underlying nonlinear theory referred to the two scalars ϕ, θ, the linearization transformation is non invertible, not all the standard electromagnetic fields being admissible in this model.

4. Its solutions are classified in homotopy classes $C_{n,m}$. As n and m are the linking number above explained, the solutions belonging to class $C_{n,m}$ can appropriately be called electromagnetic knots of class (n, m).

5. It turns out that

$$\int d^3k\, (a_R^*(\mathbf{k})a_R(\mathbf{k}) - a_L^*(\mathbf{k})a_L(\mathbf{k})) = \frac{\phi\!\!\!/}{2\hbar}(m+n). \tag{29}$$

As the left hand side is the classical limit of $N_R - N_L$, the difference between the numbers of right-handed and left-handed photons, a quantity which must take only integer values and must be allowed to take any integer value, it follows that the right normalization of the model is $\phi\!\!\!/ = 2\hbar$. With this value the magnetic and electric helicities are equal to $2n\hbar$ and $2m\hbar$, respectively. This seems to suggest that this model gives the classical limit of the quantum theory of electromagnetism with the right normalization and that $N_R - N_L$ has a topological sense.

6. With this normalization ($\phi\!\!\!/ = 2\hbar$), the model predicts the existence of a fundamental charge equal to $\sqrt{2}$, in natural units, a value close to four thirds the electron charge (up to 1.2%).

All this seems encouraging and worth of further research. First of all, the model must be generalized to general fields which are not of radiation type. This is not difficult and will be done in a forecoming paper.

It must be stressed that, once an approach based in the idea of force line is taken—as it is done in this paper and in previous work—, one falls, almost unavoidably, over the the necessity of an action constant and over the topological quantization of the electric charge. This is important and could be expressed in this way. Let us indulge in history fiction and assume, just for a moment, that the development of electromagnetism in the XIX century would have proceeded—as it almost happened—by giving primacy to the concept of force line over the idea of vector. In that case, the arguments of this paper suggest that it is conceivable that the existence of an action constant and the quantization of the electric charge could have been understood before the Planck hypothesis and the actual discovery of the electron, respectively.

A final warning seems necessary. The fact that this model predicts a topological quantization while making use of c-number fields must be stressed, but this should not be understood as an argument in favour of the reduction of quantum physics to classical terms. We must consider that there are two kind of quantum effects (Rañada 1991): the discretization of dynamical variables, as the energy of an atom, and the fact that the probabilities do not obey the Laplace rules because there is interference of amplitudes. The first are not very different from similar properties of such classical quantities as the frequencies of a drum. On the contrary, everything indicates that the probability effects are genuinely quantum, irreducible to classical physics. It is a contradiction that the very name of the theory, "quantum physics", comes from the less revolutionary change brought about. This model suggests that at least some discretization effects—as is the case with $N_R - N_L$—could be explained by a theory which uses only classical variables and could thus help to clarify the nature of the relation between classical and quantum mechanics. But it can say nothing about the probabilities, neither about the wave-particle dualism.

REFERENCES

Atiyah M 1990, *The Geometry and Physics of Knots* (Cambridge: The University Press).

Bott R and Tu L W 1982, *Differential Forms in Algebraic Topology* (New York: Springer).

Hopf H 1931, *Math. Ann.* **104** 637-665.

Kuznetsov E A and Mikhailov A V 1980, *Phys. Lett.* **77A**, 37-38.

Moffatt H K 1969 *J. Fluid Mech.* **35**, 117-129; 1981, *J. Fluid Mech.* **106**, 27-47.

Nash Ch and Sen S 1982, *Topology and Geometry for Physicists* (London: Academic).

Rañada A F 1989, *Lett. Math. Phys.* **18**, 97-106; 1990a, in *Solitons and Applications*, V G Makhankov, V K Fedyanin, and O K Pashaev, eds. (Singapore: World Scientific), pp. 180-194; 1990b, *J. Phys. A:Math. Gen.* **23**, L815-820; 1991, *Fundamental problems in Classical and quantum dynamics*, J A Ellison and H Ürberall, eds. (London: Gordon and Breach), pp. 95-117; 1992a, *J. Phys. A: Math Gen* **25**, 1621-1641; 1992b, *Eur. J. Phys* **13**, 70-76;

1993, to be published.
Rañada A F and Trueba J L 1994, to be published.
Tait P G 1911, *Scientific Papers*, Vol. I (Cambridge: The University Press), pp. 136-150.
Whittaker E 1910, *A History of the Theories of Aether and Electricity* (reprint 1973, New York: Humanities Press).

POSTSELECTION AND SQUEEZING IN NEUTRON INTERFEROMETRY AND EPR-EXPERIMENTS

H. Rauch

Atominstitut der Österreichischen Universitäten
Schüttelstrasse 115, A-1020 Wien, Austria

Recent neutron interference experiments, based on the post-selection of narrow momentum states behind the interferometer, have shown that coherent wave packets remain totally coupled in configuration space, even when they become separated in ordinary space. Such non-overlapping wave packets in real space (Schrödinger-cat-like states) show a characteristical spectral modulation and squeezing indicating its non-classical features. This shows analogies to EPR- experiments, with consequences for the debate about the non-locality phenomena of quantum mechanics.

Keywords: neutron optics, quantum optics, squeezing.

1. INTRODUCTION

Widely separated coherent beams exist inside a perfect crystal neutron interferometer which overlap at the third crystal plate where the resulting interference pattern exhibits the coherence properties of the beams (e.g., [1-3]). In stationary situations, the related coherence function is given by the auto-correlation function of the overlapping wave functions which have be to be taken as three-dimensional wave packets:

$$\Gamma(\Delta) = \langle \psi^*(0)\psi(\Delta) \rangle,$$

$$\psi(\mathbf{r}) \propto \int a(\mathbf{k})\, e^{i\mathbf{k}.\mathbf{r}}\, d\mathbf{k}\,. \qquad (1)$$

This shows that the coherence function is related to the Fourier-transformation of the momentum distribution of the beams $g(\mathbf{k}) = |a(\mathbf{k})|^2$, by

M. Ferrero and A. van der Merwe (eds.), Fundamental Problems in Quantum Physics, 279–289.

$$<|\Gamma(\Delta)|> \propto \left|\int g(\mathbf{k})\, e^{i\mathbf{k}.\Delta}\, d\mathbf{k}\right|, \tag{2}$$

with the normalization <|G(0)|> = 1. In practice, <|G(0)|> can be different from unity due to an average over different beam paths with slightly different phase shifts, due to unavoidable imperfections of the crystal and the phase shifter, and due to residual vibrations. In the following, G(**D**) denotes G(**D**) = <|G(**D**)|>/<|G(0)|>. **D** represents the spatial shift of the overlapping wave trains which can be calculated for a phase shifter with an index of refraction n, a thickness D_0 and oriented with a surface normal $\hat{s}$ as (see Fig. 1), where **K** and **k** represent the wave vectors inside and outside the phase shifter, respectively. For Gaussian wave packets which describe Gaussian momentum-distributed beams with characteristic widths dk_i (i = x, y, z), one obtains

$$\Delta = \frac{(\mathbf{K}-\mathbf{k})}{k} D_0,$$

$$\mathbf{K}-\mathbf{k} = \frac{(1-n)k\hat{\mathbf{s}}}{(\hat{\mathbf{k}}.\hat{\mathbf{s}})} \tag{3}$$

$$|G(\mathbf{D})| = \prod_{i=x,y,z} \exp\left[-(\Delta_i \delta k_i)^2/2\right], \tag{4}$$

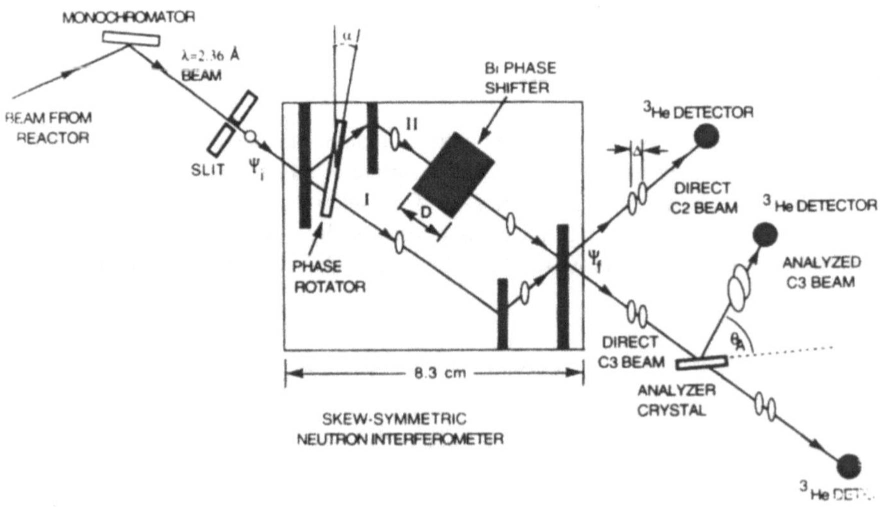

Fig. 1 Scheme of a perfect crystal neutron interferometer with indications of the shapes of the wave packets in the case of large phase shifts.

which shows that the characteristic length - the coherence length Δ_c^i - is determined by the momentum distribution of the beam in each direction. For Gaussian distributions, the relation $\Delta_c^i = 1/(2dk_i)$ holds, which is the minimum uncertainty relation.

The interference pattern behind the interferometer is given by the wave functions originating from beam paths I and II. For the forward direction (0) and the empty interferometer they are equal in amplitude and phase because the related beams are transmitted-reflected-reflected (TRR) and reflected-reflected-transmitted (RRT), respectively. A phase shifter with an effective thickness $D_{eff} = D_0/(\mathbf{k}.\mathbf{s})$ changes the related wave function by a phase factor $\exp[i(n-1)kD_{eff}] = \exp(i\mathbf{D}.\mathbf{k})$. The equation

$$I(\Delta) \propto \left|\psi_0^{I} + \psi_0^{II}\right|^2 \propto 1 + |\Gamma(\mathbf{D})| \cos(\mathbf{D}.\mathbf{k}) \qquad (5)$$

shows the direct connection between interference pattern and the coherence function (Eq.(1)). For the longitudinal direction (x) the spatial shift is given as $D = 2pNb_cD_0/k^2$. This formula indicates that the interference pattern disappears when the spatial phase shift approaches the coherence length of the beam in a certain direction $D_i \geq D_i^c = 1/(2dk_i)$, a phenomenon which has been verified experimentally by several investigations [4-6]. Thus, higher-order interference fringes can be observed for a higher monochromaticity of the incident beam, i.e., a better preselection of states, and/or a phase shift in the direction of a high internal momentum selection (y direction). This happens in the perfect crystal interferometer due to dynamic diffraction, which produces a sharp momentum selection in the y-direction providing the possibility of a nondispersive phase shifter position [6]. The loss of interference fringes at high order can be understood as a loss of wave packet overlap at the exit crystal plate of the interferometer, but it should be mentioned that this loss of interference fringes does not indicate a loss of coherence, because the interference fringes can be restored - at least to a large extent - by proper phase-echo methods [5-8]. The question is whether the interference fringes can also be restored by experimental means behind the interferometer, i.e., by a postselection procedure.

2. POSTSELECTION MEASUREMENTS

In optical interferometry, some parts of the interference pattern have been restored behind the interferometer when an additional filtering process has been applied [9]. In more recent publications, some additional effects causing spectral changes in low resolution double-slit experiments have been discussed [10-12]. Related neutron interference experiments showed that interference phenomena can be restored if an additional monochromatization is applied behind the interferometer (Fig.2) [9,13,14]. The related set-up corresponds to a double crystal arrangement between the interferometer crystal (mosaic spread $h_i \rightarrow 0$, Bragg angle θ_i) and the analyser crystal (mosaic spread η, Bragg angle θ_A) whose momentum spread is given as [13]:

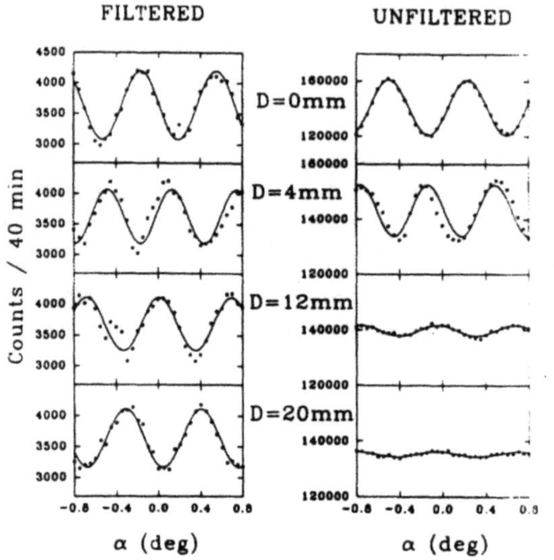

Fig. 2 Results of a recent neutron interference post-selection experiment where the enhanced visibility of the filtered (mono-chromatized) beam (left) in comparison with the overall beam (right) [13].

$$\left[\frac{\delta k'}{\delta k}\right]^2 = \frac{(\delta k_A)^2}{(\delta k)^2 + (\delta k_A)^2},$$

$$\left[\frac{\delta k_A}{k_0}\right]^2 = \eta^2 \left|\frac{\cos\theta_i \cos\theta_A}{\sin(\theta_i \pm \theta_A)}\right|^2, \qquad (6)$$

where the positive sign corresponds to the parallel setting and the minus sign for the anti-parallel configuration. Figure 2 shows the results obtained with a pressed silicon analyser crystal (h_i = 0.0035 rad) in the anti-parallel position (dk' = $1.87\text{x}10^{-2}$ A^{-1}, where dk = $3.2\text{x}10^{-2}$ A^{-1}).

Even less attenuation of the interference pattern at high order occurred with a nearly perfect crystal post-analyser, also in the anti-parallel position (h_A = 0.00015 rad, dk' = $8.0\text{x}10^{-4}$ A^{-1}). The related amplitude of the interference pattern is shown for both cases in Fig.3 [13] as a function of the phase shift which represents the coherence function defined in Eqs.(2) and (4). In a first approximation, one can assume that the wave packet does not change in the x-direction when going through the interferometer, because the tangential components of the momentum do not change at all during diffraction from the lattice planes. If one further assumes that the analyser crystals cut out a narrower band width causing a more extended wave packet, one ends up with a mutual overlap of wave functions from both beam paths, as indicated in Fig.1. This describes only partly the physical situation. At high order, the momentum dependence of the interference pattern becomes distinctly influential [15,16]. In this case, we have to rewrite Eq.(5) in a purely k-dependent form:

$$I(k) \propto |a(k)|^2 \left[1+\cos(\Delta.k)\right]$$

$$\propto \exp\left[-(k-k_0)^2/2\delta k^2\right]\left\{1+\cos\left(\chi_0 \frac{k_0}{k}\right)\right\}, \qquad (7)$$

where the mean phase shift is introduced ($c_0 = D_0.k_0 = -Nb_c l_0 D_{eff}$). Formula (7) can also be obtained from the Fourier-transformation of the two spatially (by $|D_x|$) separated (Gaussian) wave functions. This function exhibits, at high order, a marked oscillating structure in momentum space sweeping intensity between the forward and deviated beam (Fig.4). Thus, the intensity behind the interferometer is by no means the intensity sum arising from both beam paths. This indicates that each momentum component has a very distinct behavior just determined at the exit of the interferometer

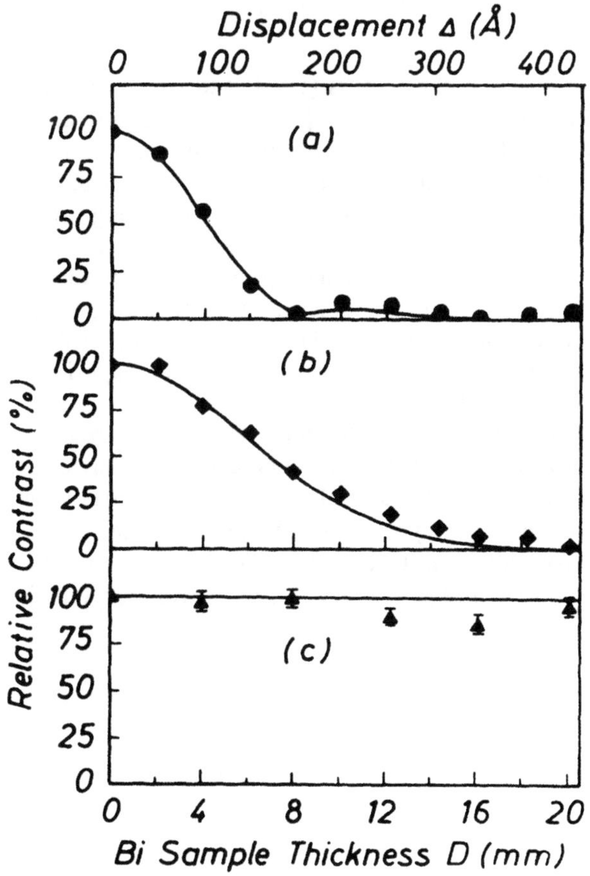

Fig. 3 Reduction of the interference contrast (= coherence function) at high order for different degrees of postselection of momentum states [13]. (a) direct beam C3, (b) pressed silicon (111)-monochro-mator in anti-parallel position, (c) nearly perfect silicon (111)-monochromator in anti-parallel position.

where the wave functions do not overlap. This underlines that the plane-wave components create a real action even outside the packet. Thus, complete interference properties can be restored in both out-going beams when proper measurement methods are applied. The vanishing interference pattern at high interference order is accomplishied with the appearance of spectral modelling in momentum space, indicating the persistent coupling in configuration space where the whole information about the history of both coherent beams remains stored, even in cases where the narrow oscillation cannot be resolved experimentally anymore. This spectral modelling effect might also have practical use for light and particle optics in achieving a spectral narrowing by interferometric means only.

The spatial distribution of the packets arising from beam paths I and II can also be obtained from Eq.(5) or, equivalently, from the Fourier- transform of the momentum distribution (Eq.(7)).

$$I_0(x) \propto \exp\left[-x^2/2\delta x^2\right] + \exp\left[-(x+\Delta)^2/2\delta x^2\right]$$

$$+ 2\exp\left[-x^2/4\delta x^2\right]\exp\left[-(x+\Delta)^2/4\delta x^2\right]\cos\chi_0 \,, \qquad (8)$$

which exhibits two separate peaks for large phase shifts (Fig.4) and which represents the amplitude function of the related wave function (e.g. [17]). For Gaussian packets, dx corresponds to the coherence length D_c and fulfills the minimum uncertainty relation dxdk = 1/2. For an appropriate large displace-ment (D >> D_c), the related state can be interpreted as a superposition state of two macroscopically distinguishable states, that is, a stationary Schrödinger-cat-like state [18,19], but here the first time for massive particles. These states - separated in ordinary space and oscillating in momentum space - seem to be notoriously fragile and sensitive to dephasing effects [20-23]. Spatially separated packets remain entangled in phase space and nonlocality appears as a result of this entanglement. The analogy with optical experiments performed in the time-frequency domain is striking [12].

The new quantum states created behind the interferometer can be analyzed with regard to their uncertainty properties. Analogies between a coherent state behavior and a free but coherently coupled particle motion inside the interferometer have been addressed previously [24]. In such cases, the dynamical conjugate variables x and p minimize the uncertainty product with identical uncertainties $(Dx)^2 = (Dk)^2 = 1/2$ (in dimensionless units). Using $I_0(k)$ and $I_0(x)$ (eqs. (7) and (8)) as distribution function we get in our case

$$<(\Delta x)^2> = <x^2> - <x>^2 = (\delta x)^2\left[1 + \frac{(\Delta_0/2\delta x)^2}{1 + e^{-(\Delta_0/2\delta x)^2/2}\cos(\Delta_0 k_0)}\right] \qquad (9)$$

and (for $\delta k/k_0 << 1$)

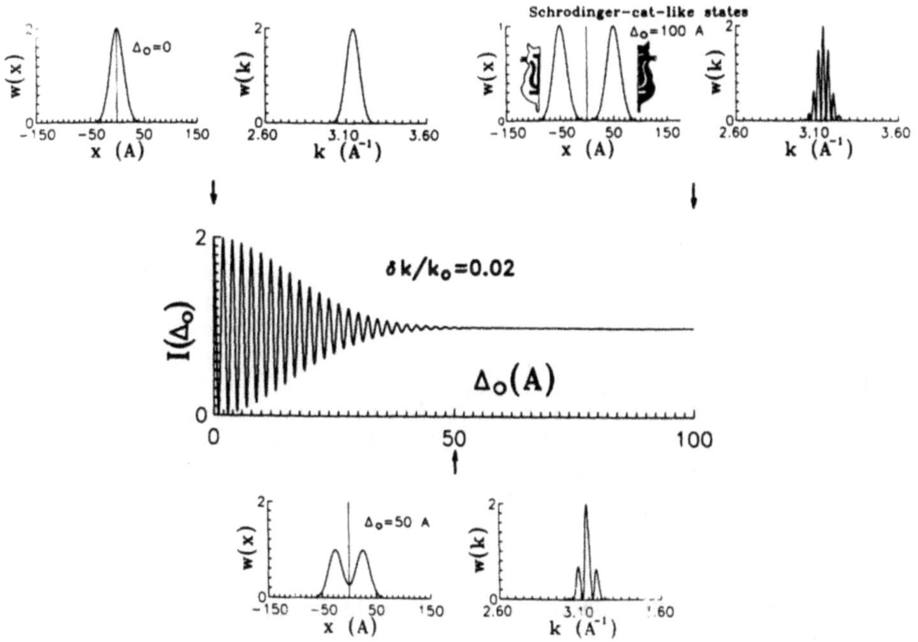

Fig. 4 Appearance of spectrum modulation when the interference pattern disappears at high order. The spectral modulation occurs in an alternative sense in both emerging beams and is sensitive to small variations of the phase shift.

$$<(\delta k)^2> = <k^2> - <k>^2$$

$$= (\delta k)^2 \left[1 + \left(\frac{\Delta_0}{2\delta x}\right)^2 \frac{e^{-(\Delta_0 2\delta x)^2}\cos(\Delta_0 k_0) + e^{-(\Delta_0/2\delta x)^2}}{1+e^{-(\Delta_0/2\delta x)^2/2}\cos(\Delta_0 k_0)} \right] \qquad (10)$$

These relations are shown in Fig.5 indicating that for $(\Delta k)^2$ a value below the coherent state value can be achieved, which in quantum optic terminology means squeezing [25-27]. One emphasizes that a single coherent state does not exhibit squeezing, but a state created by superposition of two coherent states can exhibit a considerable amount of squeezing. Thus highly nonclassical states can be made by the power of the quantum mechanical superposition principle.

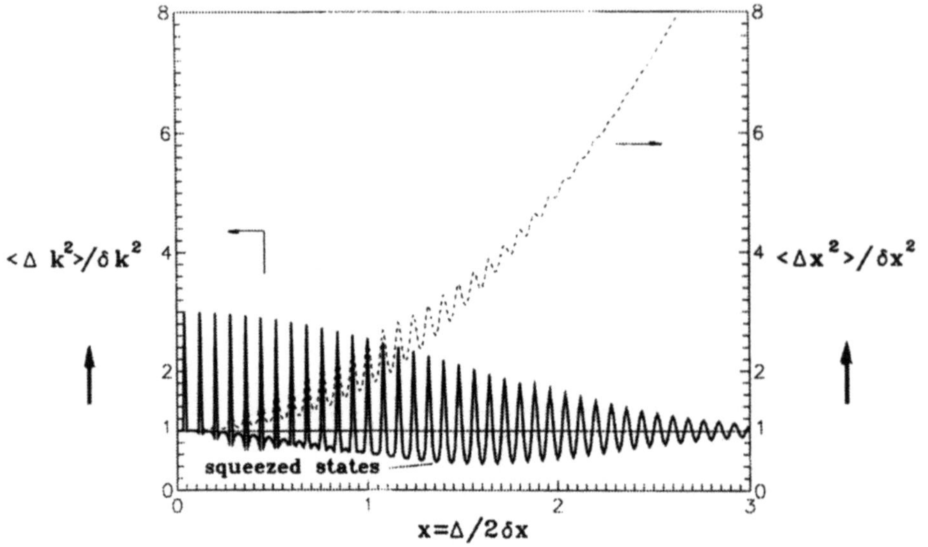

Fig. 5 Spatial and momentum uncertainties of the outgoing beams with the indication of squeezing.

3. EPR-CONFIGURATION SPACE COUPLING

In the EPR-situation [28-30,16], pairs of correlated photons (or particles) are produced which form an entangled quantum state (Fig.6):

$$\psi \propto |-k>_1 |k>_2 + |-k>_2 |k>_1 . \tag{11}$$

When the photon pairs are produced by a down conversion cascade, as shown in Fig.6, the momenta of the photons are correlated by the relation

$$k_1 + k_2 = k_{01} + k_{02} = \text{const.} \tag{12}$$

For a Gaussian distribution the wave function reads as

$$|k>_1 = \exp\left[-(k_1 - k_{01})^2 / 4\delta k^2\right] \exp\left[i(k_1 r - \omega_1 t)\right]$$

$$|k>_2 = \exp\left[-(k_2 - k_{02})^2 / 4\delta k^2\right] \exp\left[i(k_2 r - \omega_2 (t + \Delta t))\right]. \tag{13}$$

The momentum and space dependent intensity can be written in the form [16]

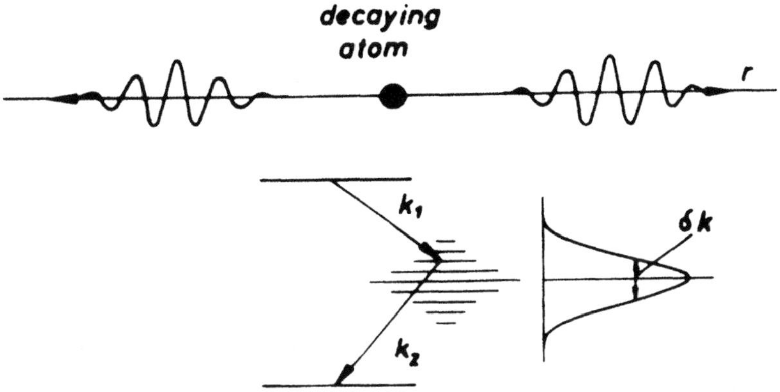

Fig. 6 Scheme of an EPR-arrangement and a simplified double photon decay scheme.

$$I(k_1,k_2,r) = \left| |-k>_1 |k>_2 + |-k>_2 |k>_1 \right|^2$$

$$= 2|a(k_1)|^2 |a(k_2)|^2 \left[1+\cos 2(k_2-k_1)r\right], \qquad (14)$$

which shows for each pair a characteristic intensity modulation (Fig.7). This formula can again be seen as Fourier-transform of spatially separated Gaussian packets, separated by 2r. It should be mentioned that the intensities $I(k_1,r)$ and $I(k_2,r)$ are smooth functions on both sides of the source determined by their distribution functions but the correlated intensity $I(k_1,k_2,r)$ exhibits marked, and for each pair characteristical, modulations. This indicates that individual |k>-states remain interacting even at arbitrarily large spatial separations of the wave packets. The coupling in configuration space persists unless a random interaction is applied to erase the fine structure of the wave function mentioned above. All EPR-experiments performed up till now were performed without momentum resolution and/or within the near field zone $r < 1/2\ dk$, thus averaging over the characteristical modulation shown in Fig.7. Therefore, it is recommended to repeat these experiments with a proper momentum resolution and to include the phase space coupling in the general discussion about EPR-phenomena. General conclusions about wave function properties can only be drawn if all accessible information about it is included. Thus the completeness of a quantum experiment has to be seen in a new light. The nonlocality phenomenon of quantum mechanics could be understood as the far reaching action of the plane wave components of the wave function as well. The band width of these plane wave components is determined by the momentum resolution of the measurement with an upper limit defined by the inverse of the source-detector distance.

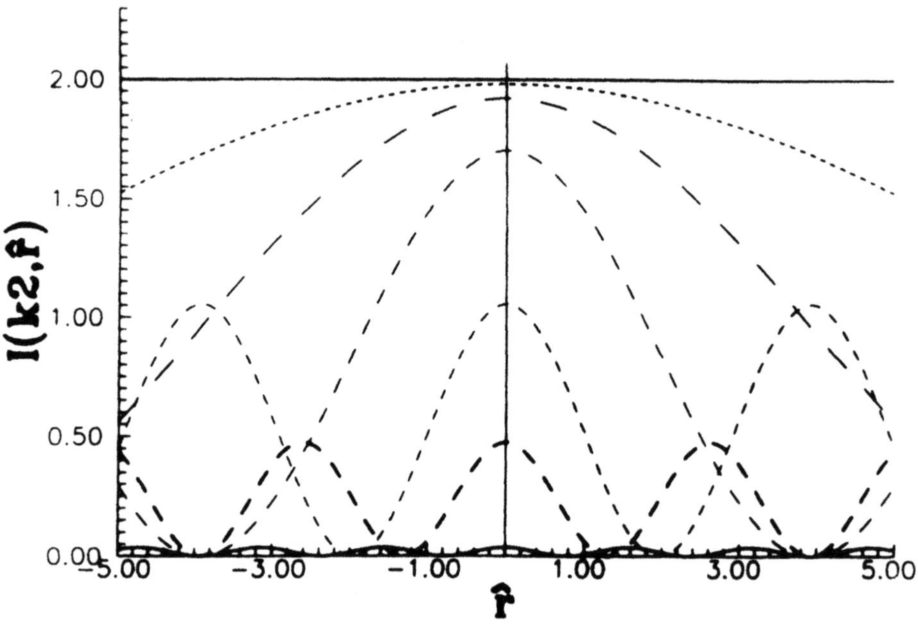

Fig. 7 Characteristical intensity modulation for different pairs of photons from a double transition decay $k_{01} = k_{02} = k_0 = 0.0015$ A^{-1} and $dk_0 = 0.05\ k_0$ for $|k_1\text{-}k_2| = 0.2\ ndk_0$ (n = 0, 0.5, 1, 2, 4, 6 and 10). $\hat{r} = r/r_c$, where $r_c = 1/2\ dk_0$ denotes the distance where the packets become separated.

4. DISCUSSION

An analogy has been shown between neutron interference (single particle) and EPR-experiments (two or more particle systems). It has been demonstrated that coherently separated wave packets remain coupled in momentum space, which can be attributed to the far reaching action of the plane-wave components constituting the wave packets. Therefore, many previously performed experiments devoted to this topic appear as incomplete experiments because not all information inherently contained in the related distribution function has been measured and used for interpretative questions. This puts a new light on the problem of non-locality measurements in quantum mechanics.

Work supported partly by Fonds zur Förderung der Wissenschaftlichen Forschung (project no. P8456).

REFERENCES

[1] U. Bonse and H. Rauch, eds., *Neutron Interferometry* (Clarendon, Oxford, 1979).
[2] A. G. Klein and S. A. Werner, *Rep. Prog. Phys.* **46** (1983) 259.
[3] G. Badurek, H. Rauch, and A. Zeilinger, eds., *Matter Wave Interferometry*, (North-Holland, Amsterdam, 1988).
[4] H. Rauch, in [1]., p. 161.
[5] H. Kaiser, S. A. Werner, and E. A. George, *Phys. Rev. Lett.* **50** (1983) 563.
[6] S. A. Werner, R. Clothier, H. Kaiser, H. Rauch, and H. Wölwitsch, *Phys. Rev. Lett.* **67** (1991) 683.
[7] H. Rauch, E. Seidl, D. Tuppinger, D. Petrascheck, and R. Scherm, *Z. Phys.* **B69** (1987) 313.
[8] R. Clothier, H. Kaiser, S. A. Werner, H. Rauch, and H. Wölwitsch, *Phys. Rev.* **A44** (1991) 5357.
[9] L. Mandel, *J. Opt. Soc. Am.* **51** (1961) 1342.
[10] D. F. V. James and E. Wolf, *Opt. Comm.* **81** (1991) 150.
[11] D. F. V. James and E. Wolf, *Phys. Lett.* **A157** (1991) 6.
[12] X. Y. Zou, T. P. Grayson, and L. Mandel, *Phys. Rev. Lett.* **69** (1992) 3041.
[13] H. Kaiser, R. Clothier, S. A. Werner, H. Rauch, and H. Wölwitsch, *Phys. Rev.* **A45** (1992) 31.
[14] D. Jacobson, S. A. Werner, and H. Rauch, *Phys. Rev.* **A49** (1994) 3196.
[15] H. Rauch, in "Vistas in Astronomy", Vol. 37, *Quantum Physics and the Universe*, M. Namiki et al., eds. (Pergamon, New York, 1993), p.157.
[16] H. Rauch, *Phys. Lett.* **A173** (1993) 240.
[17] J.-H. Levy-Leblond and F. Balibar, *Quantics* (North-Holland, Amsterdam, 1990).
[18] A. Legett, Proc. *Foundations of Quantum Mechanics*, S. Kamefuchi et al., eds. (Physical Society of Japan, Tokyo, 1984), p.74.
[19] B. Yurke, W. Schleich, and D. F. Walls, *Phys. Rev.* **A42** (1990) 1703.
[20] D. F. Walls and G. J. Milburn, *Phys. Rev.* **A31** (1985) 2403.
[21] R. J. Glauber, *New Techniques and Ideas in Quantum Measurement Theory*, D. M. Greenberger, ed. (New York Academy of Sciences, New York, 1986) p. 336.
[22] M. Namiki and S. Pascazio, *Phys. Rev.* **A44** (1991) 39.
[23] H. Zurek, *Physics Today*, Oct. 1991, p.36.
[24] H. Rauch, J. Summhammer, M. Zawisky and E. Jericha, *Phys. Rev.* **A42** (1990) 3726.
[25] W. Schleich, M. Pernigo and Fam Le Kien, *Phys. Rev.* **A44** (1991) 2172.
[26] D. F. Walls, *Nature* **306** (1983) 141.
[27] S. L. Braunstein and R. I. McLachlan, *Phys. Rev.* **A35** (1987) 1659.

MACROSCOPIC DECOHERENCE AND CLASSICAL STOCHASTIC GRAVITY

José L. Sanchez-Gomez

Departamento de Física Teórica, Univ. Autónoma de Madrid Cantoblanco, 20049-Madrid, Spain

The relevant features of a model of decoherence (previously introduced) based upon the existence of conformal stochastic fluctuations of the gravitational field are discussed and some cosmological implications of the said model are studied.

Key words: decoherence, metric fluctuations.

1. INTRODUCTION

There are essentially two ways of solving the "infamous" (*Bell dixit*) measurement problem of quantum mechanics (QM), i.e., of accomplishing the "reduction" that allows QM to be extended to the classical domain; according to Stapp [1], one should speak of: (1) "Heisenberg" reduction, in which a particular result is picked up from all the outcomes potentially present in the wave function; and (ii) "von Neumann" reduction, in which the coherence between off-diagonal elements of the density matrix—in some suitable representation—is destroyed, this fact then allowing a classical statistical interpretation of the quantum mechanical predictions (Stapp makes also distinction between subjective and objective reductions, but we are not concerned about this point here since only *objective* reductions will be contemplated in the present work).

Well-known models of type (i) are the GRW model [2] and its generalization, the Continous Spontaneous Localization (CSL) model [3]. In type (ii) are of course included all recent models of decoherence—see [4–6] and Halliwell's contribution to this Symposium.

It is possible, in my opinion, to establish a parallelism between these two different philosophies of reduction and the two ways of facing the irreversibility problem in physics. On one hand there is the idea that ir-

M. Ferrero and A. van der Merwe (eds.), Fundamental Problems in Quantum Physics, 291–297.

reversibility must be taken as a basic feature of physics and, because of it, the orthodox interpretations of both classical and quantum mechanics should be modified in order to properly incorporate that fundamental aspect of nature; this opinion is maintained, for instance, by Prigogine and his collaborators [7], who claim its origin to be traced back to the "first" Boltzmann. On the other hand, the conventional wisdom holds that irreversibility is a phenomenon caused simply by the coarse graining procedure inherent to our nature of macroscopic observers; this is the approach of Gibbs, Einstein, Ehrenfest, etc. (the parallelism is of course very rough and is intended to have only an indicative character).

In this short communication I am going to discuss the main features of a very recent model [8] in which it is assumed that the spactime metric is not totally defined since it undergoes very tiny fluctuations. Such fluctuations, albeit stochastic, are conformal and, in spite of their smallness, they are able to produce effective decoherence in the density matrix of the CM (centre of mass) of a "macroscopic" body without affecting at all the quantum-mechanical time evolution of "microscopic" systems (unitary evolution).

The general ideas of the model are introduced in Sec. 2. Sec. 3 is devoted to the analysis of important features of the model both from the physical and cosmological points of view.

2. THE MODEL

As said in the Introduction, the main idea of the model is the existence of universal fluctuations of the gravitational field (of the vacuum) which have a stochastic nature. Nevertheless, such fluctuations are not completely chaotic as they preserve the vacuum symmetries—the assumption is then that they are conformal. Hence one has

$$g_{\mu\nu}(x) = e^{\phi(x)}\langle g_{\mu\nu}(x)\rangle, \tag{1}$$

ϕ being a stochastic field ($\langle . \rangle$ stands for stochastic mean and the background metric is taken as Minkowskian for simplicity).

From the above equation, one derives the following ones:

$$\langle e^{\phi(x)}\rangle = 1, \quad \langle e^{\phi(x)}\phi_\mu(x)\rangle = 0, \tag{2}$$

where $\phi_\mu(x) = \partial_\mu\phi(x)$. Assuming the probability distribution to be Riemannian implies [9]

$$\langle \phi_\mu(x)\rangle = 0, \quad \langle \phi_\mu \phi\rangle = 0. \tag{3}$$

The preceding properties enable us to think of the field ϕ as a (approximately) gaussian stochastic process whose mean, without any loss of generality, can be set equal to zero, and having a variance

$$\langle \phi^2(x) \rangle = \sigma^2. \tag{4}$$

We will also assume the following correlation function

$$\langle \phi(x,t), \phi(x',t') \rangle = \sigma^2 e^{-(t-t')^2/\tau^2} e^{-(x,x')^2/L^2}, \tag{5}$$

where $\tau = L/c$ and L is a free (by now) parameter representing the correlation length. Note that time has been distinguished somehow from space in (5); this is really a classical problem in stochastic processes (one very difficult to deal with in the relativistic case). It will not be discussed here any longer; however we should just recall that we are in fact dealing with non-relativistic ("low velocity") processes, so that Eq. (5) is probably an adequate expression for our purposes here.

Now, just by using dimensional arguments, we arrive [8] at an approximate expression relating the parameters L and σ to each other (see also [10]).

$$\sigma^2 \sim \hbar G/(L^2 c^3), \tag{6}$$

G being the Newtonian gravitational constant. We are then left with a single free parameter: the correlation length L. It can be easily seen that for all reasonable values of L, $\sigma^2 \ll 1$, so that we are actually in the weak field case; all gravitational effects are accounted for by a stochastic Newtonian potential

$$V(x,t) = \frac{1}{2} mc^2 \phi(x,t). \tag{7}$$

Then, by using Feynman-Vernon's method [11] (see also [12]), we obtain the following equation for the time evolution of the density matrix of a particle in the coordinate representation (technical details are given in [8]):

$$\frac{\partial}{\partial t}\rho(x,x';t) = \mathcal{L}[\rho] - g(x,x')\rho(x,x';t), \tag{8}$$

where the operator $\mathcal{L}$ representing the purely quantum (unitary) evolution is given by (also in coordinate representation)

$$\mathcal{L}[\rho] = -\frac{i\hbar}{2m}\left[\Delta_x - \Delta_{x'}\right]\rho - \frac{i}{\hbar}\left[U(x,t) - U(x',t)\right]\rho. \tag{9}$$

U is any external, "deterministic" potential (of course, for a free particle $U = 0$), and the "decoherence part" in Eq. (8) is

$$g(x,x') = \frac{\tau}{\hbar^2}\left[\langle V^2 \rangle - \langle V(x,t)V(x',t) \rangle\right]. \tag{10}$$

To derive Eq. (10), one has to assume that the correlation time τ is much smaller than the relevant quantum times of the system. As we have already commented, and will be seen in more detail below, L should lie in the range $10^{-5} - 10^{-3}$cm. Now if, for definiteness, we take $L = 10^{-4}$cm then $\tau \cong 3 \times 10^{-15}$s, while the corresponding quantum time is $\cong mL^2/\hbar$, which for a proton, say, is $\cong 10^{-9}$ s., hence the above condition clearly holds.

An explicit expression for g can be obtained by inserting in (10) the correlation function given in (5) (taking (6) into account). Just to get simple—but accurate enough—expressions, we approximate the space part in (12) as follows:

$$\begin{aligned} &\sigma^2 \left[1 - (x, x')^2/L^2\right] \ , |x - x'| < L. \\ &0 \ , \qquad\qquad |x - x'| > L. \end{aligned} \tag{11}$$

This must be a good approximation but, perhaps, for $|x - x'| \cong L$.

By using (11), Eq. (8) can be written as

$$\begin{aligned} \frac{\partial}{\partial t}\rho(x, x'; t) &= \mathcal{L}[\rho] - \frac{Gm^2}{4\hbar L^3}(x - x')^2 \rho(x, x'; t), \ |x - x'| < L \\ &= \mathcal{L}[\rho] - \frac{Gm^2}{4\hbar L}\rho(x, x'; t), \qquad |x - x'| > L. \end{aligned} \tag{12}$$

One can see that the nonunitary term in (12) is negligible in the "microscopic" case. For instance, with $L = 10^{-4}$cm, and, taking such a system to be a proton, we would obtain

$$\lambda m = \frac{Gm^2}{4\hbar L^3} \cong 10^{-17}\text{cm.}^{-2}\text{s.}^{-1}. \tag{13}$$

Hence the relative coherence between parts of the proton wavepacket as separated as 1 cm would survive up to $\sim 10^{24}$s. This means that the time evolution is practically unitary (quantum). Let us go now to the—more interesting in the present context—macroscopic case.

First of all it should be pointed out that this work will be concerned with the microscopic systems with "solid" structure, that is the relative distance between any pair of constituents will not be changed by the quantum fluctuation; also—the corresponding body is taken as homogeneous and spherical. Now the treatment of the microscopic case can be straightforwardly extended to the present one simply by noting that the stochastic potential will be given by

$$V(x_i, \ldots, x_N; t) = \frac{1}{2}mc^2 \sum_i \phi(x_i, t). \tag{15}$$

In the "very" macroscopic case, i.e. when $R \gg L$, R being the radius of the body, the CM can be factored out and after a long but easy calculation one gets the following equation for the density matrix representing the CM quantum state [8], ρ_o

$$\begin{aligned} \frac{\partial}{\partial t}\rho_o(x_o, x'_o; t) &= \mathcal{L}[\rho_o] - \lambda_o(x_o - x'_o)^2 \rho_o(x_o, x'_o; t), \quad & |x_o - x'_o| < L \\ &= \mathcal{L}[\rho_o] - \gamma_o \rho_o(x_o, x'_o; t), & |x_o - x'_o| > L \end{aligned} \tag{16}$$

where

$$\lambda_o \cong GMD/\hbar, \quad \gamma_o = \lambda_o L^2, \tag{17}$$

M and D being, respectively, the mass and density of the body.

Now (16) effectively shows the decoherence between the off-diagonal terms of the CM density matrix of a macroscopic system.

Consider for instance a body of $M = 1\text{g}$, $D = 1\text{g/cm}^3$. From Eqs. (16) and (17) we see that correlations between macroscopic quantum states separated more than L (10^{-4}cm, say) would decay in about 10^{-12}s. Notice that this decoherence is produced just by stochastic fluctuations of the gravitational vacuum which are supposed to be universal, thus, in some sense, giving rise to an "effective" environment for any, however isolated, body.

3. IMPORTANT FEATURES OF THE CONFORMAL STOCHASTIC FLUCTUATIONS

First we shall discuss the energy conservation. Obviously, the non-unitary term in Eq. (16) is responsible for the change with time of the energy of the (isolated) system. In fact, in the present model such a change is quite negligible, but it has the unpleasant feature of being always positive. From (16) one easily obtains

$$\frac{d}{dt}\langle E_{\rm CM}\rangle = 2GD\hbar \cong 10^{-41} w, \tag{18}$$

which is independent of the mass.

Such a constant increase of the energy can be understood taking into account that the vacuum fluctuations "shake" the particle in a non-dissipative way—the particle's reaction on the field being disregarded. Precisely this reaction could be rather relevant in the "mesoscopic" case; however its treating in a correct mathematical way is quite complicated (some work has been already reported [13] and a more refined treatment is in progress). We think the mesoscopic case to be worth dealing with as one would expect some kind of behaviour in between the quantum and classical ones, which perhaps could give rise to some experimental tests of the present model (notice the model predicts the said mesoscopic region to lie within the range $10^{-16}\mathrm{g} < M < 10^{-8}\mathrm{g}$).

Let us now discuss some relevant cosmological points which could affect the consistency of this model.

It has been pointed out by Diósi and Lukács [14] that the so-called K-model [15] has the unpleasant (perhaps critical for the reliability of that model) feature of predicting a very large mean value of the same-time curvature—such a curvature would imply a mean energy density of the universe much larger than that of a neutron star. This unpleasant fact does not occur in the present model, due mainly to the assumed conformal nature of the stochastic metric fluctuations. It is easy to see that the mean value of the curvature is given by ($c = 1$)

$$\tilde{R} = -\frac{1}{2}\langle \exp(-2\phi)\left\{6\Box\phi + 3\left[(\partial\phi/\partial t)^2 - (\nabla\phi)^2\right]\right\}\rangle. \tag{19}$$

Now conformality plus the form of the correlation function, Eq. (5), implies

$$\langle \Box\phi \rangle = 0,$$

$$\langle (\partial\phi/\partial t)^2 \rangle = \langle (\nabla\phi)^2 \rangle = 4\sigma^2/L^2,$$

so that $\tilde{R} = 0 + 0(\phi^4)$ (recall that the background metric is Minkowskian). No problem then appears in the present model, in what concerns the mean curvature of the universe. One can also compute the *oo*-component of the energy-momentum tensor, which of course is not a Minkowski scalar. Note, however, that the relative *dynamic* velocity (i.e. cosmic expansion *not* included) between any pair of *physically relevant* systems seems to be, as a matter of fact, much smaller than the velocity of light, hence one is (probably) allowed to think that $\langle T_{oo} \rangle$—computed in the reference frame (ether?) wherein Eq. (5) holds—gives a reliable estimate of the mean energy density of the universe stored in the fluctuating vacuum. By using standard techniques, one gets

$$\langle T_{oo}/c^2 \rangle = c^2\sigma^2/(32\pi G L^2). \tag{20}$$

If it is now imposed that $\langle T_{oo}\rangle/c^2$ be less than $10^{-29}\mathrm{g/cm}^3$, as it seems reasonable that vacuum fluctuations alone cannot make the universe to be closed, and, according to Eq. (6), one takes $\sigma = \Lambda L$, where Λ is the Planck length, one can obtain a lower bound to $L : L > 10^{-3}\mathrm{cm}$. Of course this is a very rough bound, but it should give a hint—at least—of the order of magnitude of parameter L.

Finally, as to the physical origin of the—hypothetical—metrical fluctuations introduced here very little can be said by now. Perhaps one should relate such fluctuations to primordial gravity waves, originated at the end of the "Planck era"; but, as we do not have a consistent theory of quantum gravity, this is at present a mere speculation.

ACKNOWLEDGEMENTS

This work has been partially supported by C.I.C.Y.T (Spain), under contract PB91-0024.

REFERENCES

1. H. P. Stapp, *Phys. Rev. A* **46**, 78 (1992).
2. G. C. Ghirardi, A. Rimini, and T. Weber, *Phys. Rev. D* **34**, 470 (1986).
3. G. C. Ghirardi, P. Pearle, and A. Rimini, *Phys. Rev. A* **42**, 78 (1990).
4. E. Joos and H. D. Zeh, *Z. Phys. B* **59**, 223 (1985).
5. M. Gell-Mann and J. B. Hartle, in *Complexity, Entropy, and the Physics of Information*, W. Zurek, ed. (Addison-Wesley, Reading, 1990).
6. W. H. Zurek, *Phys. Today*, October 1991, p. 36.
7. I. Prigogine, *From Being to Becoming* (Freeman, New York, 1980).
8. J. L. Sánchez-Gómez, *Hadronic J. Suppl.* **8**, 129 (1993).
9. S. Roy and M. Sinha, "Einstein equation for the gravitational field and stochastic fluctuations," Calcutta preprint; see also G. Rosen, *Nuovo Cimento* **57B**, 125 (1980).
10. L. Diósi and B. Lukács, *Ann. Phys. (Leipzig)* **44**, 488 (1992).
11. R. P. Feynman and F. L. vernon, *Ann. Phys. (New York)* **24**, 118 (1963).
12. M. R. Gallis, *Phys. Rev. A* **45**, 47 (1992).
13. J. Unturbe and J. L. Sánchez-Gómez, *Nuovo Cimento B* **107**, 211 (1992).
14. L. Diósi and B. Lukács, Budapest preprint (1993).
15. F. Károlyházy, in *Sixty-two Years of Uncertainty*, A. Miller, ed. (Plenum, New York, 1990), and references therein.

DYNAMICS AND MEASUREMENT OF THE ABSOLUTE PHASE IN MACROSCOPIC QUANTUM SYSTEMS

Fernando Sols
Departamento de Física de la Materia Condensada, C-XII
Universidad Autónoma de Madrid, E-28049 Madrid, Spain

Roger A. Hegstrom
Department of Chemistry, Wake Forest University
Winston-Salem, North Carolina 27109, USA

The dynamics of the relative phase between two separate superconductors or superfluids is studied theoretically. In particular, we study the randomization of the phase after the Josephson coupling is supressed. Several time scales are identified and estimated. The phase randomizes much faster in the case of superconductors than in that of superfluids. The dynamics can however be strongly slowed down if the two superconductors are coupled to a common normal reservoir. The relation to the concept of spontaneous symmetry breaking and to the quantum measurement problem is commented upon. We analyze the question of whether one may distinguish empirically between an eigenstate of the phase and one of the particle number. We argue tentatively that such a distinction is in principle possible and notice that this claim contradicts well-established concepts in the interpretation of quantum mechanics.

Key words: macroscopic quantum mechanics, Josephson junction, gauge symmetry breaking, quantum measurement.

1. INTRODUCTION

This paper deals with some aspects of the dynamics of the absolute phase in superconductors or superfluids, which here we identify with the

M. Ferrero and A. van der Merwe (eds.), Fundamental Problems in Quantum Physics, 299–310.

relative phase between two isolated systems. This problem has bearing on the question of whether global gauge symmetry can be spontanously broken in condensed matter systems, an this in turn can be related to the debate about the quantum measurement problem in at least two ways: (i) the question of whether a physical system undergoes a spontaneous breaking of a global gauge symmetry can be rephrased as the problem of whether the corresponding order parameter can be "measured" by the environment; (ii) one may ask whether there are fundamental limitations to a direct experimental determination of the answer to the previous question. This second aspect of the relation between broken symmetry and quantum measurement will be discussed in the last part of this article.

The process of spontaneous symmetry breaking (SBS) takes place when the ground state or the equilibrium configuration of a given physical system is not invariant under the symmetry group of its original Hamiltonian (for a discussion, see Ref. 1). A special role is played by those cases involving the spontaneous breaking of a gauge symmetry (SBGS), since they are of paramount importance in the modern theory of elementary particles. Paradoxically, the concept of SBGS, which has been used pervasively in high-energy theories, was originally introduced in low temperature physics. The BCS theory of superconductivity led to the idea of broken symmetry and to the appreciation of its relevant role in elementary particle theory through what is now known as the Anderson-Higgs mechanism. Throughout this work we shall be concerned with the breaking of the Abelian $U(1)$ gauge symmetry occurring in superconductors and superfluids.

A system of charged particles is said to have gauge symmetry when its Lagrangian remains invariant under a simultaneous transformation of the electromagnetic gauge and the phase of the particle field. Thus, one can view the freedom to choose the electromagnetic gauge as the freedom to choose the phase $\varphi(\mathbf{r})$. Correspondingly, the local gauge symmetry is said to be broken if the function $\varphi(\mathbf{r})$ is uniquely determined by its value at a given point. A standard example of SBGS is provided by the BCS wave function for a conventional superconductor:

$$\begin{aligned} |\Psi_{\rm BCS}> &= \prod_k (u_k + v_k e^{-i\varphi} c^\dagger_{k\uparrow} c^\dagger_{-k\downarrow})|\text{vac}> \\ &= \sum_{N=0}^{\infty} a_N e^{-iN\varphi}|N> \simeq |\varphi>, \end{aligned} \tag{1}$$

where [2] u_k and v_k are real numbers defined by $u_k^2 = (1/2)(1+\xi_k/E_k) = 1 - v_k^2$; $\xi_k = \hbar^2 k^2/2m - \mu$ and $E_k^2 = \xi_k^2 + \Delta^2$ are, respectively, the free-electron and the quasiparticle dispersion relations, and Δ is the superconducting gap. The last approximate equality in (1) indicates that, apart from normalization factors, the BCS wave function is practically an eigenstate of the phase. This is possible due to the large uncertainty in the number of Cooper pairs, $< (\Delta N)^2 >= 4\sum_k u_k^2 v_k^2 \simeq (3\pi/2)(\Delta/E_F)N \gg 1$.

An eigenstate of the particle number operator can be obtained by projecting (1) onto its $|N>$ component:

$$|N>= (2\pi)^{-1/2} \int_0^{2\pi} d\varphi e^{iN\varphi} |\varphi> . \tag{2}$$

It is then said that, in an isolated superconductor, the *local* gauge symmetry has been broken (the phase is rigid in the sense explained above) but the *global* gauge symmetry remains unbroken, since the total particle number is conserved at the price of spreading the phase uniformly around the unit circle.

The uncertainty in the phase is clearly a fundamental consequence of the fact that the particle number is fixed. From (1) it follows that $N = -i\partial/\partial\varphi$ and, therefore, $\Delta N \Delta\varphi \geq 1/2$, but note that this inequality does not force $\Delta\varphi$ to be small when ΔN is large.

A common objection against the value of thinking about the absolute phase is that it is a meaningless concept. However, this viewpoint tends to ignore that the absolute phase is not less meaningful than more ordinary concepts like the absolute position of an isolated body in free space. It is the relative position between two bodies what has physical sense. Analogously, the relative phase between two separate superconductors has a physical meaning to the extent that these can be eventually reconnectd and display the Josephson effect. Therefore, it is legitimate to ask whether the relative phase between two separate superconductors can become spontaneously well-defined (in the sense of acquiring a small uncertainty). This question has already been raised by Anderson [1,3], who seems to propose an affirmative answer for both the case of superconductors [1] (provided the coupling to the ground is sufficiently strong) and of superfluids [3]. We shall not address this problem directly but shall try to answer related questions which lend themselves more easily to a quantitative analysis. Specifically, we will analyze theoretically the randomization of the relative phase after the Josephson coupling is supressed. We will also discuss whether it is possible in principle to learn experimentally about the nature of the quantum state of two superconductors (in particular, whether one may determine empirically if the relative phase becomes spontaneously well-defined). Sections 3 and 4 are devoted to the analysis of these two questions. In Section 2, a brief review is given of some aspects of the Josephson effect which are particularly relevant for the ensuing discussion. Part of the material of Section 3 has been presented in Refs. 4 and 5, while section 4 is essentially a summary of Ref. 6.

2. ALL YOU NEED TO KNOW ABOUT THE JOSEPHSON EFFECT

A Josephson junction is a weak link between two superconductors which permits the coherent exchange of Cooper pairs. If φ is taken to be the phase difference and N the number of transferred Cooper pairs, the Hamiltonian for the Josephson junction is

$$H_J = E_J(1 - \cos\varphi) + \frac{1}{2}E_C N^2, \tag{3}$$

where E_J and E_C are the Josephson coupling and the capacitance energies, respectively. For two superconductors, $E_C = (2e)^2/2C$, where $2e$ is the Cooper pair charge and C is the electrostatic capacitance. In the case of two superfluids, $E_C = gM_0^2/\rho A$ if the capacitance is due to gravitational forces (g is the acceleration of gravity, M_0 is twice the mass of the ^{3}He atom, ρ is the mass density, and A is the superfluid area), and $E_C = M_0^2/\rho^2 V \kappa_T$ if the volume V is held fixed (κ_T is the compressibility). The Josephson relations are

$$\begin{aligned} I(\varphi) &= \dot{N} = -\frac{1}{\hbar}\frac{\partial H_J}{\partial \varphi} = I_C \sin(\varphi), \\ \dot{\varphi} &= \frac{1}{\hbar}\frac{\partial H_J}{\partial N} = (\mu_1 - \mu_2)/\hbar, \end{aligned} \tag{4}$$

where $I_C = 2eE_J/\hbar$ is the critical current.

In a typical Josephson junction, $E_C \ll E_J$, and (3) can be approximated as

$$H \simeq \frac{1}{2}E_J\varphi^2 + \frac{1}{2}E_C N^2, \tag{5}$$

as long as $kT \ll E_J$. At equilibrium, the phase is narrowly peaked around $\varphi = 0$. In a semiclassical description (C large), φ is a c-number and the energy is given by the first term of Equation (3).

3. DYNAMICS OF THE PHASE AFTER SEPARATION

We wish to analyze quantitatively the evolution of the phase after the coupling energy E_J of a Josephson junction is suddenly made zero by, for instance, severing the contact between the two superconductors. At $t < 0$, the equilibrium average energy is $E_i = (\hbar\omega_0/2kT)\coth(\hbar\omega_0/kT)$, where $\omega_0 = (E_C E_J)^{1/2}/\hbar$ is the oscillator frequency. The initial phase uncertainty is $\Delta\varphi(0) = (E_i/E_J)^{1/2} \ll 1$. Now, let E_J become zero at $t = 0$ and let us study the subsequent evolution for $t > 0$.

A. Ballistic Randomization

In the absence of external perturbations, the evolution of the wave function at $t > 0$ is that of an initially narrow wave packet on a circle. Assuming for the moment that φ is an extended variable, the phase uncertainty at long times is

$$\Delta\varphi(t) \simeq \sqrt{E_i E_C}\, t/\hbar. \tag{6}$$

The ballistic time τ_B is defined as the time it takes $\Delta\varphi$ to be of order 2π. We get

$$\tau_B = h/\sqrt{E_i E_C}. \tag{7}$$

Typical values for a superconductor are $E_J \sim 5$ eV and $C \sim 1$ nF (a compromise value between a self-capacitance and and a plate capacitance; which one is relevant will depend on the particular physical realization), which leads, with $E_i \simeq \hbar\omega_0/2$, to $\tau_B \sim 30$ ns. For superfluid ^{3}He, $E_J \sim 1$ K. the dimensions of the baths are taken of order 1 cm (implying $E_i \simeq kT$), one obtains [4] $\tau_B \sim 4$ hours, if capacitance is caused by gravitational forces, and $\tau_B \sim 1$ minute, if it is dominated by compressibility forces. Thus, in contrast with the superconductor case, the intrinsic randomization times of superfluids are remarkably long.

B. Recurrence Time

The phase is actually an angular variable for which 0 and 2π are to be identified and this implies that N has integer eigenvalues. When $E_J = 0$, the wave function evolves as

$$\Psi(\varphi, t) = <\varphi|\Psi(t)> = \sum_N a_N \psi_N(\varphi) \exp(-iE_C N^2 t/2\hbar), \tag{8}$$

where $\psi_N(\varphi) = <\varphi|N>$ and $a_N = <N|\Psi(0)>$. From (8), we notice the existence of a recurrence time

$$\tau_R = \frac{2h}{E_C}, \tag{9}$$

after which the wave packet recombines, $\Psi(\varphi, \tau_R) = \Psi(\varphi, 0)$. The property $E_C \ll E_J$ implies $\tau_R \gg \tau_B$, so that the two time scales are well separated. This observation renders meaning to the apparently oversimplified mechanism of ballistic randomization studied above, provided one considers times much smaller than τ_R.

A detailed analysis reveals the following picture: An initially narrow wave packet begins to spread according to the law given in (6). After a time of order $0.5\tau_B$, the two tails of the wave packet begin to overlap appreciably on the opposite side of the circle and, as a result, interference on a fine scale begins to develop. By the time $t \sim 2\tau_B$, the wave packet is almost uniformly spread around the circle, appart from very fast (both in time and in angle) oscillations. At a much later time τ_R, the wave function recombines (and is broadened again) in a time scale τ_B. With the parameters given above, we find $\tau_R \sim 13\ \mu s \gg \tau_B$ for the superconductor and astronomically large values for the superfluid. The superconductor value of τ_R falls entirely within experimental reach, although it must be noted that recurrence would only be observed in quite ideal conditions (very low noise, good time resolution, etc.), perhaps not easily realizable.

C. Diffusive Randomization

So far we have neglected the possible effect of dissipation due to the environment. For brevity, we shall focus on a source of noise which, if really present, does certainly dominate the dynamics of fluctuations: we assume that the two superconductors are able to exchange particles with a normal metal reservoir.

It is useful to draw an analogy with a quantum Brownian particle (BP). The phase plays the role of the position and the capacitance is analogous to the mass. The particle number N is equivalent to the momentum p (both are conserved quantities that are exchanged with the bath). The chemical potential difference $\mu \propto \dot{\varphi}$ is the the analogue of the particle velocity v The friction coefficient η is the ratio between the the dissipative force and and the displacement of the velocity with respect to its equilibrium value $< v >= 0$, $\dot{p} = -\eta v$. Analogously, a coefficient λ may be defined relating $\dot{N}$ to a spontaneous fluctuation in the chemical potential of the superconductor with respect to the normal reservoir:

$$\dot{N} = -\lambda\mu. \tag{10}$$

(also $< \mu >= 0$). Dissipation affects the particle dynamics in a time scale $\gamma^{-1} = m/\eta$. In the superconductor case, $\gamma = \lambda E_C$, and γ^{-1} turns out to be the $R_N C$ time, where R_N is the normal state resistance between our system and the reservoir. Taking a contact area of 1 cm^2 and an average transmission probability at the Fermi level of $T_0 \sim 0.01$, we arrive at the rather small time $\gamma^{-1} \sim 7 \cdot 10^{-19}$s. The inverse temperature time is $\hbar/kT \sim$ 7 ps at $T = 1$ K.

A quantum BP which is left initially with a well-defined position and an average zero momentum in the middle of a viscous bath will increase its position uncertainty through two main mechanisms. On one hand, it will undergo a purely quantum broadening of the type described in (6) for the phase, without changing the average position and momentum. This process will result in what could be called "quantum ignorance" of the precise value of the position. On the other hand, our BP can suffer collisions with the small particles forming the viscous medium. In these microscopic processes, momentum is exchanged randomly between the particle and the bath and, as a result, "classical ignorance" about the particle position increases (we call it classical because this type of position randomization would also affect a classical particle).

Once the analogy with the BP has been established, we may borrow from the classical theory [7] and write for the evolution of the phase uncertainty

$$\Delta\varphi^2(t) = \frac{2kT}{\hbar^2\lambda}[t - \frac{1}{\gamma}(1 - e^{-\gamma t})] \tag{11}$$

At short times ($\gamma t \ll 1$), we reproduce the ballistic randomization given by (6). By contrast, at long times ($\gamma t \gg 1$), the randomization becomes diffusive,

$$\Delta\varphi(t) \simeq (\frac{2kTt}{\hbar^2\lambda})^{1/2}. \tag{12}$$

Equation (12) is a classical result. However, according to the work of Hakim and Ambegaokar [8] on a *quantum* BP with Ohmic dissipation, the diffusive law (12) applies at long times $t \gg \hbar/kT \gg \gamma^{-1}$, provided $\Delta\varphi(0) \ll 1$

From (12), we deduce that the diffusive randomization time is

$$\tau_D = \frac{\hbar^2\lambda}{2kT} \tag{13}$$

With the parameters given above, we find $\tau_D \sim 12$ minutes, a surprisingly long time if we compare it with our estimates of the ballistic time τ_B. The large disparity between the two time scales may be understood by noting that, while the fast ballistic randomization is due to the small value of the electrostatic capacitance, the diffusive dynamics (which takes over in a time scale $\gamma^{-1} \ll \tau_B$) is essentially independent of it and is determined instead by the coupling to the normal reservoir.

D. Possible Experimental Probes

One could conceivably explore the ballistic randomization time with an experimental setup consisting of a Josephson junction that at has reached equilibrium at $t < 0$ and whose coupling energy E_J is made zero at $t \geq 0$ except for brief, occasional pulses in which it recovers its original value. A plausible time dependence that would reproduce such a pattern is

$$\begin{aligned} E_J(t) &= E_J(0)\exp\{\alpha[\cos(2\pi ft) - 1]\} \\ &\simeq E_J(0)\sum_{n=0}^{\infty}\exp\{-2\pi^2\alpha(ft - n)^2\}, \end{aligned} \tag{14}$$

if $\alpha \gg 1$. For very short separation times, the phase would not have time to randomize and we would measure a zero supercurrent at each reconnection. By contrast, if the separation time is large, the wave packet is broadened and we measure a nonzero current which varies randomly from pulse to pulse. Therefore, the qualitative prediction is that $< I^2 >$ is approximately zero for $f^{-1} \ll \tau_B$, while it saturates to a nonzero value for $f^{-1} \gg \tau_B$. Under ideal conditions, one could observe the effect of recurrence: for $f^{-1} = \tau_R$, $< I^2 >$ should again become zero (it can be proved that $f^{-1} = \tau_R/4$ would suffice). The values we have estimated for τ_B and τ_R (30 ns and 13 μs, respectively) could be measured in a Josephson device modulated by a field effect transistor. These transistors may easily reach frequencies of up to 20 GHz, yielding an ample margin for exploration.

We have obtained a remarkably large value for the diffusive time τ_D. However, our estimate has been unrealistic in one specific sense: we have implicitly assumed an infinite capacitance for the normal reservoirs, so that these cannot be responsible for a further randomization of the phase. It is clear that this additional broadening would be quite fast, given the typically small values of the electrostatic capacitances. The situation is analogous to that of two quantum BP's embedded in separate viscous baths. When left in free space, an additional increase in their relative position would be caused by the baths finite mass. However, if the two particle were embedded in the same bath, the randomization of the relative position would be truly diffusive and, therefore, very slow. Analogously, one can expect the estimated value of τ_D to be relevant when the two superconductors are linked to a common normal metal reservoir. The time τ_D could be probed in a network of superconducting islands. Starting from a low temperature phase of globally coherent superconductivity, one may raise the temperature slightly and suppress the Josephson coupling between islands. A few seconds or minutes later, the system is cooled down again. If the islands are surrounded by a normal metal that acts as a common reservoir, one can expect the phases to have remained well defined during the separation period and form the same vortex configuration they had before the decoupling. If the islands have been surrounded by an insulator, or have been separated for a sufficiently long time, a new vortex configuration will be formed.

4. ON THE MEASUREMENT OF THE PHASE

In this section, we will try to answer the following question: is it possible in principle to determine experimentally whether the relative phase between the two superconductors is well- or ill-defined? More specifically, could one distinguish experimentally between $|N>$ or $|\varphi>$? We shall see that these questions are fundamentally connected to the quantum measurement problem, since they deal with the possibility of distinguishing between eigenstates of canonically conjugate varaibles.

Let us consider two separate superconductors or superfluids. In the presence of a dynamic environment, we describe the system with a density matrix. For simplicity, we focus on two extreme cases, although intermediate situations are of course possible. The density matrix may correspond to an incoherent mixture of particle eigenstates (case 1) or to an incoherent mixture of relative phase eigenstates (case 2). We write, respectively,

$$\rho^{(1)} = \sum_N \rho_N |N><N| \tag{15}$$

$$\rho^{(2)} = \int d\varphi \rho(\varphi) |\varphi><\varphi| \tag{16}$$

The matrix $\rho^{(1)}$ ($\rho^{(2)}$) describes an ensemble of systems each of which is in an eigenstate of the particle number (phase) according to a probability distribution given by ρ_N ($\rho(\varphi)$). Global gauge symmetry is spontaneously broken in case 2. We ask whether one could distinguish experimentally between cases 1 and 2. To sharpen the discussion, let us focus on the more challenging case in which both ρ_N and $\rho(\varphi)$ are approximately constant, so that

$$\rho^{(1)} \simeq \rho^{(2)} \simeq \mathbf{I}. \tag{17}$$

We face here a well-known interpretational problem. The description based on density matrices adds a "classical ignorance" to the quantum knowledge contained in a pure quantum state. For example, in Equation (15), the distribution ρ_N introduces uncertainty in the knowledge of N. In the limit (17) of total classical ignorance, we see that different physical scenarios may be described by the same density matrix and, according to the statistical interpretation of quantum mechanics, they are intrinsically indistinguishable (see, however, the discussion by D'Espagnat [9] based on finite statistics in a Stern-Gerlach context). The whole formalism of quantum statistical mechanics is based on this close interweaving of quantum and statistical averages.

The intrinsic indistiguishibility of cases 1 and 2 can be easily understood in terms of ideal measurements. Consider case 1. At a given physical realization, the system is in an eigenstate $|\varphi>$ of the phase. We connect the two superconductors, measure the supercurrent and obtain a generally nonzero value $I_C \sin\varphi$ which will vary randomly from run to run, φ following a uniform distribution. In case 2, the system is in an eigenstate $|N>$ of the particle number. We measure the supercurrent and the system collapses to an eigenstate of the phase according to a uniform probability distribution. Therefore, the predictions for cases 1 and 2 are identical. There seems to be no possible distinction as long as we deal with system ensembles, or consider single systems but perform ideal measurements. The question which naturally arises is whether one could in principle perform an experiment on a *single* quantum system that would provide enough information to distinguish between cases 1 and 2, even in the limit (17)? Below we argue (not conclusively) that the answer is yes, that it is possible to design a weak coupling experiment which would allow for such a distinction. The technical details of the argument will be published elsewhere [6].

Let us assume that we reconnect the two separate superconductors and that, before the newly formed junction relaxes to the ground state, we perform a measurement of the current by a rather unconventional (and, surely, not very practical) method which however lends itself to a convenient theoretical treatment: we send a beam of spin polarized neutrons that are deflected by the magnetic field created by the supercurrent; by measuring the deflection of the beam on a detector screen we can in principle know about the value of the supercurrent. In analogy to the Stern-Gerlach experiment, the vector position of the neutron may be viewed as a pointer coordinate that can be treated quantum-mechanically. The Hamiltonian for such a system may be written $H = H_P + H_J + V$, where $H_P = p^2/2M$ describes the neutron, H_J is given by (3), and the interaction is $V = -\mu \cdot \mathbf{B}(\mathbf{r}, \varphi) \propto I(\varphi)v(\mathbf{r})$, $v(\mathbf{r})$ being an effective interaction potential. Here, both $\mathbf{r}$ and φ (and, hence, I and $\mathbf{B}$) are treated as quantum mechanical variables. Let us analyze the interaction neutron-junction as a time-dependent scattering process and let us assume that the coupling is sufficiently weak for the Born approximation to be valid. If $\Phi_{\text{in}}(\mathbf{r}, t)$ is the wave packet that describes the incident neutron at $t \to -\infty$ and the unscattered component of the outgoing wave at at $t \to +\infty$, one can show that the probability density for the neutron at $t \to \infty$ is

$$\begin{aligned} P(\mathbf{r}, t) &= |\Phi_{in}(\mathbf{r}, t)|^2 \\ &\quad + \text{cnst.} \times < I(\varphi) > \text{Re}\{\Phi^*_{in}(\mathbf{r}, t)\Phi_{in}(\mathbf{r}\hat{e}_y, t) < \mathbf{k}_1|v(\mathbf{r})|\mathbf{k}_0 >\} \\ &\quad + O(V^2), \qquad (18) \end{aligned}$$

where $\hat{e}_y$ is the unit vector in the incident direction y, and $\mathbf{k}_0$ and $\mathbf{k}_1$ are, respectively, the neutron initial and final wawevectors. In (18), the first term is the unscattered beam and $O(V^2)$ stands for the first order fully scattered wave yielding the Born cross section. The second term of (18) comes from the interference between the unscattered wave and the first order scattered wave. It is proportional to $< I(\varphi) >$ and thus probes directly the quantum expectation value of the current. If the junction is initially in a state $|N>$, this term is zero (see Equation (2)). If it is in a state $|\varphi>$, then it is nonzero but would average to zero in an ensemble with a uniform distribution of phases φ. However, if we were able to perform measurements on a single quantum system, we would be able to notice that the interference term is nonzero. The problem is that, if we only send one neutron, we do not have enough information to reproduce $P(\mathbf{r}, t)$ and compare with (18). We obviously have to send many neutrons, each of which couples sufficiently weakly for (18) to remain valid, even after taking into account the cooperative effect of all the neutrons distorting the system wave function. The "recipe" to distinguish between $\rho^{(1)}$ and $\rho^{(2)}$ in the limit (17) could then be formulated in the following way: (i) send in *many* neutrons (to have statistics) that (ii) couple *weakly* (not to disturb the system excessively) onto (iii) a *single* junction (to avoid ensemble averaging) and (iv) analyse the *interference* term between the scattered and the incident wave. Is this recipe feasible? Adherents of the orthodox Copenhagen interpretation would probably answer no, since that would allow for a distinction between two physical systems described by the same density matrix. Here we claim that a distinction based on the recipe given above is in principle possible. It might turn out however that the orthodox view is correct, i.e., that it is impossible to reach a compromise between conditions (i) and (ii) (the requirement of sending a large enough number of neutrons which couple strongly enough to permit a meaningful resolution might necessarily lead to an excessive distortion of the single system wave function that would ultimately blur the crucial information). An exact calculation based on a semi-realistic model would be more conclusive.

5. CONCLUSIONS

We have studied some aspects of the dynamics of the macroscopic phase of an isolated superconductor or superfluid and have discussed how it may be fundamentally related to the quantum measurement problem. Specifically, we have analysed the randomization of the relative phase after suppression of the Josephson coupling. Our numerical estimates of the ballistic randomization times suggest that the phase of a superfluid behaves like a fairly classical variable while the phase of an isolated superconductor has a marked quantum character. The dissipative effect of a common normal reservoir can considerably slow down the randomization of the relative phase between two superconductors, to the point of acquiring observable time scales. We have also raised the question of whether it would be possible in principle to distinguish between physical systems prepared differently but described (as an ensemble) by the same density matrix. By considering a gedanken experiment, we argue, albeit not conclusively, that it is indeed possible to perform such a distinction, provided that the interference term in a weak coupling experiment is appropiately analysed.

ACKNOWLEDGMENTS

We would like to thank Anthony J. Leggett, Dale van Harlingen, and Ivar Zapata for stimulating discussions. The work of F.S. has been supported by CICyT, Project no. MAT91-0905, and by the Human and Capital Mobility Programme of the European Union.

REFERENCES

1. P.W. Anderson, *Basic Notions of Condensed Matter Physics* (Benjamin-Cummings, Menlo Park, 1984), Chap. 2.
2. M. Tinkham, *Introduction to Superconductivity* (Krieger, Malabar, 1975), Chap. 2.
3. P.W. Anderson, in *The Lesson of Quantum Theory*, J. de Boer, E. Dal, and O. Ulfbeck, eds. (Elservier, Amsterdam, 1986).
4. A. J. Leggett and F. Sols, *Found. Phys.* **21**, 353 (1991).
5. F. Sols, *Physica B* **194-196**, 1389 (1994).
6. R. A. Hegstrom and F. Sols, *Found. Phys.*, to be published.
7. F. Reif, *Foundations of Statistical and Thermal Physics* (McGraw-Hill, New York, 1965).
8. V. Hakim and V. Ambegaokar, *Phys. Rev. B* **32**, 423 (1984).
9. B. D'Espagnat, *Conceptual Foundations of Quantum Mechanics* (Addison-Wesley, Menlo Park, 1989).

REALISTIC QUANTUM THEORY AND RELATIVITY

Euan J. Squires

Department of Mathematical Science
University of Durham
Durham City, DH1 3LE, United Kingdom

It is argued that the Bohm model gives a very satisfactory, realistic, version of non-relativistic quantum theory, in spite of some recent arguments to the contrary. A local version, consistent with relativity, is proposed and the degree to which it violates orthodox predictions is calculated in a simple model experiment.

Key words: quantum theory, Bohm model, Lorentz-invariance.

1. INTRODUCTION

The physics of this century has been mainly about waves. Even those of us who have worked in elementary *particle* physics, have actually calculated the interactions of waves, because, as John Bell reminded us, quantum theory is a theory of waves.

Now, given that the purpose of physics is to explain the observed world, this is somewhat strange. For the world that we observe is not a world of waves; rather it is a world of objects with positions that change with time. What we have to explain is not waves - it is trajectories! And in our attempts to understand our experience, we have almost refused to talk about the things we experience.

The explanation of this is complex, and is a fascinating topic for historians, but one reason is presumably that von Neumann (and later other people) told us that we could not, consistent with the known results of quantum theory, actually talk about trajectories. This remains a persistent myth, so it is important to emphasise that these arguments are all wrong. They are wrong because measurements are "contextual." This is

M. Ferrero and A. van der Merwe (eds.), Fundamental Problems in Quantum Physics, 311–319.

a fancy way of expressing the rather obvious truth that *the answer depends upon the question.* When we add this to the fact that, as Niels Bohr emphasised, in quantum theory a proper definition of the "question" requires a more complete specification of the experimental context than in classical mechanics, we can see why the "no-go" theorems are irrelevant.

All this was made clear in 1952 by David Bohm[1], who showed that quantum phenomena could be understood in terms of particles moving along trajectories which could be calculated from the wave function. The model was simple (see below) and, with some restrictions, unique. It is a measure of how Bohm's work was neglected that a book[2] could be published in 1987 devoted almost entirely to the contextuality issue, without ever mentioning it.

The most recent claim that the Bohm model is, in some way, incorrect is due to Englert et al.[3] , who argue that in certain circumstances the model gives the "wrong" trajectories. Although I do not believe this is the proper interpretation of their results, these results themselves are interesting because they show a very nice feature of the contextuality of quantum measurements. In the next section we discuss a rather simpler situation[4] which demonstrates the same effect.

2. HOW TO FOOL A DETECTOR

We first suppose that we have a detector which consists of a particle, d, bound in some potential well and initially in the ground state, ϕ_0. This detector is designed to detect electrons, so when a wave packet containing an electron passes through the detector it is excited to the state ϕ_1. We suppose the dynamics is such that there is no significant effect on the electron. In order that the detector actually enables us to "see" that it has recorded an electron, we arrange the excited state to be unstable, so that its wavefunction, ϕ_1, spreads in space, and the detector particle, d, moves away from its initial position. We can regard it as a "pointer" indicating that an electron has been seen.

Now we want to show that we can "fool" this detector into believing that it has detected an electron, and showing this by movement of the pointer, even though no electron has passed through it.

The experiment we consider contains a Mach-Zender type interferometer (for electrons) without the final beam-splitter, as shown in fig.1. We denote the wavefunction of the electron moving along path 1(2) by ψ_1 (ψ_2), and along path $a(b)$ by ψ_a (ψ_b). Then, in the absence of the detector D, the unitary evolution of the wavefunction can be represented as

$$\Psi = 2^{-1/2}(\psi_1 + \psi_2) \qquad (1)$$

$$\rightarrow 2^{-1/2}(\psi_a + \psi_b). \tag{2}$$

Of course in orthodox quantum theory it makes no sense to ask which way the particle went; these equations are all that can be said.

The Bohm model, however, supplements this information with a position, $\mathbf{x}$, for the particle. This follows a trajectory whose form depends on the position, $\mathbf{x}_0$, relative to the wave packet, at some initial time prior to the wave packet reaching the beamsplitter[5]. For some range of $\mathbf{x}_0$ (towards the rear of the wavepacket) it will follow the path $P_1 \rightarrow P_a$; for the other values it will follow $P_2 \rightarrow P_b$.

There is actually no need to calculate in detail that the trajectories follow these paths since we know they must, at least in our two-dimensional model, as trajectories cannot cross. However, it is important to note that the change of direction that occurs in the interference region I arises because in the expression for the velocity of the particle,

$$\dot{\mathbf{x}} = Re\frac{\mathbf{p}_x(\psi_1 + \psi_2)}{m_e(\psi_1 + \psi_2)}, \tag{3}$$

both ψ_1 and ψ_2 contribute (both are non-zero for $\mathbf{x}$ in this region).

Now we consider the case where the detector is in place on path 1 as shown in fig.1. Then, instead of eqs. (1) and (2) we will have

$$\Psi = 2^{-1/2}(\psi_1 + \psi_2)\phi_0 \tag{4}$$

$$\rightarrow 2^{-1/2}(\psi_1\phi_E + \psi_2\phi_0) \tag{5}$$

$$\rightarrow 2^{-1/2}(\psi_b\phi_E + \psi_a\phi_0). \tag{6}$$

Note that the last result here is derived from the linearity of quantum mechanics together with the fact that the state $\psi_1\phi_E$ clearly evolves into $\psi_b\phi_E$, and similarly $\psi_2\phi_0$ evolves into $\psi_a\phi_0$.

Since the last equation correlates the detector state with the path of the electron, it follows that if we measured the state of the detector and the position of the electron we would find, for example, that with the electron on the path P_b the detector would always be excited, i.e., "detect" the electron. If the measurement of the state of the detector had been made prior to the measurement of the electron position we would tend to describe this as saying that the electron had passed through the detector and then gone straight on to path b. This is a rather loose way of speaking of course because orthodox quantum theory does not have particles moving along trajectories. Note also that if we measured the state of the detector *after* we had measured the position of the electron, whereas we could, if we wished, describe the situation in the same way, we would not know that an earlier measurement of the state of the detector might not have produced different results.

We now discuss the Bohm model in the presence of the detector. To do this we need an additional boundary condition, namely, the initial position of the detector particle d, say $\mathbf{y} = \mathbf{y}_0$. The path of the electron is not altered by the presence of the detector at least until it reaches the interference region I (given the above assumptions on the dynamics of the interaction). For example, if it is following the path P_2, with the wavefunction as in eq.(5), its velocity will be given by

$$\dot{\mathbf{x}} = Re\frac{\mathbf{p}_x(\psi_1\phi_E + \psi_2\phi_0)}{m_e(\psi_1\phi_E + \psi_2\phi_0)}$$

$$= Re\frac{\mathbf{p}_x\psi_2}{m_e\psi_2}, \tag{7}$$

since ψ_1 is zero when the electron is on path P_2, exactly as if there were no detector.

When the electron reaches the interference region, however, this is no longer true because then both $\psi_1(\mathbf{x})$ and $\psi_2(\mathbf{x})$ are non-zero. What happens depends on the values of $\phi_E(\mathbf{y})$ and $\phi_0(\mathbf{y})$ at the appropriate value of $\mathbf{y}$. In fact this will be still equal to $\mathbf{y}_0$, since its motion will have been governed entirely by the state ϕ_0 (as we can see from eq.(5) and the knowledge that the electron is on path P_2).

There are now two separate cases to consider. First, we suppose that the lifetime of the excited state of the detector is large compared to the time taken by the electron to travel through the interferometer. Then the state ϕ_E will still essentially be confined to the same region of space as the bound state, ϕ_0, until after the electron has passed through the region I. (This means that the "measurement" is made after the electron has passed through the interferometer.) In this case it is clearly possible to choose a value of $\mathbf{y}_0$ such that $\phi_E(\mathbf{y}_0) = \phi_0(\mathbf{y}_0)$. It is reasonable to assume that the argument below works when this is only approximately true, but detailed calculations are necessary to establish the precise regions (see [4]). Then we can cancel the ϕ states in eq.(6), thereby obtaining eq.(3), and showing that the electron moves as if the detector were not present, i.e., it will again follow the path $P_2 \rightarrow P_b$. The subsequent path of the detector particle will then be determined by

$$\dot{\mathbf{y}} = Re\frac{\mathbf{p}_y(\psi_b\phi_E + \psi_a\phi_0)}{m_d(\psi_b\phi_E + \psi_a\phi_0)} \tag{8}$$

$$= Re\frac{\mathbf{p}_y\phi_E}{m_d\phi_E}, \tag{9}$$

where we have used the fact that $\psi_a = 0$ for the path P_b.

In other words the detector particle is moving as if it were in the excited state, even though the electron did not pass through the detector. The particle guided by eq. (9) will move away from the detector, which will be seen to have "fired". Thus, the detector appears to have seen the electron, even though the electron has not passed through it. Englert et al. claim that this shows the Bohm trajectories are not real.

On the contrary we believe that this is yet another illustration of the contextuality of measurements; the detector in the interferometer, as in fig.1, does not behave in the same way as an isolated detector, and it can fire when the electron passes on the other arm of the interferometer. This is the effect of the quantum potential, which here knocks the detector particle out of the detector. It is a simple matter of *calculation* to see that the fact that the electron goes the other way does not imply that the detector is unaffected. (There is nothing particularly remarkable about this; even in classical physics theoretical input is required to interpret the readings of a measuring device.) Other examples of the effect of, so-called, empty waves have recently been given[6]. They are a calculable consequence of the Bohm model.

Finally we note that in the alternative situation where the lifetime of the detector excited state is small, so that the measurement is made prior to the electron reaching the interference region, then in the case above, $\phi_E(\mathbf{y}_0)$ will be zero when the electron reaches this region, so that only the ψ_2 part of the electron state will be effective and the electron will follow path P_a. Consequently the detector will not be seen to have fired, and the curious effect of its firing without an electron will not occur.

3. A RELATIVISTICALLY INVARIANT BOHM MODEL

The fact that the Bohm model is explicitly non-local and also non-Lorentz-invariant is often used as an argument against it. Certainly for those of us trained to believe that the fundamental equations of physics have to be invariant under Lorentz transformations, this argument carries some force, even though, as John Bell showed us[7], the non-locality is a necessary feature of *any* hidden-variable model (for a similar result relating to the lack of Lorentz-invariance see Hardy[8] and Hardy and Squires[9]). Nevertheless it is important to realise that the non-locality is intrinsic to quantum theory, independent of any particular interpretation. A realistic interpretation *exposes* this fact rather than creates it. As is well known the non-locality has its origin in the fact the a quantum wavefunction lives in configuration space, i.e., depends on many positions.

The standard reaction to the clash between quantum theory and locality, encouraged by experiments like those of Aspect et al.[10], has been to abandon locality and preserve the results of quantum theory. However, as we shall see, the Bohm model rather naturally suggests the opposite procedure of preserving locality (and Lorentz invariance) at the expense of all the predictions of quantum theory. Such a procedure is not ruled out by the experiments unless it can be shown that these really do discriminate between the two possibilities. The calculation given below suggests that present experiments fail to do this.

Our starting point is that in the Bohm model the wavefunction is something that gives a potential. Then there is nothing particularly noteworthy in the fact that it is defined in configuration space; the same thing is true for example of the potential of classical (non-relativistic) electrostatics and gravitation. In the Bohm model, as in these classical theories, we are interested in the values of the functions only at the positions of particles. The effect of special relativity on the potential of classical electromagnetism is to replace the positions of particles at some common time by the positions at a retarded time. Here we shall use exactly the same procedure in the Bohm model. Thus, in the standard velocity equation of the Bohm model

$$\dot{\mathbf{x}}_i(t) = Re\frac{\mathbf{p}_1\psi(\mathbf{x}_1(t_1),\mathbf{x}_2(t_2),..)}{m_1\psi(\mathbf{x}_1(t_1),\mathbf{x}_2(t_2),..)}, \tag{10}$$

instead of putting $t_1 = t_2 = ... = t$, we use retarded times defined by

$$t_k = t_i - \frac{|\mathbf{x}_i(t_i) - \mathbf{x}_k(t_k)|}{c}. \tag{11}$$

In order to see how the resulting model disagrees with quantum theory we consider a very idealised experiment in which a "photon" is emitted from some source and travels to a pair of detectors. We work in one space dimension and suppose that the detectors are on opposite sides of the source. Clearly when a photon is emitted it should, according to orthodox quantum theory, be detected by one, and only one, detector. We follow the standard procedure in the Bohm model of assuming that there is no trajectory (hidden variable) for the photon, so that the choice of which detector records it is determined by the exact positions of the detector particles, as we shall see below.

We suppose that the detectors consist simply of free particles, initially at rest with gaussian wavepackets. If a photon reaches one of the detectors it causes it to acquire a momentum in the direction of the photon, which is then absorbed. Thus the photon is detected when the detector particle moves. Explicitly, we can consider a wavepacket initially centered at $x = L$. After an interaction with a photon this becomes a moving wavepacket which we can write as

$$\psi(x,t) = \exp\Big[-\frac{i}{\hbar}\big(xp - \frac{p^2}{2m}t\big) - \frac{a}{2}\big(x - L - \frac{p}{m}t\big)\big(1 - \frac{i\hbar at}{m}\big)^{-1}\Big], \quad (12)$$

where p is the imparted momentum, which we take to be positive, and where the interaction is assumed to take place at $t = 0$. Here, and elsewhere, we omit the overall normalisation factor in the wavefunction. Since the detector particle is assumed to be macroscopic, we shall ignore the quantum factor $\left(\frac{i\hbar at}{m}\right)^{-1}$.

The velocity of the particle in this packet is easily seen to be given by

$$\dot{x} = \frac{p}{m}. \quad (13)$$

It is this velocity which is "seen" and corresponds to the photon having been detected.

Now suppose we have the situation of a photon wave in two parts, one travelling to the left and the other to the right. The state is a linear superposition of the two wave packets, which for simplicity we assume to have the same magnitude. We also suppose there are two detectors, a right-hand detector, as above, initially at position L, and an identical left-hand detector, initially centered at $-L$. After the photon wave has passed through the detectors, and the photon has been absorbed, the state of the detectors will be a linear superposition:

(left-hand detector still at rest and right-hand detector with a momentum p) plus (left-hand detector with momentum $-p$ and right hand detector at rest).

When we use the Bohm equation to give the velocities of the detector particles, we find (see [11] for details):

$$\dot{x}_1 = p/m,\ \dot{x}_2 = 0,\ \text{if } (x_1 - L) > -(x_2 + L), \quad (14)$$

$$\dot{x}_2 = -p/m,\ \dot{x}_1 = 0,\ \text{if } (x_1 - L) < -(x_2 + L). \quad (15)$$

Thus, one and only one, detector records the photon; which one depending on the relative initial positions of the detector particles. Note that this clearly shows that anticoincidence does not need actual photon trajectories. It also demonstrates the contextuality and non-locality of measurements in the Bohm model.

Now we consider what happens when we use the retarded positions in the velocity equations, i.e., as in eqs. (10) and (11). Initially both detector particles start to move (or not) exactly as they would if there was only one detector. Eventually however, the two detectors affect each other, and we find again that one and only one detector records the photon unless

$$|(x-1-L)+(x_2+L)| < \frac{pT}{m}, \tag{16}$$

where T is the time for a light signal to travel from one detector to the other.

Thus there is a (small) region of starting values (where both particles have similar initial positions relative to the center of their wavepackets) for which "wrong" results occur. We now need to use this expression to see to what proportion of wrong results it corresponds.

We suppose that a measurement is completed when the pointer has moved a distance K times the original position uncertainty in its wave packet, which is of the order of $a^{-1/2}$. We might expect a value of K around 10 to be reasonable. The time this takes is the time, T_M, for the measurement to be complete, and is given by

$$T_M = \frac{mK}{pa^{1/2}}, \tag{17}$$

where we have used eq.(13) to give the velocity. Then, from (16), it follows that wrong results occur whenever

$$|(x-1-L)+(x_2+L)| < \frac{KT}{a^{1/2}T_M}. \tag{18}$$

Now typical values of the right-hand side of this inequality are $a^{-1/2}$, so the condition for very few wrong results to occur is that

$$\frac{T}{T_M} \ll K. \tag{19}$$

Given that the distances between the detectors in the Aspect, et al., experiments were of the order of a few metres, it is clear that very few wrong results would have been seen unless the macroscopic detection "particle", i.e., whatever plays the role of the pointer, had moved significantly in times less than about 10^{-7} sec. Hence it is reasonable to claim that the experiments are unlikely to have been able to distinguish between the standard non-local Bohm model and our local version given in eqs. (10) and (11). As with Lorentz invariant versions of collapse models [12], experiments carried out over distances one or two orders of magnitude larger than presently used seem to be called for.

Two critisisms of the above calculation should be noted. First, we have conveniently dropped the photon field function on the assumption that the photon is absorbed by the detectors. However, it should be possible to analyse the situation with only one detector present. Then the photon wave in the other direction would not be absorbed, so there would be no unique equation for the particle trajectory. This seems to be a probelm with the Bohm model if photon trajectories are not allowed. The problem would not arise if a model experiment involving a spin measurement were made, and it would be interesting to try the retarded model in such a situation.

Secondly, the explicit time dependence of the wavefunction was omitted from eq. (10), and it is not obvious how it should be included so that the Lorentz invariance is preserved. The answer may lie in the fact that at the fundamental level there is no explicit time dependence, as would be suggested for example by the Wheeler-deWitt equation.

To conclude it is worth mentioning an interesting application of these ideas to the decay of black-holes. Recently there has been considerable interest associated with the apparant loss of information associated with one of a pair of "pair-produced" particles falling into a black hole, with the result that (so it is claimed) a pure state turns into a mixture, i.e., that corresponding to radiation from a black body at a given temperature. I am not aware of any calculations of this effect using the Bohm model, but such calculations would be interesting. The model is deterministic, so it is hard to see how there can be any loss of information. This is particularly the case if the above Lorentz invariant from, since one particle would see the other on its backward light-cone, and it would never actually disappear into the black-hole. Calculations of this effect are in progress.

REFERENCES

1. D.Bohm, *Phys. Rev.* **85** (1952) 166, 180.
2. M.L.G.Redhead, *Incompleteness, Non-locality and Realism* (Clarendon, Oxford, 1989).
3. B.Englert, M.O.Scully, G.Sussman, and H.Walthur, *Z.Naturforsch* **47a** (1992) 1175.
4. C.Dewdney, L.Hardy and E.J.Squires, *Phys. Lett.* **A184** (1993) 6.
5. C.Dewdney and B.J.Hiley, *Found. Phys.* **12** (1982) 27.
6. L.Hardy, *Phys. Lett.* **A167** (1992) 11.
7. J.S.Bell, *Physics* **1** (1964) 195.
8. L.Hardy, *Phys. Rev. Lett.* **68** (1992) 2981.
9. L.Hardy and E.J.Squires, Phys. Letters, **A168** (1992) 169.
10. A.Aspect, P.Grangier, and G.Roger, *Phys. Rev. Lett.* **49** (1982) 91.
11. E.J.Squires, *Phys. Lett.* **A178** (1993) 22.
12. E.J.Squires, *Phys. Lett.* **A145** (1990) 297.

ON THE EMPIRICAL LAW OF EPISTEMOLOGY: PHYSICS AS AN ARTIFACT OF MATHEMATICS

Nikos A. Tambakis

Interdisciplinary Research Group
Department of Physics, University of Athens
Athens, Greece

The philosophy of physics and the research of its foundations usually takes for granted that physics is written and created by mathematics. Why, though, should mathematics correctly represent the physical world? And what if this does not really occur? These questions were dealt with by Eugene Wigner, who thus formulated the empirical law of epistemology (ELE). However, excessive reliance on the effectiveness of mathematics may give rise to some ideologies in quantum mechanics (QM), such as that a theory of everything is possible. Against this supposed completeness of any physical theory an argument is given in two steps. First, the actual development of physics suggests that any theory refers only to a part of a greater structure. Its completeness is then refuted for either a finite or an infinite universe from a set theoretical or Gödelian point of view, respectively. The absence of a one-to-one correspondence between physical and mathematical entities also calls for some reservations with respect to the ontological status of physico-mathematical structures and the natural philosophy of the Copenhagen interpretation of QM. For example, the "projection postulate" recast in the density operators' representation, rather than that of state vectors, loses its appearance as a "postulate."

Key words: empirical law of epistemology, ideologies in quantum mechanics, completeness, projection postulate.

1. INTRODUCTION

"...The miracle of the appropriateness of the language of maths for the formulation of the laws of physics is a wonderful gift which we neither understand nor deserve ...." [1]

M. Ferrero and A. van der Merwe (eds.), Fundamental Problems in Quantum Physics, 321–327.

The above well-known quotation is from E. Wigner's famous paper "The unreasonable effectiveness of mathematics in physical science," in which he proceeds to give the "wonderful gift" the name "empirical law of epistemology" [ELE] [2]. At this point it is quite possible that the non-philosophically inclined could pose some naive questions, such as:

(a) Is this "effectiveness of mathematics" really a *gift*? (And if so, from whom?)

(b) Is it possible that this "effectiveness" is *reasonable* instead of unreasonable? In other words, is it possible that the ELE is not at all empirical, but is rather based on some deeper unknown causes?

In this short paper it is not my intention to deal with the above questions, although I regard them as important [3]. I shall try, instead, to concentrate only on certain negative aspects of ELE, since the positive ones—marvellous predictions and retrodictions—have so often been exalted by Wigner and many others [4]. In thinking, however, about the negative aspects of ELE, I shall not go as far as to ask, together with Alfred Whitehead, whether perhaps the whole of mathematical physics is in fact a "huge fairy tale"! [5] I shall try only to point out that excessive reliance on the "effectiveness" of mathematics in quantum mechanics (QM) may give rise to some ideologies, i.e., the traditional double one, that nature is in its heart both probabilistic and complementary, and more recently, that a theory of everything is possible.

I shall deal with the most recent one first.

2. ON THE INCOMPLETENESS OF QM AND PHYSICS IN GENERAL

2.1 Descriptions and Explanations

The strong tendency to build physics as a pure mathematical science has been explicitly expressed in a famous programme in Paris in 1900. David Hilbert suggested as the sixth of 23 problems "to treat those physical sciences, in which mathematics play an important role, by means of axioms, like geometry."

It is, of course, well known that Hilbert's programme partially collapsed in 1934 because of Gödel's theorem. This fact was quickly assimilated by mathematicians, new terms like Gödelisation, Gödel numbers, etc., entered into mathematical jargon, and the Gödelian view has had a profound effect on the subsequent development of mathematics [6]. I think, however, that, more than half a century later, physicists are rather unwilling to take it up. This is implied at least when it is said that we "approach the end of physics" or that some "theory of everything" is within our reach [7]. There is a feeling of completeness of physics, and especially of QM, against which I shall argue from two different points of view.

The first is a human view of the development of physics, according to which the history of physics consists, very roughly, of the interplay between *description* and *explanation*.

Think, for example of the world view of the earth carried by an elephant, and the subsequent world view *explanation* in which a turtle is added underneath the elephant. Translating this zoo into modern terms, we may say that a transition from description to explanation follows the scheme:

$$\text{description } (a, r) \rightarrow \text{explanation } (a, a', r, r'),$$

where physical entities a and their relations r are supplemented by new ones, a' and r', under the condition: $r' \rightarrow r$ (r's enjoy the name of "theoretical laws.")

A typical example of such a process is represented by the transition from Keplerian astronomy to Newtonian, which included such impressive new physical entities as the sun's mass and, of course, the Newtonian laws imply the Keplerian. Turning to the question of whether, from today's point of view, the Newtonian gravitational law is indeed an explanation, let me quote R. Feynman: "Previously the angels pushed the planets towards the center, now along the trajectory!" [8]

The moral seems to be that every description turns into a wider explanatory scheme, which later appears as a description itself looking for a new explanation, and so on. Of course, the whole picture I give corresponds to the old traditional scheme of the development of science as an ever-expanding knowledge aiming towards some absolute truth. It certainly misses the more abrupt and erratic evolution of science which does not follow the above simpler scheme of successive inclusion [9]. My point, however, is that in any scheme of either normal or revolutionary science, every world view theory T_w refers admittedly [10] only to a *part w of a greater structure U*, absolute or not.

We are now ready to confront at a more abstract level the question of completeness of a tentative universal theory T_u.

2.2 Every Physical Theory is Incomplete

Let us, then, suppose that T_u is complete. The claim that we possess complete knowledge of an *unobservable* universal structure implies that the theory T_u is *isomorphic* to some theory T_w, which is the best and broadest theory that we have already mastered. Hence, T_w should also be a complete theory and its object should perfectly reflect every aspect of the universal structure U. In a historical context, W would then be a "Leibnizian unit" mirroring the whole universe. We now go on to sketch a refutation of the Leibnizian universe.

Let us first consider the easier case of a finite U. It is then a well known fact of set theory that there cannot be a complete structure U isomorphic to any of the (complete) proper substructures.

However the infinite U case is not that simple at all. The modern theory of categories establishes that under certain circumstances an infinite mathematical structure U can indeed be arranged in Leibnizian units. In the case of a *physical* infinite system U, however, we can argue against the completeness of its theory T_u from a second point of view, a Gödelian one: *Every consistent theory is incomplete.* A corollary of this proposition is even more disturbing for our physical theories: For an infinite consistent theory there are non-isomorphic models.

Summarizing, then, our argument against completeness, we note that a theory of everything, with a quantum one as its best candidate, is highly improbable. Such attempts are wonderful creations not of physics but of "*physics as an artifact of mathematics.*" [11]

At this point it is of interest to note that the incompleteness of all physical theories has been used as an argument in favour of an *inherent indeterminism* of the quantum world. Herman Bondi, for example, used the idea of incompleteness to support the probabilistic Copenhagen interpretation of QM [12]. It seems to me, however, that this line of argument confuses methodology of science, where the interpretation questions belong, with natural philosophy, which has to do with the indeterminism problem. Such arguments, which do not come from quantum theory itself, attempt to support one branch of traditional ideology surrounding the Copenhagen interpretation, namely, the inherent determinism of nature.

Let us then take a look at the whole tree of Copenhagen ideology.

3. THE "REDUCTION OF THE WAVE PACKET" AS AN ARTIFACT OF MATHEMATICS

Complementarity enters the Copenhagen philosophy as a third constituent in order to adequately modify, or perhaps to bedevil even more, its two main views, namely:

(1) that nature is inherently probabilistic and
(2) that the common sense spacetime description is impossible.

Shortly after its appearance, the complementarity principle acquired an autonomous existence and invaded other disciplines, scientific as well as popular. There is a famous topic of the standard QM where the above three constituents of Copenhagen philosphy make a common appearance. This is known as the "collapse of the state under Ψ," "reduction of the wave packet," or "projection postulate."

To begin with, it seems to me that the traditional teaching of QM goes along the same lines and principles settled more than half a century ago, though an almost magical aura envelops the concept of Ψ collapse. (And one might say that even the very word "collapse" suggests that something dramatic takes place in the dynamics.) This magic perpetuates through generations, since every student stumbles upon a variety of possible reasons for the Ψ collapse. In the best case, they learn that there is a "projection postulate" of von Neumann, which presents the reduction as an axiom of QM.

It is, however, well known that there exists by now some explanation and almost proof of this so-called "postulate," even in the framework of orthodox QM. I say "almost proof" since it works easily only for a particular class of measurements, the so-called "ideal." [13] This kind of measurement (already discussed by von Neumann) is, in spite of its name, met in such common situations as the partial polarisation of spins when a Stern-Gerlach apparatus is used. What is necessary for this restricted proof of the projection postulate is some extension of the traditional concept of quantum state and its replacement, at least for some problems, wtih the notion of *density operator*.

In this approach, a "state" does not concern our information on an individual system but rather that on a statistical ensemble of systems. More precisely, a quantum state is *defined* as the set of the expectation values $<\hat{A}>$ of all observables-operators $\hat{A}$. From this definition it follows that states may be adequately represented by density operators which become a principal concept, whereas states are now derived from them instead of the traditional converse way. The proof of the projection postulate is based on the fact that for an ideal measurement, the off-diagonal blocks of the density operator $\hat{\rho}$ become zero. This follows formally from some rather natural assumptions, e.g., that the variance of an observable is nonnegative and that two commuting observables can be measured together [14].

An explanation of the wave-packet collapse that is based on the vanishing of the off-diagonal blocks of $\hat{\rho}$ is also found in quantum statistical mechanics, whereas this vanishing is based on the interaction of the quantum system with the measurement apparatus, which has an infinite number of degrees of freedom. Alternatively, the cancellation of the off-diagonal blocks of $\hat{\rho}$ may be shown by averaging phase factors in standard QM based on the state concept.

I'm not going to delve into the advantages of density operators instead of state vectors, but I would like to remind you that in the mathematical formulation which uses state vectors only, two systems that have interacted and correlated in the past can, in general, no longer be considered separately. It is, then, worthwhile to notice that Landau, as early as 1927, suggested representing a state by a density matrix. Much more recently, in 1963, Wigner emphasized the difficulties which arise in the case of partial polarisation of spins as in a Stern-Gerlach experiment. This kind of measurement also needs a purely probabilistic formulation as expressed by density operators [15].

However, I would like to stress that, in many cases, the suggested use of $\hat{\rho}$ instead of Ψ should not be considered as an exclusive answer to the question of a suitable description of quantum effects in general. Density operators simply offer a *representation* of quantum phenomena (and to my mind a better one in most cases than state vectors) but of course, there are other representations as well, for example, the "Wigner representation," the "polarisation vector representation" and so on, *each one more suitable for some problems*. I think that keeping in mind the possibility of different representations may help us relax the somehow metaphysical picture of Hilbert space as the land of all state vectors (a picture which is subconsciously well grounded and feeds each successive generation of students). Different representations are simply possible because we use the mathematical notions of completeness and of a vector space, of which Hilbert space is just one of the possible manifestations.

4. CONCLUSIONS

Of course the above remarks are not restricted to the notorious case of Ψ reduction. More generally, there are no unique mathematical representations of physical phenomena, and of physical entities in particular. There is, instead, a polymorphism of physics to mathematics and the same physical entity accepts many different representations. There is also, of course, the reverse polymorphism of mathematics to physics. This is more trivial; for example, a vector may represent a static force, an e.m. field or a wave function. The conclusion is that there is no one-to-one correspondence between physical and mathematical entities. I think that this fact on its own calls for some reservations in our human tendency of rushing to give ontological status to our physico-mathematical structures; it may be that some of them are simply artifacts of our mathematics.

As a concluding remark, I would like to apologise for the rather crude term "ideologies of quantum theory" that I have sometimes used earlier. I intended this to be an antidote to the severe treatment imposed on many of us, i.e., that we have to renounce our basic categories of thought in order to understand physics! As far as I can see there is no empirical support for this kind of treatment [16]. Only indirect support is offered, that of the efficiency of the mathematical tools being used, i.e., that of successful predictions and retrodictions.

The point of this paper is precisely to suggest that this indirect evidence is vulnerable even from a quite general point of view. *The effectiveness of mathematics in the physical sciences does not always guarantee their ontologies and even more so, their natural philosophy.*

NOTES AND REFERENCES

1. E. Wigner, "The unreasonable effectiveness of mathematics in the natural sciences," in *Symmetries and Reflections* (M.I.T. Press, Cambridge, 1970), p. 237.
2. Ref. 1, p. 233.
3. N. A. Tambakis, "The empirical law of epistemology," unpublished thesis (in Greek), University of Athens, 1987.
4. See for example, R. L. Oldershaw, *Am. J. Phys.* **56**, 12 (1988).
5. A. N. Whitehead, *An Introduction to Mathematics* (Oxford University Press, Oxford, 1982) p. 33.
6. See, for instance, Morris Kline, *Mathematics, the Loss of Certainty* (Oxford University Press, Oxford, 1982), Chap. XII; E. W. Beth, *The Foundations of Mathematics* (North-Holland, Amsterdam, 1959).
7. See, for example, S. W. Hawking, *Is the End in Sight for Theoretical Physics?* (Cambridge University Press, Cambridge, 1980).
8. R. Feynman, *The Character of Physical Law*, (M.I.T. Press, Cambridge, 1967).
9. As described in the classic of T.S. Kuhn, *The Structure of Scientific Revolution* (Chicago University Press, Chicago, 1962).
10. The supposed existence of "black matter" points out that, perhaps, we are, at the moment, touching only 10% of the elephant!
11. Such a view of physics is presented by N. Cartwright, *How the Laws of Physics Lie* (Clarendon, New York, 1983).
12. H. Bondi, "The lure of completeness," in *The Encylopedia of Ignorance*, R. Duncan et al., eds. (Pergamon, Toronto, 1977).
13. Defined by *repeatability* of measurements of the same quantity and *compatibility* of simultaneous measurements of commuting observables. See R. Balian, *Am. J. Phys.* **57**, 111 (1989).
14. See Ref. 14.
15. E. Wigner, *Am. J. Phys.* **31**, 6 (1963).
16. F. Selleri, "Quantum paradoxes and physical reality," *Philosophia Naturalis* **27** (1) (1990); M. Ferrero et al., *Am. J. Phys.* **58**, 7 (1990).

15. E. Wigner, Am. J. Phys. 31, 6 (1963).
16. F. Selleri, "Quantum paradoxes and physical reality,"

SEARCH OF A FIRST PRINCIPLE FOR QUANTUM PHYSICS

Alberto Clemente de la Torre[1]

Departamento de Física
Universidad Nacional de Mar Del Plata
Funes 3350, 7600 Mar Del Plata, Argentina

A classification of physical systems is presented, suggesting the adoption of the law that imposes a lower bound to the action, as a first principle for quantum theory. Many of the essential features of quantum theory can be traced back to that principle, but a systematic generation of the formalism remains to be found. In this approach, a formalism for quantum theory, characterized by a continuous transition to classical physics, allowing thereby an unambiguous interpretation, is sought.

Key words: action, classical-limit, uncertainty, contextuality, nonseparability, spin.

1. INTRODUCTION

One of the reasons why quantum mechanics has not yet found a satisfactory interpretation, is that no "first principle" has been proposed from which the whole theory could be derived. A theory derived from a first principle poses no serious problem in its interpretation because, once the principle is accepted, he is made responsible for all logical consequences that it may have. This is, for instance, the case of special relativity. Perhaps the best way in the attempt to find an interpretation of quantum mechanics is to propose some first principle from which the theory could be derived with a smooth transition to classical mechanics.

2. A CLASSIFICATION SCHEME

Classifying every known physical system according to their characteristic velocity and action, we find what seems to be, two fundamental laws

M. Ferrero and A. van der Merwe (eds.), Fundamental Problems in Quantum Physics, 329–340.

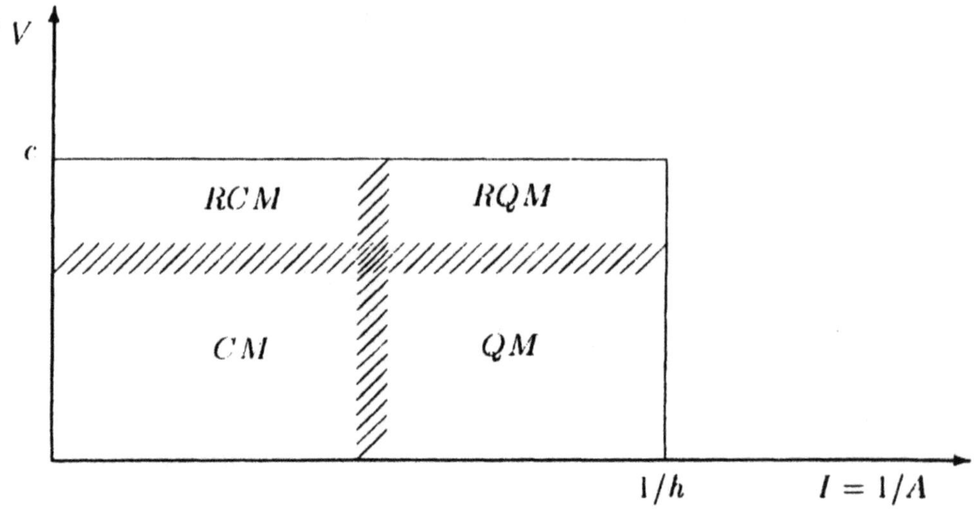

Fig. 1. The velocity - inaction, $(V - I)$ diagram.

of nature: $V \leq c$ and $A \geq \hbar$. Representing each system by a point, or region, in a diagram spanned by two axes corresponding to velocity and to the inverse of the action (inaction?), we find that all known physical systems fall within the rectangle limited by the axes, by $V = c$ and

$1/A = 1/\hbar$, as shown in Fig. 1. Physicists have the dream, or prejudice, that a theory will sometime be developed, appropriate to study every existing physical system. We are far from the realization of such a dream, but we have very good approximations to it, valid in some limited regions shown in the diagram. Physical systems with characteristic action much bigger than $\hbar$ and velocities much smaller than c, that is, those located in the region marked by CM in the diagram, are well described by classical mechanics. Similarly, we find the regions corresponding to quantum mechanics: QM, relativistic classical mechanics: RCM, and relativistic quantum mechanics: RQM. These regions have marked differences in the formalism and interpretations of the theories invoked. The "classical" sector, relativistic or not, is characterized by a closed formalism possessing a well defined interpretation. On the contrary, the quantum sector is lacking a definite interpretation for a formalism; well defined in the nonrelativistic case but still problematic in the relativistic case. Another important difference is that whereas CM and RCM are connected in a "smooth" manner, there is an abrupt discontinuity between CM and QM. The smooth connection between CM and RCM, represented by $\lim_{c \to \infty}(RCM) = CM$, means that, taking the limit for $c \to \infty$ in any formula of RCM, we obtain a valid relation in CM, and similarly, every concept of mass, energy, space and time, etc., are smoothly connected between both theories. Nothing of the sort happens between CM and QM. Even though their experimental predictions are related by Ehrenfest theorem, both theories are not continuously connected. This abrupt transition is precisely the reason why we have not been able to produce an interpretation of quantum mechanics in terms of classical concepts, or of any conceptual frame connected "in the limit" with classical concepts.

3. A FIRST PRINCIPLE

Is it possible to mock the smooth transition from CM to RCM going "up" in the diagram to develop, by going "right" in the diagram, a quantum theory that could be continuously connected with CM? Is it possible to postulate some type of specular symmetry in the diagram across the diagonal? Could $A \geq \hbar$ play, for quantum physics, the same rôle that $V \leq c$ has played for special relativity? in other words, is $A \geq \hbar$ a first principle for quantum physics?

Let us take then $A \geq \hbar$ as a first principle and let us find out, what else we *must* accept in order to avoid the apparently trivial ways to violate the principle. Before we do that, we must extend the concept of *action* such that it may be understood as some property of the physical system. we want to consider the action for a given interval $[t_1, t_2]$ as a dynamic property of the system determining its evolution, in the same way that velocity was taken to be an intrinsic quality of the system, generating its relativistic kinematic. We are then allowed to say that the system *has* some value of action, as we say that a particle *has* some velocity. Using standard notation, the classical action defined by,

$$A = \int_{t_1}^{t_2} \mathcal{L}(q_k, \dot{q}_k, t)\, dt \ , \tag{1}$$

can be written, for N noninteracting degrees of freedom, that is, for a hamiltonian $H = \sum_{k=1}^{N} p_k^2/2m$, as

$$A = \sum_{k=1}^{N} A_k \ , \text{with } A_k = \frac{1}{2}\int_{q_{k1}}^{q_{k2}} p_k dq_k. \tag{2}$$

4. UNCERTAINTY RELATIONS

How can the absolute value of the classical action be made vanishing small? In the equations above we can recognize two ways in which the classical action can vanish. The first way appears when we choose vanishing intervals of integration. In fact, the classical action vanishes if we simply take $t_1 \to t_2$ or $q_{k,1} \to q_{k,2}$. In a second way, the classical action may vanish due to canceling contributions from different degrees of freedom. If we want to forbid the vanishing of the action, then we must make, at any cost, the two ways mentioned above, impossible. This will certainly lead to some consequences. In order to avoid the first way of vanishing, let us consider the contribution to the action due to one degree of freedom, as shown in Eq. (2). We may here postulate that the limit $q_{k,1} \rightarrow q_{k,2}$ is physically meaningless. That is, there is a lower bound for any relevant space interval which we denote by Δq_k. Therefore the contribution of the degree of freedom to the action will be at least $\frac{1}{2}p_k\Delta q_k$. Now p_k should not be allowed to take an arbitrarily small value because this could violate $A_k \geq \hbar$. If we accept that there is a minimum interval for the coordinate q_k, then there must also be a minimum interval for the momentum p_k, since there is a canonical transformation $q \rightarrow p$ and $p \rightarrow -q$ where the rôles of coordinates and momentum are permuted. Let this minimum momentum interval be Δp_k. Placing then this minimum value in the expression above, and requiring that the so defined minimal contribution of a degree of freedom to the action should not be less than $\hbar$, we obtain

$$\Delta q_k \Delta p_k \geq \hbar \ . \tag{3}$$

The impossibility to define an arbitrarily small interval for a coordinate or momentum, implies that the numerical value assigned to it can not have a larger precision than the interval itself. This value must be then *uncertain* within this interval, giving to the expression above a similar meaning as the usual uncertainty relations in quantum mechanics. There are however differences with some interpretations that ought to be mentioned. Here, the uncertainty relations do not arise as a consequence of some perturbation of the system due to some observation. The uncertainties do not arise from some statistical dispersion but are instead, a consequence of an inherent indefiniteness in the properties of the system. The first consequence, then, of assuming $A_k \geq \hbar$ is that, in contrast with classical mechanics where the observables are assigned precise numerical values, quantum observables are assigned not only a numerical value but also an uncertainty, or an interval of indefiniteness. In a similar way as above, a time-energy uncertainty relation can be derived from the nonvanishing of the action.

5. NONSEPARABILITY AND CONTEXTUALITY

Besides the correlations between coordinates and their conjugate momenta, characterized by Heisenberg uncertainty relation, there is another type of quantum correlation involving nonconjugate pairs of observables. They are characterized by the concepts of nonseparability and contextuality that, although they were "discovered" in the 60's, they were already hidden in Schrödinger's version of the uncertainty relations [1,2,3]. We will now see that this second type of correlations can also be traced back to the nonvanishing of the action. In order to illustrate this, let us consider a system with two noninteracting degrees of freedom. Let q_1, q_2, p_1, p_2 be the generalized coordinates and momenta. Let us imagine now another set of coordinates Q_1, Q_2, P_1, P_2 related to the previous ones by

$$Q_1 = q_1 - q_2 \,, \qquad P_1 = (p_1 - p_2)/2 \,, \qquad Q_2 = (q_1 + q_2)/2 \,, \qquad P_2 = p_1 + p_2 \,.$$

Let us imagine that at some time t, the system is prepared in a state characterized by some precise values for two of these last coordinates. That is, by $Q_1 = D$ and $P_2 = P$ with uncertainties $\Delta Q_1 = \epsilon$, and $\Delta P_2 = \delta$. These uncertainties are small and numerically of the same order of magnitude. We have no theory to relate precisely these uncertainties in the new coordinates with the corresponding ones in the old coordinates. However it is reasonable to assume that they are related quadratically, as if they were statistical errors. Therefore, we have

$$\Delta^2 q_1 + \Delta^2 q_2 = \epsilon^2 \,, \qquad \Delta^2 p_1 + \Delta^2 p_2 = \delta^2 \,.$$

As we have already seen, the uncertainties in the coordinates and momenta are constrained by an inequality, avoiding the vanishing of the action associated to the degree of freedom in question. Then, squaring the uncertainty relation, we find

$$\Delta^2 q_1 \Delta^2 p_1 \geq \hbar^2 \,, \qquad \Delta^2 q_2 \Delta^2 p_2 \geq \hbar^2 \,.$$

We can now combine these two inequalities, consequence of the nonzero-action principle, with the equations above, consequence of the particular state in which the system is prepared:

$$\delta^2 = \Delta^2 p_1 + \Delta^2 p_2 \geq \frac{\hbar^2}{\Delta^2 q_1} + \frac{\hbar^2}{\Delta^2 q_2} = \hbar^2 \frac{\Delta^2 q_1 + \Delta^2 q_2}{\Delta^2 q_1 \Delta^2 q_2} = \frac{\hbar^2 \epsilon^2}{\Delta^2 q_1 \Delta^2 q_2} \,.$$

Therefore, we obtain

$$\Delta q_1 \Delta q_2 \geq \hbar \frac{\epsilon}{\delta} \approx \hbar \,. \tag{4}$$

We see here that the nonvanishing of the action imposes a classically unexpected correlation between two degrees of freedom. These correlations are different from the correlations found before, between canonical conjugate variables, in the sense that they appear for some particularly prepared states. Identifying the two degrees of freedom q_1 and q_2 with the coordinates of two particles moving in a line, we obtain precisely the system and preparation used in the EPR argument. That is, two particles in a state with relative position D and total momentum P. The argument presented, becomes then the EPR argument and here again, the system leads to the surprising nonseparability. If, instead, we identify the two degrees of freedom q_1 and q_2 with the orthogonal coordinates of one particle moving in a plane, we are led to the, less dramatic but as surprising, concept of contextuality. The equivalence of these two identifications has been exploited [4] in order to relate nonseparability with contextuality.

6. SPIN AND ORBITAL ANGULAR MOMENTUM

What about spin? Can the principle $A \geq \hbar$ somehow account for this essential feature in the quantum mechanical description of particles? Let us consider a particle in its rest frame. Well, we have already seen that a quantum particle at "rest" is far from being "quiet". In fact, $A \geq \hbar$ implies that the location of the particle must be uncertain within a region ΔR, meaning that for the particle there is no "here" and "there" within this region. Similarly, the particle can not be considered at a sharp instant of time. Within a time measure, the particle has no "past" and "future" but has, instead, a "present" of duration ΔT. Momentum and energy are also affected by an inherent uncertainty combined with the other uncertainties in a way that the action is, at least, $\hbar$. A classical particle at rest becomes schizophrenic in the quantum description. What is now the rotational state of this particle? Choosing an axis, in an arbitrary direction, we estimate with equation (2) the contribution to the action, due to the rotational degree of freedom, $dA = \frac{1}{2}S\, d\theta$. Where S is the intrinsic angular momentum (spin) and $d\theta$ is an angular sector. Let us find the least possible value for S, compatible with the least possible action $\hbar$. This will be obtained, taking the largest possible angular sector. Since the angle is taken "modulo 2π," the largest possible interval is $d\theta = \pi$. Therefore, we obtain

$$S = \frac{2\hbar}{d\theta} = \frac{2\hbar}{\pi} \,. \tag{5}$$

This is close to the value of the spin "1/2" of the usual formalism of quantum mechanics. Of course, in these imprecise arguments we can not claim exact results. This result, obtained for an arbitrary axis, should be valid for all axis because we require a nonvanishing rotational action in any direction, in a clear departure from the classical case, where the rotation around an axis, implies zero angular momentum with respect to another axis orthogonal to it. Here we must have $S = 2\hbar/\pi$ with respect to *all* axis. This mixture, of a classical rotational state with a (nonclassical) requirement $A \geq \hbar$, allows us to think about spin in a way where its coupling with the external rotational state, that is, orbital angular momentum is conceivable. In other interpretations, where a classical image is impossible, spin is taken as an intrinsic quantum number, like mass, electric charge or lepton number. But then, the coupling or addition of this "internal" quantum number with an "external" variable, like orbital angular momentum, becomes very mysterious. Now that we have established a rotational state along an axis, we may attempt to cancel this rotation, and thereby violate the proposed principle, by kinematic means. That is, rotating the reference frame by an appropriate angular velocity, or equivalently, imparting an orbital angular momentum to the particle. In order to avoid this canceling of the action, we must impose that the minimal possible, nonzero, orbital angular momentum must be *twice* the amount of intrinsic angular momentum, or spin. Thereby the spin will flip direction for the rotating observer, and the action will not vanish. In this manner, orbital angular momentum will never exactly cancel the intrinsic angular momentum of a particle. Therefore, the requirement $A \geq \hbar$ also explains the important feature that spin takes half integer values and orbital angular momentum takes integer values. We had no troubles explaining spin, but we may run into difficulties if we want to understand the case $S = 0$. The theory doesn't have scalar particles. After all, this might be no problem for the theory, because all presently known elementary (noncomposite) particles, quarks, leptons and gauge vector bosons, have nonzero spin. The standard theory of elementary particles predicts the existence of scalar Higgs particles but, in spite of intensive search, they have not been found. If they are found, then the present argument, anyway a nonrigorous one, will have to be revised in order to accommodate for them. There are nonelementary particles, that is, composed particles, like pions for instance (or maybe the Higgs), whose total angular momentum, arising from the addition of the spins of the quarks and gluons and their relative orbital motion, add to zero. These are quite complicated compound states, and perhaps the vanishing of the total angular momentum is not equivalent with the vanishing of the action. In any case, the treatment of such a complicated system requires the complete derivation of the theory. Intermediate vector bosons have integer spin and one may wonder if an orbital rotation, also an integer value, could cancel the action. Here again, a complete and detailed theory is needed to clarify the issue. In fact, we may speculate that for massive bosons, that is nonstable ones, their state can be considered to be a compound state of leptons or of quarks (their decay products) in a complicated state. For the massless photon an orbital canceling of the spin is kinematically inaccessible. Anyway, the photon could also be considered as a compound state of the annihilated particle-antiparticle pair that created it.

7. NONCOMMUTATIVITY

If $A \geq \hbar$ is a first principle from which quantum theory should be derived, and if the usual Hilbert space formalism of quantum mechanics is correct, then this formalism must somehow contain the above principle. Where is it? We will see that it is to be found in the noncommutativity of the algebra inherent to the Hilbert space formalism of quantum mechanics. In order to illustrate this, consider a simple system with just one noninteracting degree of freedom. Taking a short time interval dt, the action is given by Eq. (2) as

$$dA = \frac{1}{2} p \, dq = \frac{1}{2} p \dot{q} \, dt \; .$$

Writing this as a quantum mechanical operator equation, we obtain, using the time evolution equation for $\dot{q}$ given by the hamiltonian $H = p^2/2m$,

$$dA = \frac{1}{4} \frac{-i}{\hbar} \frac{1}{2m} \left(p^2 [q,p] + 2p[q,p]p + [q,p]p^2 \right) dt \; . \tag{6}$$

In this last expression we can see that the noncommutativity of p and q avoids the identically vanishing of the action dA. Furthermore, if the commutator $[q, p]$ is given precisely the value $i\hbar$, usually postulated as a quantization condition, then we recover the expected expression $dA = H \, dt$. The postulate $A \geq \hbar$ provides then a physical meaning to the noncommutativity of the formalism, an otherwise rather mysterious mathematical feature.

8. TOWARDS A QUANTUM THEORY

In all the heuristic arguments presented above, we have used the classical action and, quite brutally, imposed a lower bound for it. The conclusions drawn, support the idea that the nonvanishing of the action could play a significant rôle as a first principle for quantum theory. The rigorous formalization of this idea remains to be done but we can already imagine some of its features. Most probably an adequate mathematical tool will have to be developed in order to obtain a definition for the action, consistent with the results obtained before. In fact, if neither spatial and temporal coordinates nor energy and momentum, can be defined with arbitrary precision, then all action integrals are not mathematically well defined. Perhaps a new integration has to be defined, related to the usual one in a similar way as *fuzzy* set theory is related to set theory.

Regardless on how do we arrive at a precise definition for the quantum action, we may concentrate our attention at a particular feature of the classical action and question whether it may remain in the quantum action. Let us assume a physical system S that can be decomposed into two subsystems S_1 and S_2. For instance we can think on a system with N degrees of freedom where N_1 of them build the subsystem S_1 and the remaining N_2 degrees of freedom belong to S_2. Another important case of decomposition is to consider two contiguous intervals of evolution for the same system. That is, if S is a physical system in its evolution from t_1 to t_2 then S_1 could be the system in its evolution from t_1 to t_0 and S_2 the same in its evolution from t_0 to t_2. A similar decomposition is obtained considering intervals of evolution in the generalized coordinates. In all these cases, if we denote by A_1 and A_2 the corresponding action of the subsystems S_1 and S_2, respectively, then the *classical* action A of the compound system is such that

$$A = A_1 + A_2 \ .$$

This property should not remain valid for the quantum action because it could otherwise be used to violate the proposed principle $A \geq \hbar$. What is then the composition law for the quantum action? The requirement of a nonvanishing action implies that the action of a composite quantum system is not simply the addition of the corresponding action of the subsystems, as is classically the case. In the classical limit, however, we should recover additivity for the action. Here a similarity with special relativity is useful. Besides the velocity V, we define in relativity a quantity, the rapidity Y, that behaves like the classical velocity. That is, it is unbounded and additive like usual numbers. They are related by (with $c = 1$)

$$V = \tanh(Y) \ , \qquad Y = \frac{1}{2}\ln\left(\frac{1+V}{1-V}\right) \ . \tag{7}$$

Adding two rapidities $Y = Y_1 + Y_2$, we end up at the usual relativistic composition law for the velocity. Similarly, let us define a quantity W, the "activity", that should behave like the classical action. It is additive and it may vanish. It is related to the action A by

$$A = F(W) \ , \qquad W = G(A) \ . \tag{8}$$

In Fig. 2 we can see the qualitative relation of velocity versus rapidity and action versus activity. Now let us consider a system composed of two subsystems $S = (S_1, S_2)$ with their corresponding activities such that $W = W_1 + W_2$. Therefore $G(A) = G(A_1) + G(A_2)$. This equation is solved for A in order to obtain $A = H(A_1, A_2)$. Of course, without the exact knowledge of the function $F(W)$ and its inverse $G(A)$, it is impossible to find out how the action is precisely given by the action of its subsystems. Presently we don't have any reason to choose among the many possible functions with the qualitative behavior shown in Fig. 2. As illustration we could consider, for the positive branch $W \geq 0$, the functions $A = F(W) = W + \hbar\exp(-\frac{W}{\hbar})$,

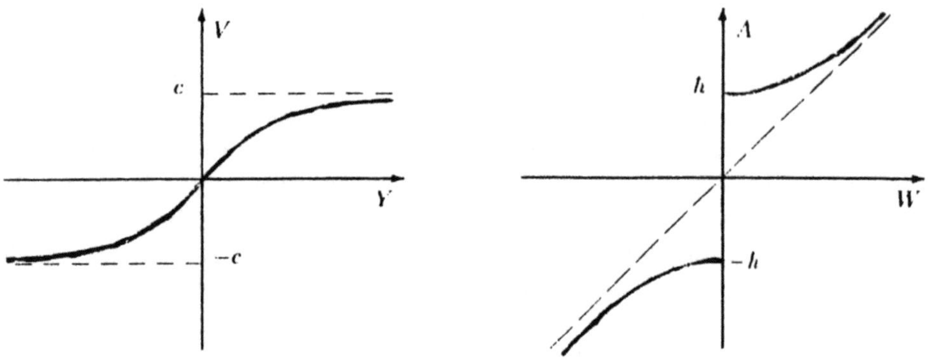

Fig. 2. Qualitative relation of velocity vs rapidity and action vs activity.

or $A = (W^n + \hbar^n)^{1/n}$ with $n = 1, 2, \ldots$. The negative branch being obtained by antisymmetric extension. In general, we expect

$$A = A_1 + A_2 + \hbar\, f(A_1, A_2)\ , \tag{9}$$

with some function $f(A_1, A_2)$ impeding the additivity of the action. An example, perhaps more interesting, for a relation between action and activity is suggested by the observation that, if we plot the inverse of the action (Inaction) versus the inverse of the activity (Inactivity?), then we obtain a figure with a shape like the left part of Fig. 2. This suggests the following relation:

$$\frac{\hbar}{A} = \tanh\left(\frac{\hbar}{W}\right), \qquad \frac{\hbar}{W} = \frac{1}{2}\ln\left(\frac{A+\hbar}{A-\hbar}\right) . \tag{10}$$

Writing $W = W_1 + W_2$ in terms of A, A_1, and A_2, we obtain, after some algebra,

$$\frac{A}{\hbar} = \coth\left(\frac{1}{2}\frac{\ell_1 \ell_2}{\ell_1 + \ell_2}\right) \quad \text{with } \ell_i = \ln\left(\frac{A_i + \hbar}{A_i - \hbar}\right) . \tag{11}$$

This composition law for the action of a system, in terms of the action of two of its subsystems has the required classical limit: When both subsystems are classical, that is $A_1 \gg \hbar$ and $A_2 \gg \hbar$, then we get the classical result $A = A_1 + A_2$. If one subsystem is classical $A_1 \gg \hbar$ and the other has $A_2 \approx \hbar$ then $A = A_1$. Furthermore, the composition law has also the necessary feature of associativity. That is, if we have a system composed by N subsystems, then the order in which we combine them to build the total action is irrelevant. In fact, in this case we obtain

$$\frac{A}{\hbar} = \coth\left(\frac{1}{2}\frac{\prod_{i=1}^{N} \ell_i}{\sum_{k=1}^{N}\prod_{i\neq k}^{N} \ell_i}\right) . \tag{12}$$

With this last expression, we can investigate the limit for large N. Let us assume that we have a system composed by a very large number of *quantum* subsystems. Large N means, not only large number of degrees of freedom but also, possibly, intervals of evolution decomposed in large number of subintervals. If all the subsystems are quantum ones, then their action is close to $\hbar$. Let us take $A_i \approx n\hbar$ where n is some small number, say $1 < n < 5$. With this, $\ell_i \approx x$, where x is some number close to 1. Then, replacing above we get:

$$\frac{A}{\hbar} \approx \coth\left(\frac{1}{2}\frac{x^N}{Nx^{N-1}}\right) \approx \coth\left(\frac{1}{2}\frac{x}{N}\right) \approx \frac{2}{x}N \ . \tag{13}$$

Therefore, for large N, the action of the system goes like $A \approx N\hbar$. We see here that a system becomes classic ($A \gg \hbar$) when it has a large number of subsystems. This is in agreement with some results indicating that the classical limit is obtained by large number of degrees of freedom [5]. As said before, there is no strong reason to prefer this composition law. However we may find this choice esthetically appealing because it can be motivated by a novel formal symmetry. In fact we may imagine a kinematic-dynamic transformation that changes Velocity with Inaction with their corresponding bounds, c and $1/\hbar$. Let us dare to speculate. Under this transformation RCM becomes QM in Figure 1 and, assuming symmetry, we could attempt to develop a quantum theory just by copying and translating special relativity. How is the space-time geometry of special relativity imaged under this transformation? What is the geometry of dynamics?

9. COMMENTS AND CONCLUSIONS

In all these examples, and in particular in equation (9), we see that $\hbar$ plays the rôle of a coupling constant between the subsystems in a composite system. This is an essential feature of quantum theory implying that a quantum system can not be simply considered as the sum of its parts, but must necessarily be taken as a whole (quantum holism). Since the subsystems above may correspond, not only to different degrees of freedom but also, to different intervals of evolution, $\hbar$ plays a double rôle as a coupling constant. In one case, when the subsystems correspond to different degrees of freedom, the coupling results in contextuality or nonseparability, whereas in the other case, when the subsystems correspond to different intervals of evolution, the coupling results in uncertainty relations between conjugate variables.

We have seen that $A \geq \hbar$, taken as a first principle, can play a significant rôle in the generation of a quantum theory. In fact we have seen that many essential features of quantum theory can be traced back to the principle of nonvanishing action. The nonzero uncertainty product between canonical conjugate variables and the contextuality, or nonseparability, for some states, can be deduced from the principle. An intrinsic angular rotation state, spin, and its relation with orbital rotation, quite naturally arise from such a principle. Furthermore, we have seen that the conventional formalism of quantum mechanics contains the principle, hidden in its noncommutative algebra. All these nonrigorous arguments give us confidence in attempting to develop quantum theory, based on this principle.

I would like to thank J. L. Iguain for comments and H. Mártin for enlightening discussions and suggestions.

1. E. Schrödinger, "Zum Heisenbergschen Unschärfeprinzip," *Berliner Ber.* 296-303 (1930).
2. A. C. de la Torre, P. Catuogno, and S. Ferrando, "Uncertainty and nonseparability," *Found. Phys. Lett.* **2**, 235-244 (1991).
3. A. C. de la Torre, P. Catuogno, and S. Ferrando, "Nonseparability and noncommutativity in quantum systems," *Found. Phys. Lett.* **4**, 49-59 (1991).
4. A. C. de la Torre, "Contextuality in quantum systems," MARDP-DLT6 preprint, submitted to *Am. J. Phys.*
5. J. L. Sánchez-Gómez, private communication.

NOTE

1. Member of "Consejo Nacional de Investigaciones Científicas y Técnicas" (CONICET), Argentina.

DECOHERENCE IN AN ISOLATED MACROSCOPIC QUANTUM SYSTEM: A PARAMETER-FREE MODEL INVOLVING GRAVITY

Jesus Unturbe

Departamento de Fisica Teorica
Universidad Autonoma de Madrid
Cantoblanca, 28049 Madrid, Spain

We will present here a proposal about how gravitation can contribute to produce the appropriate decoherence. Our model is restricted to the range of weak gravitational fields and non-relativistic velocities, in a somehow semiclassical limit of a "hypothetical" quantum gravity. We so elude all the huge difficulties entailed in superstrings, superoperators, subquantum world, quantum cosmology, etc. Despite the simplicity of the model, one can extract some important conclusions about the emergence of classical and quantum behaviour, the measurement problem, the question of mesoscopic bodies, etc.

Key words: decoherence, macroscopic quantum systems.

1. A RADICAL INTERPRETATION OF QUANTUM MECHANICS

The basic concepts of the model lie, heuristically, upon a radical interpretation of quantum mechanics (QM), namely, (i) the wave function Ψ represents an *individual* system in an exhaustive manner, and (ii) the probability $|\Psi|^2$ predicted by QM is *objective*, that is, it cannot be reduced to lack of knowledge of a more fundamental deterministic theory (i.e., in the line with (i), hidden variables are excluded).

Since a complete theory cannot be, at the same time, deterministic and stochastic, then we must go beyond the current theory, and this fact leads us just to gravitation, as we shall see immediately. The model presented here has no free parameters, which distinguish it from other models also going beyond QM (see [1]).

M. Ferrero and A. van der Merwe (eds.), Fundamental Problems in Quantum Physics, 341–345.

2. QUALITATIVE DESCRIPTION OF THE MODEL

Let us imagine a quantum particle of mass m. According to QM this particle has not a definite position, being, say, in the region Δx. Because of its mass, the particle produces a gravitational field. Now, since its position is not definite, that gravitational field (i.e., the metric) will acquire a certain blurriness $\Delta g^{\mu\nu}$ in each point of the space-time; note that it is a *radical application of the equivalence principle*: this is the way gravitation is introduced in the model.

Now, the particle wave function itself is immerse in this *blurry metric*. The phase is especially sensitive to this blurriness and, in some cases—e.g., the CM of a macroscopic solid—the phase can be irreversibly lost producing the decoherence and therefore giving rise to the classical limit.

A microscopic system—having very minute mass—is less sensitive to distortion of the metric and the latter *does not produce decoherence at all* (quantum limit), except when a measurement is performed (then this micro-system interacts with, is reduced by, a macro-system.

In contrast to other approaches (see [1,2]), the model also predicts a new behaviour—neither classical nor quantum—in the systems with intermediate masses ($\approx 10^{-13}$g.); the so-called mesoscopic bodies.

3. HEURISTIC APPROXIMATION TO THE CONCEPT OF BLURRY METRIC

Step 1. Point corpuscle with probability distribution $|\Psi|^2 dx^3$. The metric has mean value $\langle g^{\mu\nu}\rangle = \int |\Psi|^2 g^{\mu\nu} dx^3$ and dispersion $\Delta g^{\mu\nu} = \left(\langle (g^{\mu\nu})^2\rangle - \langle g^{\mu\nu}\rangle^2\right)^{1/2}$.

Step 2. Now, the point corpuscle *does not have a definite position* (QM). Hence, $g^{\mu\nu}$ will not have a definite value, either. (So it is somehow "semiclassical.") One has, with $r_s := 2Gm/c^2$,

$$\begin{aligned} g^{oo} = 1 - r_s/r, \quad & \delta g^{oo} \approx r_s \delta r/r^2, \\ & \delta g^{oo} \approx r_s \Delta x/r^2, \quad r \gg \Delta x, \\ & \delta g^{oo} \approx r_s/\Delta x, \quad r \ll \Delta x. \end{aligned} \tag{1}$$

Step 3. A body with a great number of constituents: Ψ has many arguments, so one has to make some methodological simplifications, *but there is* **only one** *wave function and* **only one** *metric*, and the former produces certain blurriness in the latter.

4. DECOHERENCE BY THE BLURRINESS OF METRIC

Step 4. Starting from the Klein-Gordon general covariant equation (spin should be irrelevant here) and introducing a variation, δg^{μ}, in the metric we easily arrive [3] at the expression of the indefiniteness of phase $\delta\phi$, in a first approximation:

$$\delta\phi = \frac{mc^2}{2\hbar} t\delta g^{oo} + 0\left[(\delta g^{oo})^2\right] \tag{2}$$

5. MEANING OF THE BLURRY METRIC IN COMPUNDS SYSTEMS ([4], [5])

Of course, it is very hard to calculate the blurry metric from a wave function with many arguments. Obviously, this shows a lack in the theory and, therefore, we restrict the treatment to the level of a simple model. In our approach, Ψ factorizes in such a way that one can extract the relevant aspects—wave function of CM—and so the problem is *almost* reduced to the one-particle case. The way to do this can be found in [3], and with more detail in [4]. In any case, whether the system is individual or compound, our model involves the existence of **only one metric**—and nothing like a superposition of metrics. Of course the metric is blurry, so we refer to it as an "average," but this is only a word and does not mean an ensemble of metrics at all.

6. MAIN RESULTS OF THE MODEL ([3], [4])

(1) All elementary particles, atoms, etc. fulfill $\Delta\phi \ll 1$ for all t. All the predictions of QM are recovered here.

(2) The CM wave function of a compound system changes its behaviour when the corresponding mass is of the order of the critical mass:

$$M_o = \left(\frac{m_e e^2}{G}\right)^{1/3} \approx 10^{-13}\text{g}; \tag{3}$$

here m_e denotes the electron mass, e the electron charge, and G the Newton gravitational constant.

(3) The CM wave function of a system with mass M well larger than the critical mass M_o (say $M > 10^{-11}$g.) has a Gaussian shape with a constant (*non-spreading*) width given by $a_o^{CM} = \hbar^2 N^{1/6}/GM_o^2 M$, where N is the number of constituents. For $M = 1$gm, $a_o^{CM} = 10^{-17}$cm. *This precludes the superposition of macroscopic states.*

(4) The metric indefiniteness is always very small: $\Delta g^{\mu\nu} \ll 1$. (More exactly, $\Delta g^{oo} < 10^{-36}$; other tensor components are even smaller.)

7. GOOD ASPECTS

(1) There is a natural emerging of *macroscopic properties.*

(2) The relative (i.e., depending on the relative coordinates) wave function *is not affected by the blurriness of the metric*—even in a macroscopic body.

(3) The concept of blurry metric might help solving the problem of the contradiction between *"semiclassical" General Relativity* and *QM.*

(4) The energy needed to keep in a steady state the CM packet of macroscopic bodies has a *gravitational origin*, being *the minimum possible*—this implies the stability of the wave packet.

(5) As already said, *there are no free parameters.*

(6) *In the microscopic case*, the metric is "blind" with respect to the quantum basis and hence, *there is no preferred basis*—QM is *well* recovered.

(7) *The privileged role that the observer has in QM is suppressed.*

(8) At mesoscopic level ($\approx 10^{-13}$g.), in suitable conditions, self-reductions are possible, which, as already commented, could entail a new, peculiar behaviour of mesoscopic systems.

8. FINAL COMMENTS [5]

Perhaps, in a more general theory transcending QM, either the projection postulate or the *decoherence* inherent to the coarse graining procedure should be "carried out" not by the observer but by the metric. In our model, the information of quantum correlations is not transferred to the metric (there are no superposed metrics). But this loss of information is actually apparent only when matter systems are sufficiently sensitive to the blurriness of metric, i.e. for macroscopic systems. Hence, the microscopic ones, not being affected by such a loss of information, closely follow quantum behaviour.

Now, is the blurry metric a new fundamental concept beyond quantum and classical theories, or is it simply an aspect of the metric in the "semiclassical" approximation at low energies of quantum theory of gravity?

REFERENCES

1. G. C. Ghirardi, A. Rimini, and T. Weber, *Phys. Rev. D* **34**, 470 (1986);

ibid. **36**, 3287 (1987); G. C. Ghirardi, P. Pearle, and A. Rimini, *Phys. Rev. A* **42**, 78 (1990).

2. F. Károlyházy, A. Frenkel, and B. Lukács; in *Physics as Natural Philosophy*, A. Shimony and H. Feschbach, eds. (MIT Press, Cambridge, 1982), p. 204.
3. J. Unturbe and J. L. Sánchez-Gómez, *Nuovo Cimento B* **107**, 211 (1992).
4. J. Unturbe, *Sobre el papel de la Gravedad en la Mecanica Cuantica de los Sistemas Macroscopicos*, Doctoral Thesis, Universidad Autonoma de Madrid, 1992.
5. J. Unturbe and J. L. Sánchez-Gómez, "The role of gravity in the wave function of macroscopic systems," to appear in *Physics Essays*, about Dec. 1993; FTUAM 93/15.

NONLOCAL MEASUREMENTS AND TELEPORTATION OF QUANTUM STATES

Lev Vaidman

School of Physics and Astronomy
Raymond and Beverly Sackler Faculty of Exact Sciences
Tel-Aviv University, Tel-Aviv, 69978 Israel

Recent result of Bennett *et al.* on teleportation of an unknown quantum state is obtained in the framework of nonlocal measurements proposed by Aharonov and Albert. The latter method is generalized to the teleportation of a quantum state of a system with continuous variables.

Key words: nonlocality, teleportation, many-worlds interpretation.

1. TELEPORTATION

When I first heard the word "teleportation," I imagined heroes of "Star Trek" entering a transmitting cabin in their star ship Enterprise: In a few seconds they disappear and immediately appear on a distant planet. It sounds, however, like bad science fiction: Too many laws are broken in this picture. For example, we tend to believe that the center of mass of a closed system should not move. Thus, when the heroes move far away, the spaceship has to move a little in the opposite direction, but this is not shown in the movies. Also, this story requires huge currents to fulfill the continuity equation, and it is very hard to understand how these currents can be created.

However, quantum theory teaches us that the essential feature of an object is not the *matter* out which it is made, but its shape. Indeed, all objects are "made" out of identical elementary particles, and what distinguishes one object from another is the state of these particles. Thus, we can consider a different kind of machine for teleportation. The receiving teleportation chamber is not empty before the transmission, but it contains elementary particles in a number equal to the number of particles of the object to be transmitted. Then, transmission results in building the Star Trek heroes out of these particles, while the heroes in the cabin on the Enterprise revert to an unstructured set of elementary particles.

One might be tempted to build the heroes in several locations, i.e., to

M. Ferrero and A. van der Merwe (eds.), Fundamental Problems in Quantum Physics, 347–356.

produce several copies. This is certainly not teleportation. But the unitarity of quantum theory prevents this possibility. It is impossible to clone an unknown quantum state. It is also impossible to identify an unknown quantum state without significantly changing it. Therefore, the only option quantum mechanics leaves is destruction of the heroes in one place while creating them in another.

The procedure for such an operation was discovered recently. Bennett, Brassard, Crepeau, Jozsa, Peres, and Wootters (BBCJPW) [1] have shown how to teleport an unknown quantum state. They found a method to transmit an unknown quantum state of a spin-1/2 particle to another distant spin-1/2 particle without actually moving it from one place to another. We will show how this can be achieved using another method [2] based on the nonlocal measurements of Aharonov and Albert [3].

In the next section we will review the method of nonlocal measurements. Section 3 is devoted to teleportation of the state of a spin-1/2 particle. In Sec. 4 we generalize the method to systems with continuous variables. Section 5 deals with philosophical aspects of teleportation. A brief summary concludes the paper in Sec. 6.

2. NONLOCAL MEASUREMENTS

We call a measurement *nonlocal* if it cannot be reduced to a set of local measurements. An example is a measurement of a sum of variables A_1 and A_2 related to two separate locations 1 and 2. The method of Aharonov and Albert for nonlocal measurements uses only local interactions. The measurement is described by interaction Hamiltonian

$$H = g(t)P_1A_1 + g(t)P_2A_2, \tag{1}$$

where $g(t)$ is normalized function with a compact support at the time of the measurement; P_1, P_2 are conjugate momenta of the pointer variables of the two parts of measuring device which locally interact at locations 1 and 2. In order to perform a nonlocal measurement (and not two local measurements), the initial state of the measuring device has to be

$$Q_1 + Q_2 = 0, \qquad P_1 - P_2 = 0. \tag{2}$$

After the interaction is completed,

$$A_1 + A_2 = Q_1 + Q_2, \tag{3}$$

and local measurements of Q_1 and Q_2, therefore, yield the value of $A_1 + A_2$.

A set of measurements of nonlocal variables can serve as a verification of a nonlocal (entangled) state. The EPR-Bohm state of two spin-1/2 particles (completely anticorrelated state) can be verified using two consecutive measurements: first of $\sigma_{1x} + \sigma_{2x}$ and then of $\sigma_{1y} + \sigma_{2y}$; see Fig. 1. Here and below we will use the units of $\hbar/2$ for the spin components, such that each component can have values ± 1. If the outcomes are

$$\sigma_{1x}(t_1) + \sigma_{2x}(t_1) = 0, \qquad \sigma_{1y}(t_2) + \sigma_{2y}(t_2) = 0, \tag{4}$$

where $t_2 > t_1$, then, after time t_2, the system is in the EPR-Bohm state. Had we started at time $t < t_1$ with the EPR-Bohm state, we would be certain to obtain the outcomes (4).

The method of Aharonov and Albert is applicable also for measurements which are nonlocal not only in space but also in time. The interaction Hamiltonian has to be modified such that local interactions in separate locations will take place at different times. For example, for a measurement of the sum $A_1(t_1) + A_2(t_2)$, the Hamiltonian is

$$H = g(t - t_1)P_1 A_1 + g(t - t_2)P_2 A_2, \tag{5}$$

where $g(t)$ has compact support around zero. The sums and also modular

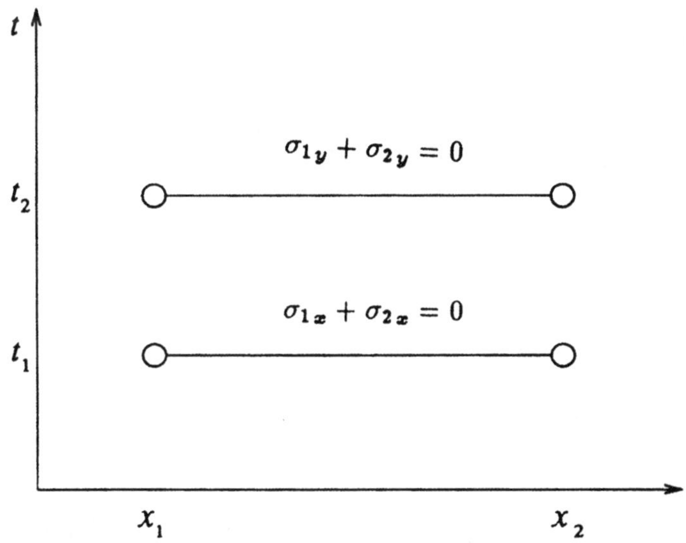

Fig. 1. If the system is in the EPR-Bohm state then the outcomes of the nonlocal measurements have to be as shown in the figure. Conversely, if these are the outcomes of the nonlocal measurements then, after the measurements, the system is in the EPR-Bohm state.

sums of local variables are measurable. For measuring a sum modulo a the measuring device has to be set in the following initial state:

$$(Q_1 + Q_2)\text{mod}a = 0, \qquad P_1 - P_2 = 0, \qquad P_1 \text{mod}\frac{2\pi\hbar}{a} = 0. \tag{6}$$

For more details see Ref. 4.

3. TELEPORTATION OF A STATE OF A SPIN-1/2 PARTICLE

Let us assume that the state of a spin-1/2 particle 1 is $\alpha|\uparrow\rangle + \beta|\downarrow\rangle$ and we have to teleport it to particle 2. To this end, consider the "crossed" measurements of $\sigma_{1x}(t_1) - \sigma_{2x}(t_2)$ and $\sigma_{1y}(t_2) - \sigma_{2y}(t_1)$, see Fig. 2. If the outcomes are

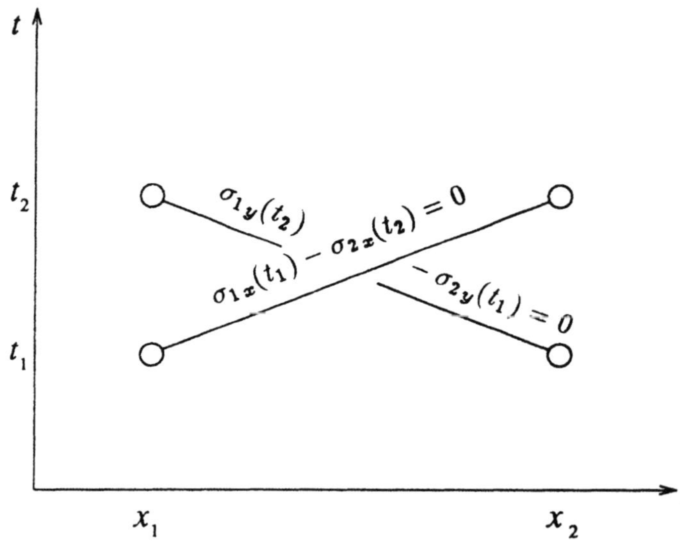

Fig. 2. Teleportation via "crossed" space-time nonlocal measurements. The measurements with the outcomes in the figure cause the state of particle 2 after t_2 to be the state of particle 1 before t_1 (and the state of particle 1 after t_2 to be the state of particle 2 before t_1). For *reliable* teleportation the nonlocal measurements to be performed are the measurements of the differences of the spin components modulo 4 accompanied by appropriate local rotations.

$$\sigma_{1x}(t_1) - \sigma_{2x}(t_2) = 0, \qquad \sigma_{1y}(t_2) - \sigma_{2y}(t_1) = 0, \tag{7}$$

then, taking into account the measurement interaction (5) with $A_1 = \sigma_{1x}$ ($A_1 = \sigma_{1y}$) and $A_2 = -\sigma_{2x}$ ($A_2 = -\sigma_{2y}$), the initial state of the measuring device (2), and the outcomes of the local readings of the measuring device (which are also described by Eq. (2)), we obtain, after straightforward calculation, that the final state of particle 2 is $\alpha|\uparrow\rangle + \beta|\downarrow\rangle$, i.e., we have succeeded in teleporting the state of particle 1 to particle 2.

However, this procedure is not good enough, since the nonlocal measurements might not yield the specific outcomes (7). The difference between the spin components might equal ± 2 and in that case we destroy the state without teleporting it. In order to obtain *reliable* teleportation (such as the one suggested by BBCJPW), we must measure, instead, the following nonlocal observables:

$$\Big(\sigma_{1x}(t_1) - \sigma_{2x}(t_2)\Big)\mathrm{mod}4, \qquad \Big(\sigma_{1y}(t_2) - \sigma_{2y}(t_1)\Big)\mathrm{mod}4. \tag{8}$$

A null outcome reduces to the previous case. If, however, the outcome of one of the above is 2, then we can convert it to 0 by appropriate rotation of the coordinate frame of the second particle (for example, $\sigma_{2x} = -\sigma_{2x'}$, for $\hat{x}' = -\hat{x}$). Thus, for any set of outcomes of the nonlocal measurements (8) the spin state is teleported; in some cases the state is rotated, but the resulting rotation can be inferred from the nonlocal measurements. We can complete, then, the teleportation by the following transformations. In the case of two null outcomes no additional transformation is needed; in three other cases a transformation of rotation by the angle π is necessary: the rotation around the y axis for the outcome (2,0), around the x axis for (0,2), and around the z axis for the outcome (2,2).

The Aharonov-Albert method for nonlocal measurement contains the following elements: (i) a preparation of an entangled state of the measuring device, (ii) local interactions with separate parts of the system, (iii) local readings of the separate parts of the measuring device resulting in a set of numbers obtained in the respective space-time locations of the parts of the system. These numbers represent classical information which must be transmitted for completing the teleportation. (In our example, the information tells us which rotation must be performed). The initial entanglement of the measuring device, which is the core of the method, may employ pairs of spin-1/2 particles in the EPR-Bohm state (see Sec.IV of Ref. 4), making this method very similar to the BBCJPW proposal. In this case we also need to transmit just two bits of classical information, which is the minimal information for teleportation of a spin state, as has been proven in Ref. 1. The number of "nonlocal channels" in our method is two instead of just one in the BBCJPW method. This is because we have accomplished two-way teleportation. (Obviously, for teleporting also the state of particle

2 to particle 1 we need to send another two bits of classical information from the site 2 to the site 1.)

The BBCJPW method can be presented in our language as in Fig. 3. The EPR-Bohm pair which is employed by BBCJPW can be created via two (successful) measurements (4). The measurement "in the Bell operator basis" (Eqs. 1 and 2 of Ref. 1) at the location 1, performed on the composite system consisting of particle 1 and one member of the EPR-Bohm pair, is equivalent to two consecutive measurements of the modular sums:

$$\Big(\sigma_{1x}(t_1) + \sigma_{2x}(t_1)\Big)\mathrm{mod}4, \qquad \Big(\sigma_{1y}(t_2) + \sigma_{2y}(t_2)\Big)\mathrm{mod}4. \tag{9}$$

The four different combinations of the outcomes of the nonlocal measurements (9) correspond to the four outcomes of local measurement of BBCJPW. The procedure of teleportation is completed by appropriate rotation according to these results. After the teleportation, particle 1 is in a mixed state and contains no information. This is in contrast with the "crossed" measurements method in which (after the appropriate rotation) the the final state of particle 1 is the initial state of particle 2.

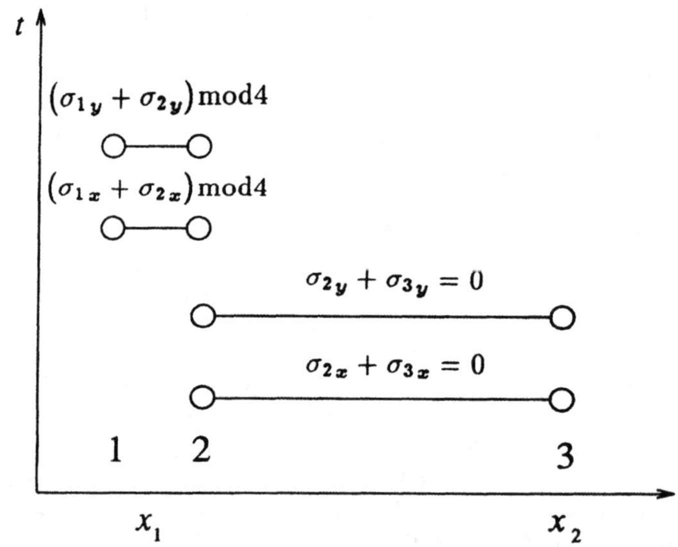

Fig. 3. Nonlocal measurements of the BBCJPW teleportation scheme. The state of particle 1 is teleported to particle 3. To this end the particle 3 is prepared in the EPR-Bohm state with particle 2 located near particle 1 using nonlocal measurements. Then the measurements of the composite system consisting of particles 1 and 2 causes the teleportation (possibly, with rotation).

4. TELEPORTATION OF A CONTINUOUS VARIABLE STATE

In the framework of nonlocal measurements there is a natural way of extending the teleportation scheme to the systems with continues variables. Consider two similar systems located far away from each other and described by continuous variables q_1, q_2 with corresponding conjugate momenta p_1 and p_2. In order to teleport a quantum state $\Psi(q_1)$, we perform the following "crossed" nonlocal measurements (see Fig. 4*a*), obtaining the outcomes a and b:

$$q_1(t_1) - q_2(t_2) = a, \qquad p_1(t_2) - p_2(t_1) = b. \tag{10}$$

Straightforward calculation shows that these nonlocal "crossed" measurements correlate the state of particle 1 before t_1 and the state of particle 2 after t_2, thus teleporting the quantum state to the second particle up to a shift of $-a$ in q and $-b$ in p. These shifts are known (after the results of local measurements have been transmitted), and can easily be corrected by appropriate back shifts even if the state is unknown, thus completing a reliable teleportation of the state $\Psi(q_1)$ to $\Psi(q_2)$.

A generalization of the BBCJPW scheme to the case of continuous variables is also possible, see Fig. 4*b*. The method contains the following stages: first, the preparation of the EPR state of particles 2 and 3,

$$q_2 + q_3 = 0, \qquad p_2 - p_3 = 0, \tag{11}$$

second, the consecutive measurements performed on particles 1 and 2, yielding the outcomes a and b:

$$q_1 + q_2 = a, \qquad p_1 - p_2 = b. \tag{12}$$

Each pair of measurements (11), (12) causes an anticorrelation, thus the anticorrelation between particles 2 and 3 together with the anticorrelation between particles 1 and 2 lead to a correlation between particles 1 and 3. The only difference between the states is due to the shifts both in q and in p:

$$q_3 = q_1 - a, \qquad p_3 = p_1 - b. \tag{13}$$

If the initial state of particle 1 is $\Psi(q_1)$, then the state of particle 3, after the measurements (11) and (12) have been performed, is $e^{ibq_3}\Psi(q_3 + a)$, which is exactly the state obtained after the "crossed" measurements (10). The final stage of teleportation are the appropriate back shifts of the state in p and q. Note again that while the crossed measurements yield two-way teleportation, we have obtained now only a one-way teleportation: from particle 1 to particle 3.

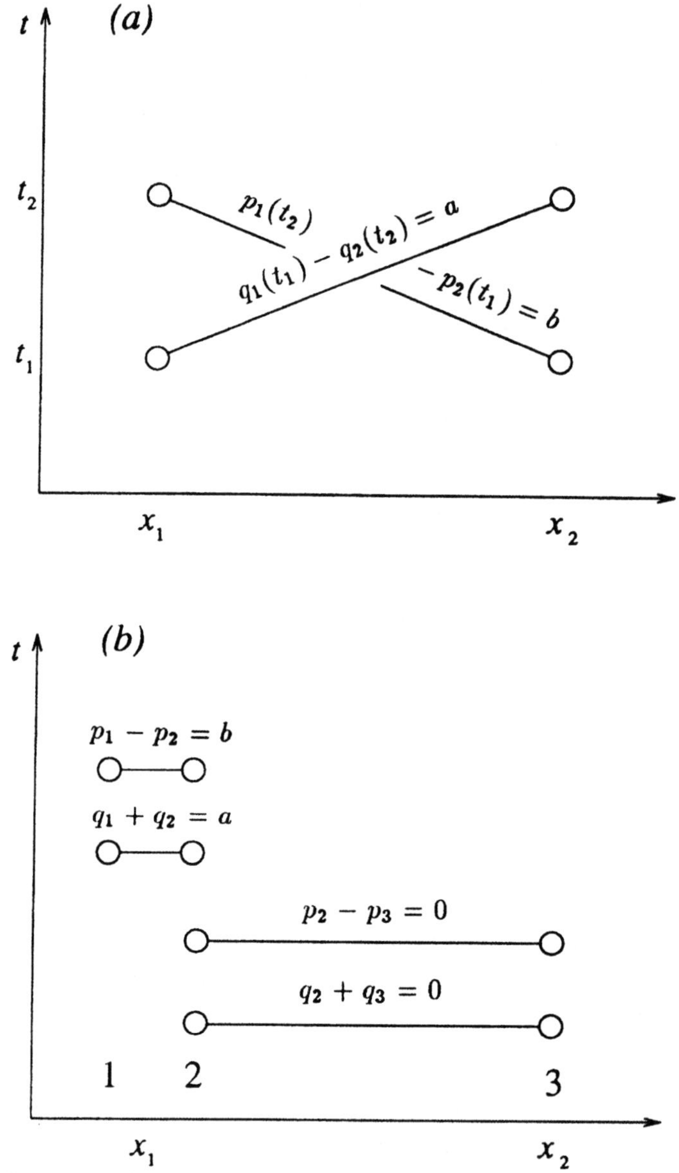

Fig. 4. Teleportation of an unknown quantum state of a system with continuous variables: (a) the method of "crossed" space-time nonlocal measurements, (b) the method analogues to the BBCJPW proposal. In both cases the final state (before the back shifts) is $e^{ibq}\Psi(q+a)$.

5. IS THERE A PARADOX WITH TELEPORTATION?

Consider teleportation, say in the BBCJPW scheme. We perform some action in one place and the state is *immediately* teleported (sometimes also rotated and/or shifted) to an arbitrary distant location. But relativity theory teaches us that anything which is physically significant cannot move faster than light. If we decide that an unknown rotation of the state is important, then we have to transmit classical information (which cannot be done with superluminal velocity) about the kind of back rotation to be performed for obtaining, after teleportation, exactly the same state. The classical information about the rotation is very small, but it is the only thing which is not transmitted immediately. Thus, it seems, that this classical information is the only essential part of the quantum state. Is the essence of a state of a spin-1/2 particle just 2 bits? I tend to attach a lot of physical meaning to a quantum state, not only because Aharonov and I found a method for measuring the state of a single quantum system [5], but also because I am a proponent of the many-worlds interpretation of quantum theory [6]. For me everything is a quantum state. But I also believe in relativistic invariance, so the only entities that cannot move faster than light have physical reality. Thus, teleportation poses a serious problem to my attitude.

A similar and even simpler example of this difficulty is a single spin measurement of one particle of an EPR-Bohm pair. It can also be be considered as some kind of teleportation. The state of the second particle of the pair immediately after the measurement is (up to inversion) the state of the first one. According to standard quantum mechanics there were no pure states of the particles before the measurement. The local measurement *created* the spin state of the first particle, and it was immediately transmitted and reversed (without destroying the state of the first particle) to a distant particle. This is a kind of superluminal teleportation.

I resolve this paradoxical situation in my interpretation of the many-worlds interpretation [7]. Anything superluminal is forbidden in a physical world (or universe as I call it). The universe incorporates all our worlds as members of a superposition which is the quantum wave (i.e., the universe). The act of measurement on one particle changes no physical property of the second particle from the point of view of an observer who can see the whole Universe, i.e. all worlds together. If, before the measurement, the second spin was correlated to the states of the first spin, then after the measurement, it will be also correlated to the measuring device and the observer of the first particle, but its density matrix remains unchanged. Thus, there is no change in any measurable property of the second particle. Why, then, do we believe that the state of the second particle has changed? Because our measurement locally splits the world of the observer of the first particle. When he (in one of the new worlds) later will look or ask about the second particle he will find it in the (teleported) state. But this is more the property of his world rather than an objective property of the second particle.

6. CONCLUSIONS

We have shown how nonlocal measurements can be used for teleportation of unknown quantum states, first for a spin state, and then for a system with a continuous variable. We presented a "crossed measurement" method which yields two-way teleportation. The method avoids the necessity of measuring Bell's type operator which currently poses a serious experimental challenge. But our formalism also suggests a way of splitting the Bell operator measurement into two apparently simpler measurements. Finally, we resolved some philosophical difficulties posed by teleportation to a believer in the many-worlds interpretation of quantum theory.

We have not presented here an exact prescription for an engineer who is going to build a teleportation camera in a spaceship of tomorrow. But we have shown the type of teleportation which does not contradict physics laws known today, and we pointed out the direction for the beginning of the experimental work.

ACKNOWLEDGMENTS

It is a pleasure to thank Daniel Rohrlich for very helpful suggestions. The research was supported in part by grant 425/91-1 of the the Basic Research Foundation (administered by the Israel Academy of Sciences and Humanities).

6. REFERENCES

1. C.H. Bennett, G. Brassard, C. Crepeau, R. Jozsa, A. Peres, and W.K. Wootters, *Phys. Rev. Lett.* **70**, 1895 (1993).
2. L. Vaidman, *Phys. Rev. A* **49**, 1473 (1994).
3. Y. Aharonov and D. Albert, *Phys. Rev. D* **21**, 3316 (1980); **24**, 359 (1981).
4. Y. Aharonov, D. Albert, and L. Vaidman, *Phys. Rev. D* **34**, 1805 (1986).
5. Y. Aharonov and L. Vaidman, *Phys. Lett. A* **178**, 38 (1993).
6. L. Vaidman, Tel-Aviv preprint TAUP 2058-90 (1990).
7. L. Vaidman, *PSA 1994*, Volume 1, D. Hull, M. Forbes, and R.M. Burian, eds. (PSA, East Lansing, MI, 1994).

QUANTUM NOISE IN OPTICAL PHOTON DETECTORS

Antonio Vidiella-Barranco and Emilio Santos

Departamento de Física Moderna, Universidad de Cantabria
39005 Santander, Spain

We consider a simple model of a photon detector which consists of a set of N three-level atoms. Absorption takes place when a transition happens from the initial, metastable state $|1\rangle$ to the upper state $|2\rangle$. Irreversible amplification takes place when the atom spontaneously decays (from level $|2\rangle$) to a state $|3\rangle$ lower than $|1\rangle$. We show that the product of the dark rate with the dead time is bounded from below by a number which increases linearly with the frequency width (characteristic of the incoming beam) times the detector area.

Key words: photodetection, quantum noise, theory of measurement.

Detectors of optical photons are essentially characterized by four parameters: the quantum efficiency η, the dark rate R, the dead time τ, and the area A of the sensitive region. It is well known that there are constraints on the values of those parameters, and that the ideal situation $\eta \to 1$, $R \to 0$, $\tau \to 0$ cannot be achieved by a number of reasons. Some of the causes of non-ideality can be removed, or at least diminished. For instance, thermal fluctuations can in principle be eliminated by lowering the temperature. However, there are also fundamental constraints, due to quantum noise, which can never be eliminated. As we are going to see, this will cause severe limitations on the detector's performance, and therefore this also constitutes a problem of practical significance. Our purpose here is to derive these fundamental constraints by means of a solvable model. Solvable models are very useful in quantum theory of measurement, because they often provide

M. Ferrero and A. van der Merwe (eds.), Fundamental Problems in Quantum Physics, 357–361.

physicists with a clear understanding of the phenomena involved [1].

In theoretical studies of the photon detection process [2], it is usual to treat the detector as a two-level atom (or a collection of such atoms). It is assumed that the essential part of the detection ends when an atom initially in the ground state, say $|1\rangle$, is excited to the higher level $|2\rangle$. Within this two-level model it is possible to treat ideal detectors, having 100% efficiency and no noise. Nevertheless, it is well known that any real detection process should combine the absorption of the signal with an (irreversible) amplification, and the detection is considered to be achieved only when the second step has taken place. Thus we propose the inclusion of the amplification process by considering three-level atoms. Besides the initial state $|1\rangle$ and the upper state $|2\rangle$, we assume the existence of another state, namely $|3\rangle$, with energy lower than $|1\rangle$. The detection would consist of an absorption ($|1\rangle \to |2\rangle$), followed by a spontaneous emission process ($|2\rangle \to |3\rangle$) which gives rise to a new photon, of energy greater than the incoming one. Further amplification might be obtained by the repetition of the process or by other means, but in our simplified approach we shall consider that the detection finishes when the second photon is emitted.

In a real photomultiplier [3], the detection process is similar to what we have here described. The main difference is that an electron, instead of a photon, is liberated after the detection. Again we have a kind of three-level system, the role of the excited state $|2\rangle$ being played now by a continuum of states for the free electron, and the final state $|3\rangle$ corresponding to a number of electron states in the deep potential produced by the applied voltage. Thus we hope that our simple model captures all the essential features of real detectors.

Our detector consists of a cylinder of area A and width x, which contains n three-level atoms per unit of volume. When the detector is ready, all (or most of) the atoms will be found in level $|1\rangle$, which we assume is metastable, i.e., the atoms cannot decay to the lowest state $|3\rangle$ by spontaneous emission (as a first order process). Once a photon of energy $\hbar\omega$ is absorbed, the atom may spontaneously decay either to the third state $|3\rangle$, or back to the initial state $|1\rangle$, with probabilities per unit time γ_{23} and γ_{21}, respectively.

The calculation of the efficiency is now straightforward. If $\sigma(\omega)$ is the total absorption cross-section of a photon with energy $\hbar\omega$ (regarding the atomic transition $|1\rangle \to |2\rangle$), then a beam of photons with frequency ω will be reduced in intensity by a factor

$$I_{out}/I_{in} = \exp[-n\sigma(\omega)x] = \exp[-N\sigma(\omega)/A], \qquad (1)$$

where $N = nAx$ is the total number of atoms in the detector. It is assumed that the beam is normally incident to the surface of the detector (of area A).

As we have already mentioned, not all absorbed photons give rise to a count, due to the back decay $|2\rangle \rightarrow |1\rangle$. If we assume, for simplicity, that the photon produced in the back decay is emitted in the same direction as the incoming beam, the transition $|1\rangle \rightarrow |2\rangle$ followed by $|2\rangle \rightarrow |1\rangle$ will not correspond to an interaction, and therefore we can take this effect into account simply by multiplying the cross-section $\sigma(\omega)$ by the factor $\gamma_{23}/(\gamma_{23}+\gamma_{21})$. Thus the efficiency of the detector can be written as

$$\eta = 1 - \exp[-n\sigma x\, \gamma_{23}/(\gamma_{23}+\gamma_{21})] = 1 - \exp\left(-N\sigma\, \gamma_{23}/[A(\gamma_{23}+\gamma_{21})]\right). \tag{2}$$

Hence it follows that, for any values of n, σ and γ_{ij}, the ideal efficiency $\eta \rightarrow 1$ can be achieved just by increasing the thickness x of the detector (or the number N of atoms for a given area A). However, we shall see that this will also increase the non-ideal behavior.

In actual photomultipliers, the so-called dead time is the time required by the detector to process an event, and depends on the several elements constituting the detector [3]. In our model, we shall assume that until the second photon is emitted, the detector is not ready for a new detection, so that the equivalent to the dead time τ will be simply

$$\tau = 1/\gamma_{23}. \tag{3}$$

This gives a reasonable lower limit for the detector's dead time, which means that if two incoming photons are absorbed within a time shorter than τ, we may assume that a single count is recorded. Of course, the dead time will be usually much longer than γ_{23}^{-1}, because of a number of practical reasons.

We also have that quantum fluctuations may give rise to direct passage from $|1\rangle$ to $|3\rangle$, by a second order effect involving the intermediate (virtual) state $|2\rangle$, with the simultaneous emission of two photons. We shall assume that any process of this type, with a probability per unit time denoted by γ_{13} and which liberates the energy $\hbar\omega_{13}$, will give rise to a detection (dark count). Our argument for this assumption is that the total energy liberated by a detection event is $\hbar(\omega_{23} - \omega_{12}) = \hbar\omega_{13}$, which is just the same released in the second-order process leading to a transition $|1\rangle \rightarrow |3\rangle$. As there are N atoms which may experience that process, we conclude that the detector's minimal dark rate R is just

$$R = N\,\gamma_{13}. \tag{4}$$

Again, thermal fluctuations and other causes may greatily increase the dark rate.

The calculation of the quantities involved in our model, namely, γ_{ij} and σ, is straightforward if we know the atomic matrix elements of the atomic electric dipole $e\mathbf{D}_{ij}$ (where e is the electron charge). If we sum over the final states and average over the orientation of the atomic dipoles, the results are:

$$\gamma_{2j} = \frac{e^2|\mathbf{D}_{j2}|^2}{3\,\pi\,\hbar\,c^3}\,\omega_{j2}^3, \qquad j = 1,3, \tag{5}$$

$$\gamma_{13} = \frac{e^4|\mathbf{D}_{12}|^2|\mathbf{D}_{23}|^2}{360\,\pi^3\hbar^2c^6}\,\omega_{13}^5, \tag{6}$$

and

$$\sigma = \frac{\pi e^2|\mathbf{D}_{12}|^2}{3\,\hbar\,c}\,\frac{\omega_{12}}{\Delta\omega}, \tag{7}$$

where $\Delta\omega$ is the frequency width of the incoming beam and σ has been averaged over the frequency range $\Delta\omega$.

It is easy to eliminate the model's parameters γ_{21}, γ_{23}, γ_{13}, σ, N, $|\mathbf{D}_{12}|$, $|\mathbf{D}_{23}|$ (remember that we assumed $\mathbf{D}_{13} = 0$) using eqs. (2) to (7), in order to obtain a relationship involving only the detector's measurable parameters η, R, τ, A, as well as ω_{12} and $\Delta\omega$. Although straightforward, the result is somewhat involved. We shall then write only a simplified relation valid if the following relations, which are natural in our model, hold true:

$$\omega_{13} \simeq \omega_{23} \gg \omega_{12}, \qquad \gamma_{23} \gg \gamma_{21}. \tag{8}$$

In this case the back decay may be neglected, and we obtain

$$R\,\tau > \frac{A}{40\,\pi^3c^2}[-\ln(1-\eta)]\,\frac{\Delta\omega\,\omega_{13}^2}{\omega_{12}} \gg \frac{A\,\Delta\omega\;\omega_{12}\,\eta}{40\,\pi^3c^2}. \tag{9}$$

This is the main result of this paper.

We see that the "noise" $R\,\tau$ in a photon detector may be reduced by narrowing the range $\Delta\omega$ of frequencies where it responds, and by reducing the sensitive area A. Both features are well known ones. However, neither A nor $\Delta\omega$ can be lowered indefinitely, because they are constrained by the obvious inequalities:

$$A \gtrsim \lambda^2 = \frac{4\pi^2c^2}{\omega_{12}}, \qquad \Delta\omega \gtrsim 2\pi\,\gamma_{13} = \frac{2\pi}{\tau}. \tag{10}$$

Substituting (10) into (9), we obtain

$$R\,\tau^2\omega_{12} \gtrsim \frac{1}{10\pi}[-\ln(1-\eta)]\left(\frac{\omega_{13}}{\omega_{12}}\right)^2 \gg 0.03\,[-\ln(1-\eta)], \tag{11}$$

which reduces to $R\tau^2\omega_{12} \gg 0.03\,\eta$ if $\eta \ll 1$.

We may test this inequalities with the characteristic values of a typical detector of optical photons, namely:

$$\eta \simeq 0.3, \quad R \simeq 10^2 sec.^{-1}, \quad \tau \simeq 2\,nsec., \quad \omega_{12} \simeq 4 \times 10^{15} Hz. \tag{12}$$

We can easily see that these values are consistent with the limit imposed by (11) if $\omega_{13}/\omega_{12} \simeq 10$, which is quite a reasonable value.

We have analized an extremely simple model of a photon detector which, nevertheless, provides an accurate description of the non-ideal behavior. According to our model, the fundamental quantum noise does not depend dramatically on the detector's efficiency, except if this is very close to unity. In any case, the dark rate, the dead time and the frequency of the incoming light fulfill the simple inequality (11), and the quantum noise increases dramatically if $A \gg \lambda^2$ and/or $\Delta\omega \gg \tau^{-1}$.

We would like to point out that all our work refers to optical photons only, that is, we do not consider the detection of X-rays or gamma radiation. In this case the amplification part of the detection process here discussed may not be necessary because every high energy photon contains already enough energy to excite many atoms (of the order of $E_\gamma/10\,eV$). On the other hand, for low energy photons (say microwaves), ω_{12} is so small that equation (11) forbids any detection of individual photons. Of course, if we do not intend to detect individual photons, but only the total energy of a beam in a given frequency range, the restrictions here discussed do not exist, and a quite good efficiency with negligible noise could in principle be achieved.

ACKNOWLEDGMENTS

We acknowledge financial support from DGICYT, Project Number PB92-0507 (Spain). One of the authors (A.V.B.) would like to acknowledge the hospitality of the Universidad de Cantabria.

REFERENCES

[1] For a model of particle detection, see, e.g., H. Nakazato and S. Pascazio, *Phys. Rev. Lett.* **70**, 1 (1993).

[2] R. Loudon, *The Quantum Theory of Light* (Oxford University Press, Oxford, 1983).

[3] W.R. Leo, *Techniques for Nuclear and Particle Physics Experiments* (Springer, Berlin, 1987).

EXTENSIONS OF BELL THEOREM: EXPERIMENT INVOLVING INDEPENDENT SOURCES IN "EVENT-READY" CONFIGURATION

M. Żukowski,[a,b] A. Zeilinger,[a] M. A. Horne,[c] and A. Ekert[a,d]

[a] *Institut fuer Experimentalphysik, Unversitaet Innsbruck, A-6020, Austria*
[b] *Instut Fizyki Teoretycznej i Astrofizyki, Uniwersytet Gdanski, PL-80-952, Poland*
[c] *Stonehill College, North Easton, MA 02357*
[d] *Merton College, Oxford University, OX1 4JD, United Kingdom*

With two independent parametric down-converters one can realize an "event-ready" Bell-EPR experiment. Monitoring for sharply coincident idlers results in entangling and pre-selecting the independent signals, which are then fed into an interferometric setup. We give the general conditions for high visibility and particle collection efficiency for Bell-type experiments involving two independent sources. These ideas can also lead to a realizable GHZ experiment.

Key words: "event ready" detectors, independent sources, entanglement swapping.

Bell's theorem is usually stated as the contradiction between local realism and quantum mechanics. The property of entanglement is at its root both in the two-particle case, where Bell's inequality is violated, and for states with higher particle number, where the GHZ Theorem applies. Usually entanglement is a consequence of some interaction of the particles in their common past. Recently, Yurke and Stoler [1] have proposed, both for the two and the three particle case, that Bell-EPR phenomena may occur in the coincidence count rates of particles originating from independent sources. We will show, that the expected phenomena only occur if the emission acts of the independent sources are monitored with time resolution much sharper than the coherence time of the

M. Ferrero and A. van der Merwe (eds.), Fundamental Problems in Quantum Physics, 363–373.

radiation fed into the interferometic setups. If this condition is met, in a special setup one can entangle photons which do not even share any common past. We call this technique *entanglement swapping*. Surprisingly, it leads to the first explicit proposal of a practical realization of the scheme of "event-ready detectors." Such a scheme, which has been called for by John Bell since 1971 [2], allows one to operationally pre-select (control, monitor) the ensemble of particle pairs before they enter the measuring apparata. This assures that the ensemble of particle pairs in the experiment does not depend upon particular parameter settings of the apparata. One has an operational definition of even the particle-non-detection events [2,3]. Previously, event-ready detectors scheme was thought to be unfeasible [4].

Consider the setup of Fig. 1. Two independent sources I and II emit one pair of entangled photons each. One can give a simplistic description of the four-photon state as

$$\frac{1}{2}\left(|a\rangle|b\rangle+|a'\rangle|b'\rangle\right)\left(|c\rangle|d\rangle+|c'\rangle|d'\rangle\right). \tag{1}$$

The scheme of Fig. 1 can experimentally be realized by parametric down-conversion (PDC) and the specific geometry can be achieved using a suitable arrangement of mirrors and apertures [6].

The photons in the beams a, a' and b, b' are not entangled with photons in the beams c, c' and d, d'. Let us call the two photons in beams a, a' and d, d' signals and the other two idlers. Suppose we register an idler in the detector i_1 in coincidence with an idler in the detector i_2. Then, using the standard rules of a (symmetric) beamsplitter one can show that the state of the signals collapses into

$$2^{-1/2}\left(|a\rangle|d'\rangle+|a'\rangle|d\rangle\right). \tag{2}$$

The entangled state (2) leads to correlations violating Bell's inequality in an extension of the setup shown by the dashed lines in Fig. 1 [7]. The new state of the signals is a consequence both of (a) the initial signal-idler entanglements (1), and (b) the coincident registration of the idlers in detectors i_1 and i_2, which projects these idler photons into

$$2^{-1/2}\left(|b\rangle|c'\rangle+|b'\rangle|c\rangle\right). \tag{1'}$$

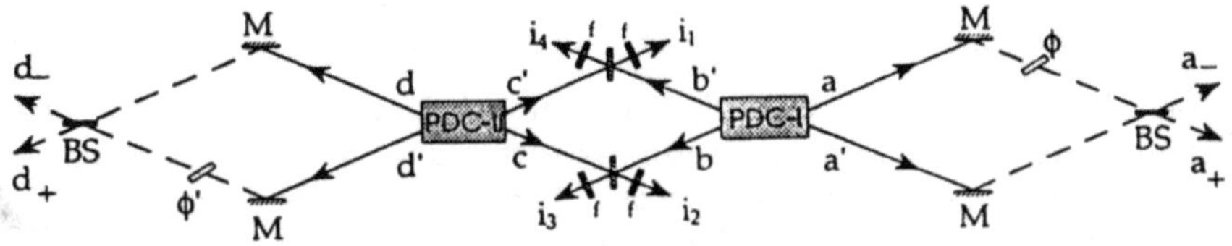

fig.1. Principle of an event ready Bell-EPR experiment [16]. Two down conversion sources PDC-I and PDC-II emit a photon pair each. The specific geometry of each source is obtainable by a suitable arrangement of mirrors and apertures. The initially independent signal photons get entangled by coincident registration of the idlers. (M mirrors, BS beamsplitters, *f* filters, ϕ, ϕ' local phase shifters.)

The resulting entanglement swapping is due to a non-destructive measurement of the signals via a destructive registration of the idlers. Note that the experiment can be arranged such that all registration events occur outside each other's light cones: *the entangled signals do not share any common past*!

We assume in our discussion that, whenever a pair of idlers is registered in coincidence at i_1 and i_2, each source contributed only one particle. Of course it is equally likely that both particles came from the same source. Fortunately in the laboratory these events can be distinguished and excluded because they lead to coincident signal counts at the same end of Fig. 1. Although our experiment depends on accidental coincidences between independent sources we assume the count rates to be low enough such that we are not troubled by triple accidental emissions. Also, since we are interested in those traits of the experiment distinguishing it from the standard ones, we assume for most of our discussion perfect detectors.

The meaning of coincident detection in such an experiment and its implications for Bell's theorem require careful analysis. Note, that even for ideal devices with perfect time resolution one always has to impose a finite time gate to *define* two counts as coincident. Furthermore, the bandwidth $\Delta\omega$ implies that the detection time of a signal is determined by the registration time of its idler up to around its coherence time $T_c \approx 1/\Delta\omega$ and *vice versa* [5]. Since down-conversion radiation is extremely broadband one defines in a practical experiment the bandwidth through apertures and filters. Because of the phase matching condition for frequency (essentially, energy conservation) the filtering of, say, the idlers also limits the bandwidth of the coincident signals. Thus we consider idler filtering only.

Consider first (**experiment A**) the intuitive choice to accept as coincident two idlers arriving at the detectors i_1 and i_2 within a time window $\tau_1 = T_c$. This implies that two signals are within the interferometric setup. We "activate" the signal detectors (in the real experiment this could be done by associating a certain pair of signal detections with given pair of registered idlers via an analysis of the arrival times of all photons, long after the actual events). Our perfect detectors will sooner or later register both signals. Now, a signal caught earlier can be thought with a higher probability to be paired with the idler registered earlier. Therefore, the larger both the time separation between the registrations of the idlers and that between the registrations of the signals, the less "fuzzy" is the information on signal path. The resulting *partial distinguishability* of the paths taken by the signal photon pair leads to a *reduction of the two-particle interference contrast.* This reduction keeps the experiment from violating Bell's inequality, if the visibility drops below 71% [2,4].

A possible improvement (**experiment B**) could be to only accept *signal* detection events within a time window much smaller than the coherence time, what we will call *ultra-coincidence*. Then the registration times of the idlers, which may still differ by up to T_c, do not provide any information on the paths taken by the signal photon pair. This implies a very high contrast of the signal fringes [9]. However, this is certainly a *post-selection* procedure. We accept a only a small subset of all the signal pairs registered (compare experiment A). This post-selection introduces analogous problems to those associated with the detection-efficiency loophole. Thus, such an experiment can only serve as a test of *Bell's inequality with the additional assumption* that the subensemble of signal-pairs registered within the ultra-coincident time gate is sufficiently independent of the phase settings.

To overcome the problems of experiments A and B we propose to impose *ultra-coincidence conditions* (**experiment C**) solely on the registration times of the *idlers*. One has high visibility: the idlers are accepted within a time window narrow enough to *erease* the signal *path information*, no matter at which time difference the signals are registered. In fact, the time gate defining coincidence of signals τ_s, as we shall see, should substantially exceed T_c, and is limited only by the requirement that it has to be by an order of magnitude smaller than the average time between two emissions of entangled pairs by the sources. Thus, the ultra-coincident registration of the idlers provides an entangled signal pair for the interferometers, and as $\tau_s > T_c$ *almost all signal pairs* thus selected take part in the buildup of the two-particle interference fringes. This is a *pre-selection* procedure based on ultra-coincident idlers. We have a valid test of local realism. Besides the imperfection of the devices, no other problems occur. In fact this is a realization of Bell's old idea of an experiment with "event-ready detectors" [1,3]: ultra-coincident registration of the idlers "activates" the signal detectors.

We now turn to a quantitative analysis of the experiments A,B, and C. The initial two photon state produced by the PDC source I, which we consider to incorporate also the filters of the idler radiation, can be described as [10]

$$\sqrt{1/2}\int d\omega_s \int d\omega_i \Delta(\omega_s+\omega_i-\omega_p) f(\omega_i;\omega_f) \Big(|\omega_s,a\rangle|\omega_i,b'\rangle + |\omega_s,a'\rangle|\omega_i,b\rangle \Big), \tag{3}$$

where, e.g. $|\omega_s,a\rangle$ ($|\omega_i,b\rangle$) describes the signal (idler) of frequency ω_s (ω_i) in beam a (b). The entanglement of the beams is a consequence of the phase matching condition for the momenta (essentially, momentum conservation), and of an appropriate selection of apertures [6]. The function $\Delta(\omega_s+\omega_i-\omega_p)$ reflects the phase matching condition for the signal and idler frequencies (within a high accuracy $\omega_s+\omega_i$ is equal to the pump frequency ω_p). It is sharply peaked at zero and in practice can be replaced by Dirac's delta function. By $f(\omega_i;\omega_f)$ we denote the response function of a filter of central frequency ω_f. A similar state describes the emission by the source II.

The detection of an idler in i_1 at, say, t_1 and of another one in i_2, at t_2, causes a wave packet collapse into an *entangled state of the signals*

$$2^{-1/2}\Big(|t_2,a\rangle|t_1,d'\rangle+|t_1,a'\rangle|t_2,d\rangle\Big), \tag{2'}$$

where e.g. $|t_1,a\rangle$ is given by

$$\int d\omega \exp[-i(\omega_p-\omega)t_1]h(\omega)|\omega,a\rangle.$$

The expression $\exp[-i(\omega_p-\omega)t_1]$ describes the amplitude to find a photon (here an idler) with frequency $\omega_p-\omega$ at time t and the function $h(\omega)$ reads $Nf(\omega_p-\omega;\omega_f)$ (N is a normalization constant). This more refined description of the state of the signals has the structure of (2).

Subsequently at some instant t we register a signal emitted by PDC-I. We check whether the signal emitted by PDC-II and detected at t' satisfies $|t-t'|<\tau_s$. The quantum prediction for the probability $p(i,j|\phi,\phi';\tau_s,\tau_i)$ of having counts both in beam a_i and d_j (i, j = $\pm$) within τ_s, *provided we have detected two idlers* in i_1 and i_2 within $|t_1-t_2|<\tau_i$, is given by

$$\frac{1}{2\tau_i}\int_{-\tau_i}^{\tau_i} d\tau'\int_{-\tau_s}^{\tau_s} d\tau\int_{-\infty}^{\infty} dt\,\Big|2^{-3/2}\Big(H(t'-t_1)H(t-t_2)$$

$$+\,ij e^{-i(\phi+\phi')}H(t-t_1)H(t'-t_2)\Big)\Big|^2, \tag{4}$$

where, $\tau=t'-t$, $\tau'=t_1-t_2$, the symbol ij equals + for $i=j$ and $-$ otherwise, and $H(t)$ denotes the normalized Fourier transform of $h(\omega)$. Let $P(\tau_s|\tau_i)$ be the conditional probability of registering the "event-ready" prepared signal pair at any two detectors within τ_s, provided both idlers were detected within τ_i, and $V(\tau_s,\tau_i)$ be the signals fringes visibility under these conditions. Then

$$V(\tau_s,\tau_i)P(\tau_s|\tau_i)$$

$$= \frac{1}{2\tau_i}\left|\int_{-\tau_i}^{\tau_i} d\tau' \int_{-\tau_s}^{\tau_s} d\tau \int_{-\infty}^{\infty} dt H(t+\tau)H(t+\tau')H^*(t)H^*(t+\tau+\tau')\right|. \qquad (5)$$

Let us introduce for $f(\omega_i;\omega_f,\Delta\omega)$ an approximated Fabry–Perot function $1/(\omega_i-\omega_f+i\Delta\omega/2)$ [11]. The Fourier integrals of the spectral functions $h(\omega)$ can then be approximated by extending their range to negative frequencies. Then,

$$H(t) = (\Delta\omega)^{1/2}\exp[-\Delta\omega t/2-i(\omega_p-\omega_f)t]\Theta(t),$$

where $\Theta(t)$ is the step function. Thus, our specific choice of f (and hence h) ensures that this approximation does not imply a loss of causality in the description of the filters. The coincidence gates can be expressed in the natural units

$$\mathcal{T}_i = \tau_i/T_c \text{ and } \mathcal{T}_s = \tau_s/T_c.$$

For finite τ_s the probability $P(\tau_s|\tau_i)$ then reads

$$1 - \sinh\mathcal{T}_i \exp(-\mathcal{T}_s)/\mathcal{T}_i \quad \text{for} \quad \tau_i<\tau_s,$$

and it is

$$[\mathcal{T}_s - \exp(-\mathcal{T}_i)\sinh\mathcal{T}_s]/\mathcal{T}_i \quad \text{for} \quad \tau_i>\tau_s.$$

In general one has

$$V(\tau_s,\tau_i)P(\tau_s|\tau_i) = \frac{[1-\exp(-\mathcal{T}_i)][1-\exp(-\mathcal{T}_s)]}{\mathcal{T}_i}. \qquad (6)$$

The actual form of the expressions is a consequence of the specific f chosen. Nevertheless, one can show that any sensible filter function gives qualitatively similar results: the expression can have high numerical value (above 0.71) only provided $\mathcal{T}_i$ is sufficiently smaller than 1, and $\mathcal{T}_s$ much bigger than 1.

Let us now recall Bell's Theorem. Consider Bell's inequality in the form

$$E_{hv}(\phi_1,\phi_2) + E_{hv}(\phi_1,\phi_2')+ E_{hv}(\phi_2,\phi_2') - E_{hv}(\phi_2,\phi_1')\leq 2, \qquad (7)$$

where, the correlation function $E_{hv}(\phi_1,\phi_1')$ is defined by

$$\sum_{ij} ijp^{hv}(i,j|\phi,\phi';\tau_s,\tau_i),$$

where in turn $p^{hv}(i,j|\phi,\phi';\tau_s,\tau_i)$ is any local realistic prediction for probability that the detectors a_i and d_j fire. Here, the definition of the correlation function must include the fact that some signals may not fit into the required time gate τ_s. Following [2,4], we let the result on each side to be ±1, if the signals were detected within τ_i. We we could simply discard the counts from our ensemble if a signal photon from an event ready pair arrived too late. However, for the analysis that follows this is equivalent to keeping the ensemble intact and assigning to such an event the value 0 (we follow this approach).

We now turn to the implications of our results (4-6). With the use of (4), we find the quantum prediction for the correlation function for the specific time gates τ_s and τ_i:

$$V(\tau_s,\tau_i)P(\tau_s|\tau_i)\cos(\phi+\phi').$$

Thus, the product $V(\tau_s,\tau_i)P(\tau_s|\tau_i)$ gives the *amplitude of the modulation of the quantum correlation function.* It is a well known fact that Bell's inequality is only violated if this amplitude of modulation exceeds $\sqrt{1/2}$. Therefore, an experiment with independent sources as discussed here *violates Bell's inequality only if* the coincidence gate time windows obey the condition

$$V(\tau_s,\tau_i)P(\tau_s|\tau_i) > \sqrt{1/2}\,. \qquad (8)$$

This inequality, together with Eq. (6) imposes conditions on

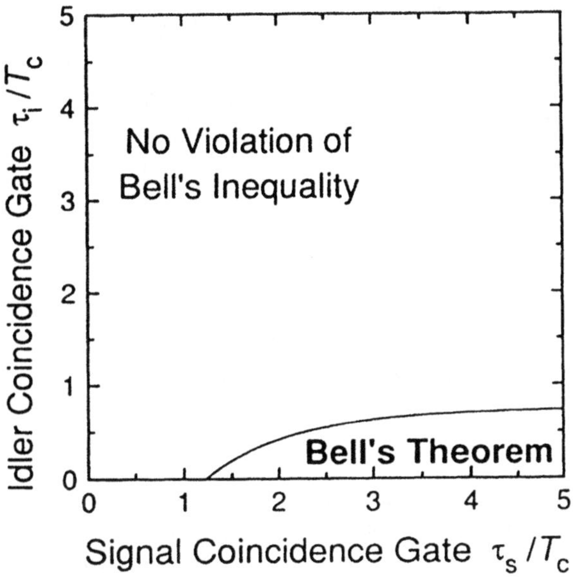

fig.2. Boundary of the experimental parameters (6) necessary to violate Bell's inequality [16] (for our model of the filter).

the time gates for an experiment to be a valid test of Bell's Theorem [10].

In Fig.2 we show that region of time gates for the specific choice of our filter model. Note that (original) Bell's inequalities can only be shown to be violated if (a) the idlers are detected in coincidence much sharper than T_c (b) the signal coincidence windows is significantly longer than T_c. This conclusion rules out in particular all experiments in which one does not care about the idlers at all. Also, those experiments where one can only monitor the emission events of one of the sources like those where single photon radiation is mixed with very weak coherent local oscillator fields fall into the same category [13]. The specific details of the dependence of the signal fringe visibility and the signal acceptance probability $P(\tau_s|\tau_i)$ can be seen in Fig.3.

Like all experiments testing local realism performed so far, the ones proposed here also would suffer from a finite detector efficiency and an abundance of instrumental imperfections which lower the visibility [3,12]. Of course, if the detector efficiency is not sufficiently high, one must resort to some auxiliary assumption such as fair sampling, as

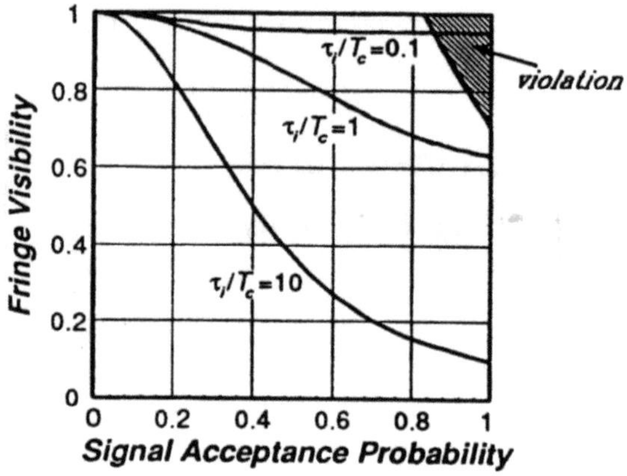

fig.3. Two-photon interference visibility versus signal acceptance probability, which is defined as the square root of the probability $P(\tau_s|\tau_i)$ to register both signals within τ_s given both idlers were registered within τ_i [16]. The three curves are obtained by sweeping τ_s from 0 to ∞ with fixed τ_i.

in the past. Nevertheless, if there are some limitations intrinsic to the setup we must work in an ultra-coincidence regime ($\tau_i \ll T_c$).

Finally we would like to add that using a similar scheme one can obtain a source exhibiting GHZ-correlations for an arbitrary number of independent particles in an event-ready configuration. Again, sharp coincidence for the idlers and long time gates for registration of the third-order interference pattern are required. Note, that in a realistic GHZ experiment one cannot test directly "Bell's theorem without inequalities" but rather resort to generalized Bell inequalities for the GHZ correlations [14].

Concluding, we dare to remark that the experiment with event-ready detection might be a step forward toward a definitive test against local realism (cf. [15]).

We would like to thank Tycho Sleator and Harald Weinfurter for useful comments and discussions. This work has been supported by the Austrian Science Foundation FWF, project No. S 6502 (Schwerpunkt Quantenoptik), the National Science Foundation, Grant No. PHY92-13964, and also University of Gdańsk Grant No. BW/5400-5-0156-4).

REFERENCES

[1] B. Yurke and D. Stoler, *Phys. Rev.A* 46, 2229 (1992); *Phys. Rev. Lett.* **68**, 1251 (1992).

[2] J.S. Bell, in *Foundations of Quantum Mechanics*, B. d'Espagnat ed. (Academic, New York, 1971); J.S. Bell *Speakable and Unspeakable in Quantum Mechanics* (Cambridge University Press, Cambridge, 1988).

[3] A. Garg and N.D. Mermin, *Phys. Rev. D* **35**, 3638 (1987).

[4] J.F. Clauser and A. Shimony, *Rep. Prog. Phys.* **41**, 1881 (1978).

[5] D.M. Greenberger, M. Horne, and A. Zeilinger, in *Bell's Theorem, Quantum Theory, and Conceptions of the Universe*, M. Kafatos ed.,(Kluwer, Dordrecht, 1989); D.M. Greenberger, M. Horne, A. Shimony, and A. Zeilinger, *Am. J. Phys.* **58**, 1131 (1990).

[6] M.A. Horne, A. Shimony, and A. Zeilinger, *Phys. Rev. Lett*, 62 2209 (1989); J.G. Rarity and P.R. Tapster, Phys. Rev. Lett. 64 2495 (1990).

[7] M.A. Horne and A. Zeilinger, in *Proceedings of the Symposium on the Foundations of Modern Physics*, P. Lahti and P. Mittelstaedt eds., (World Scientific, Singapore, 1985); M. Żukowski and J. Pykacz, *Phys. Lett.* A **127**, 1 (1988).

[8] M. A. Horne, H.J. Bernstein, D.M. Greenberger, and A. Zeilinger in *Proceedings of International Workshop on Squeezed States and Uncertainty Relations*, Y.S. Kim ed., (World Scientific, Singapore, 1992).

[9] Z.Y. Ou, *Phys. Rev. A*, 37 1607 (1988).

[10] C.K. Hong and L. Mandel, *Phys. Rev. A* 31, 2409 (1985); M. Żukowski, *Phys. Lett. A* 150, 136 (1990).

[11] J.H. Eberly and K. Wódkiewicz, *J. Opt. Soc. Am.* 67, 1252 (1977).

[12] P. H. Eberhard, *Phys.Rev. A* **47**, R747 (1993).

[13] S.M. Tan, M.J. Holland, and D.F. Walls, *Optics Comm.* 77, 285 (1990); and S.M. Tan, D.F. Walls, and M.J. Collett, *Phys. Rev. Lett.* 66, 252 (1991); B.J. Oliver and C.R. Stroud Jr., *Phys. Lett. A* **135**, 407 (1989); see also [8] and, for the GHZ case, M. Zukowski, *Phys. Lett. A* **157**, 203 (1991).

[14] N.D. Mermin, *Phys. Rev. Lett.* **65**, 1838 (1990); M. Ardehali, *Phys. Rev. A* **46**, (1992) 5375; M. Zukowski, *Phys. Lett. A* **177**, 290 (1993).

[15] E. Santos, *Phys. Rev. A* **46** 3646 (1992).

[16] M Żukowski, A. Zeilinger, M.A. Horne and A.K. Ekert, *Phys. Rev. Lett* 71, 4287 (1993).

AUTHOR INDEX

Fundamental Theories of Physics

22. A.O. Barut and A. van der Merwe (eds.): *Selected Scientific Papers of Alfred Landé. [1888-1975]*. 1988 ISBN 90-277-2594-2
23. W.T. Grandy, Jr.: *Foundations of Statistical Mechanics.* Vol. II: *Nonequilibrium Phenomena.* 1988 ISBN 90-277-2649-3
24. E.I. Bitsakis and C.A. Nicolaides (eds.): *The Concept of Probability.* Proceedings of the Delphi Conference (Delphi, Greece, 1987). 1989 ISBN 90-277-2679-5
25. A. van der Merwe, F. Selleri and G. Tarozzi (eds.): *Microphysical Reality and Quantum Formalism, Vol. 1.* Proceedings of the International Conference (Urbino, Italy, 1985). 1988 ISBN 90-277-2683-3
26. A. van der Merwe, F. Selleri and G. Tarozzi (eds.): *Microphysical Reality and Quantum Formalism, Vol. 2.* Proceedings of the International Conference (Urbino, Italy, 1985). 1988 ISBN 90-277-2684-1
27. I.D. Novikov and V.P. Frolov: *Physics of Black Holes.* 1989 ISBN 90-277-2685-X
28. G. Tarozzi and A. van der Merwe (eds.): *The Nature of Quantum Paradoxes.* Italian Studies in the Foundations and Philosophy of Modern Physics. 1988 ISBN 90-277-2703-1
29. B.R. Iyer, N. Mukunda and C.V. Vishveshwara (eds.): *Gravitation, Gauge Theories and the Early Universe.* 1989 ISBN 90-277-2710-4
30. H. Mark and L. Wood (eds.): *Energy in Physics, War and Peace.* A Festschrift celebrating Edward Teller's 80th Birthday. 1988 ISBN 90-277-2775-9
31. G.J. Erickson and C.R. Smith (eds.): *Maximum-Entropy and Bayesian Methods in Science and Engineering.* Vol. I: *Foundations.* 1988 ISBN 90-277-2793-7
32. G.J. Erickson and C.R. Smith (eds.): *Maximum-Entropy and Bayesian Methods in Science and Engineering.* Vol. II: *Applications.* 1988 ISBN 90-277-2794-5
33. M.E. Noz and Y.S. Kim (eds.): *Special Relativity and Quantum Theory.* A Collection of Papers on the Poincaré Group. 1988 ISBN 90-277-2799-6
34. I.Yu. Kobzarev and Yu.I. Manin: *Elementary Particles. Mathematics, Physics and Philosophy.* 1989 ISBN 0-7923-0098-X
35. F. Selleri: *Quantum Paradoxes and Physical Reality.* 1990 ISBN 0-7923-0253-2
36. J. Skilling (ed.): *Maximum-Entropy and Bayesian Methods.* Proceedings of the 8th International Workshop (Cambridge, UK, 1988). 1989 ISBN 0-7923-0224-9
37. M. Kafatos (ed.): *Bell's Theorem, Quantum Theory and Conceptions of the Universe.* 1989 ISBN 0-7923-0496-9
38. Yu.A. Izyumov and V.N. Syromyatnikov: *Phase Transitions and Crystal Symmetry.* 1990 ISBN 0-7923-0542-6
39. P.F. Fougère (ed.): *Maximum-Entropy and Bayesian Methods.* Proceedings of the 9th International Workshop (Dartmouth, Massachusetts, USA, 1989). 1990 ISBN 0-7923-0928-6
40. L. de Broglie: *Heisenberg's Uncertainties and the Probabilistic Interpretation of Wave Mechanics.* With Critical Notes of the Author. 1990 ISBN 0-7923-0929-4
41. W.T. Grandy, Jr.: *Relativistic Quantum Mechanics of Leptons and Fields.* 1991 ISBN 0-7923-1049-7
42. Yu.L. Klimontovich: *Turbulent Motion and the Structure of Chaos.* A New Approach to the Statistical Theory of Open Systems. 1991 ISBN 0-7923-1114-0

AUTHOR INDEX

Fundamental Theories of Physics

Series Editor: Alwyn van der Merwe, *University of Denver, USA*

1. M. Sachs: ***General Relativity and Matter***. A Spinor Field Theory from Fermis to Light-Years. With a Foreword by C. Kilmister. 1982 ISBN 90-277-1381-2
2. G.H. Duffey: ***A Development of Quantum Mechanics***. Based on Symmetry Considerations. 1985 ISBN 90-277-1587-4
3. S. Diner, D. Fargue, G. Lochak and F. Selleri (eds.): ***The Wave-Particle Dualism***. A Tribute to Louis de Broglie on his 90th Birthday. 1984 ISBN 90-277-1664-1
4. E. Prugovečki: ***Stochastic Quantum Mechanics and Quantum Spacetime***. A Consistent Unification of Relativity and Quantum Theory based on Stochastic Spaces. 1984; 2nd printing 1986 ISBN 90-277-1617-X
5. D. Hestenes and G. Sobczyk: ***Clifford Algebra to Geometric Calculus***. A Unified Language for Mathematics and Physics. 1984 ISBN 90-277-1673-0; Pb (1987) 90-277-2561-6
6. P. Exner: ***Open Quantum Systems and Feynman Integrals***. 1985 ISBN 90-277-1678-1
7. L. Mayants: ***The Enigma of Probability and Physics***. 1984 ISBN 90-277-1674-9
8. E. Tocaci: ***Relativistic Mechanics, Time and Inertia***. Translated from Romanian. Edited and with a Foreword by C.W. Kilmister. 1985 ISBN 90-277-1769-9
9. B. Bertotti, F. de Felice and A. Pascolini (eds.): ***General Relativity and Gravitation***. Proceedings of the 10th International Conference (Padova, Italy, 1983). 1984 ISBN 90-277-1819-9
10. G. Tarozzi and A. van der Merwe (eds.): ***Open Questions in Quantum Physics***. 1985 ISBN 90-277-1853-9
11. J.V. Narlikar and T. Padmanabhan: ***Gravity, Gauge Theories and Quantum Cosmology***. 1986 ISBN 90-277-1948-9
12. G.S. Asanov: ***Finsler Geometry, Relativity and Gauge Theories***. 1985 ISBN 90-277-1960-8
13. K. Namsrai: ***Nonlocal Quantum Field Theory and Stochastic Quantum Mechanics***. 1986 ISBN 90-277-2001-0
14. C. Ray Smith and W.T. Grandy, Jr. (eds.): ***Maximum-Entropy and Bayesian Methods in Inverse Problems***. Proceedings of the 1st and 2nd International Workshop (Laramie, Wyoming, USA). 1985 ISBN 90-277-2074-6
15. D. Hestenes: ***New Foundations for Classical Mechanics***. 1986 ISBN 90-277-2090-8; Pb (1987) 90-277-2526-8
16. S.J. Prokhovnik: ***Light in Einstein's Universe***. The Role of Energy in Cosmology and Relativity. 1985 ISBN 90-277-2093-2
17. Y.S. Kim and M.E. Noz: ***Theory and Applications of the Poincaré Group***. 1986 ISBN 90-277-2141-6
18. M. Sachs: ***Quantum Mechanics from General Relativity***. An Approximation for a Theory of Inertia. 1986 ISBN 90-277-2247-1
19. W.T. Grandy, Jr.: ***Foundations of Statistical Mechanics***.
    Vol. I: ***Equilibrium Theory***. 1987 ISBN 90-277-2489-X
20. H.-H von Borzeszkowski and H.-J. Treder: ***The Meaning of Quantum Gravity***. 1988 ISBN 90-277-2518-7
21. C. Ray Smith and G.J. Erickson (eds.): ***Maximum-Entropy and Bayesian Spectral Analysis and Estimation Problems***. Proceedings of the 3rd International Workshop (Laramie, Wyoming, USA, 1983). 1987 ISBN 90-277-2579-9

Fundamental Theories of Physics

22. A.O. Barut and A. van der Merwe (eds.): *Selected Scientific Papers of Alfred Landé. [1888-1975]*. 1988 ISBN 90-277-2594-2
23. W.T. Grandy, Jr.: *Foundations of Statistical Mechanics.* Vol. II: *Nonequilibrium Phenomena.* 1988 ISBN 90-277-2649-3
24. E.I. Bitsakis and C.A. Nicolaides (eds.): *The Concept of Probability.* Proceedings of the Delphi Conference (Delphi, Greece, 1987). 1989 ISBN 90-277-2679-5
25. A. van der Merwe, F. Selleri and G. Tarozzi (eds.): *Microphysical Reality and Quantum Formalism, Vol. 1.* Proceedings of the International Conference (Urbino, Italy, 1985). 1988 ISBN 90-277-2683-3
26. A. van der Merwe, F. Selleri and G. Tarozzi (eds.): *Microphysical Reality and Quantum Formalism, Vol. 2.* Proceedings of the International Conference (Urbino, Italy, 1985). 1988 ISBN 90-277-2684-1
27. I.D. Novikov and V.P. Frolov: *Physics of Black Holes.* 1989 ISBN 90-277-2685-X
28. G. Tarozzi and A. van der Merwe (eds.): *The Nature of Quantum Paradoxes.* Italian Studies in the Foundations and Philosophy of Modern Physics. 1988 ISBN 90-277-2703-1
29. B.R. Iyer, N. Mukunda and C.V. Vishveshwara (eds.): *Gravitation, Gauge Theories and the Early Universe.* 1989 ISBN 90-277-2710-4
30. H. Mark and L. Wood (eds.): *Energy in Physics, War and Peace.* A Festschrift celebrating Edward Teller's 80th Birthday. 1988 ISBN 90-277-2775-9
31. G.J. Erickson and C.R. Smith (eds.): *Maximum-Entropy and Bayesian Methods in Science and Engineering.* Vol. I: *Foundations.* 1988 ISBN 90-277-2793-7
32. G.J. Erickson and C.R. Smith (eds.): *Maximum-Entropy and Bayesian Methods in Science and Engineering.* Vol. II: *Applications.* 1988 ISBN 90-277-2794-5
33. M.E. Noz and Y.S. Kim (eds.): *Special Relativity and Quantum Theory.* A Collection of Papers on the Poincaré Group. 1988 ISBN 90-277-2799-6
34. I.Yu. Kobzarev and Yu.I. Manin: *Elementary Particles. Mathematics, Physics and Philosophy.* 1989 ISBN 0-7923-0098-X
35. F. Selleri: *Quantum Paradoxes and Physical Reality.* 1990 ISBN 0-7923-0253-2
36. J. Skilling (ed.): *Maximum-Entropy and Bayesian Methods.* Proceedings of the 8th International Workshop (Cambridge, UK, 1988). 1989 ISBN 0-7923-0224-9
37. M. Kafatos (ed.): *Bell's Theorem, Quantum Theory and Conceptions of the Universe.* 1989 ISBN 0-7923-0496-9
38. Yu.A. Izyumov and V.N. Syromyatnikov: *Phase Transitions and Crystal Symmetry.* 1990 ISBN 0-7923-0542-6
39. P.F. Fougère (ed.): *Maximum-Entropy and Bayesian Methods.* Proceedings of the 9th International Workshop (Dartmouth, Massachusetts, USA, 1989). 1990 ISBN 0-7923-0928-6
40. L. de Broglie: *Heisenberg's Uncertainties and the Probabilistic Interpretation of Wave Mechanics.* With Critical Notes of the Author. 1990 ISBN 0-7923-0929-4
41. W.T. Grandy, Jr.: *Relativistic Quantum Mechanics of Leptons and Fields.* 1991 ISBN 0-7923-1049-7
42. Yu.L. Klimontovich: *Turbulent Motion and the Structure of Chaos.* A New Approach to the Statistical Theory of Open Systems. 1991 ISBN 0-7923-1114-0

Fundamental Theories of Physics

43. W.T. Grandy, Jr. and L.H. Schick (eds.): *Maximum-Entropy and Bayesian Methods.* Proceedings of the 10th International Workshop (Laramie, Wyoming, USA, 1990). 1991 ISBN 0-7923-1140-X
44. P.Pták and S. Pulmannová: *Orthomodular Structures as Quantum Logics.* Intrinsic Properties, State Space and Probabilistic Topics. 1991 ISBN 0-7923-1207-4
45. D. Hestenes and A. Weingartshofer (eds.): *The Electron.* New Theory and Experiment. 1991 ISBN 0-7923-1356-9
46. P.P.J.M. Schram: *Kinetic Theory of Gases and Plasmas.* 1991 ISBN 0-7923-1392-5
47. A. Micali, R. Boudet and J. Helmstetter (eds.): *Clifford Algebras and their Applications in Mathematical Physics.* 1992 ISBN 0-7923-1623-1
48. E. Prugovečki: *Quantum Geometry.* A Framework for Quantum General Relativity. 1992 ISBN 0-7923-1640-1
49. M.H. Mac Gregor: *The Enigmatic Electron.* 1992 ISBN 0-7923-1982-6
50. C.R. Smith, G.J. Erickson and P.O. Neudorfer (eds.): *Maximum Entropy and Bayesian Methods.* Proceedings of the 11th International Workshop (Seattle, 1991). 1993 ISBN 0-7923-2031-X
51. D.J. Hoekzema: *The Quantum Labyrinth.* 1993 ISBN 0-7923-2066-2
52. Z. Oziewicz, B. Jancewicz and A. Borowiec (eds.): *Spinors, Twistors, Clifford Algebras and Quantum Deformations.* Proceedings of the Second Max Born Symposium (Wrocław, Poland, 1992). 1993 ISBN 0-7923-2251-7
53. A. Mohammad-Djafari and G. Demoment (eds.): *Maximum Entropy and Bayesian Methods.* Proceedings of the 12th International Workshop (Paris, France, 1992). 1993 ISBN 0-7923-2280-0
54. M. Riesz: *Clifford Numbers and Spinors* with Riesz' Private Lectures to E. Folke Bolinder and a Historical Review by Pertti Lounesto. E.F. Bolinder and P. Lounesto (eds.). 1993 ISBN 0-7923-2299-1
55. F. Brackx, R. Delanghe and H. Serras (eds.): *Clifford Algebras and their Applications in Mathematical Physics.* Proceedings of the Third Conference (Deinze, 1993) 1993 ISBN 0-7923-2347-5
56. J.R. Fanchi: *Parametrized Relativistic Quantum Theory.* 1993 ISBN 0-7923-2376-9
57. A. Peres: *Quantum Theory: Concepts and Methods.* 1993 ISBN 0-7923-2549-4
58. P.L. Antonelli, R.S. Ingarden and M. Matsumoto: *The Theory of Sprays and Finsler Spaces with Applications in Physics and Biology.* 1993 ISBN 0-7923-2577-X
59. R. Miron and M. Anastasiei: *The Geometry of Lagrange Spaces: Theory and Applications.* 1994 ISBN 0-7923-2591-5
60. G. Adomian: *Solving Frontier Problems of Physics: The Decomposition Method.* 1994 ISBN 0-7923-2644-X
61 B.S. Kerner and V.V. Osipov: *Autosolitons.* A New Approach to Problems of Self-Organization and Turbulence. 1994 ISBN 0-7923-2816-7
62. A. Heidbreder (ed.): *Maximum Entropy and Bayesian Methods.* Proceedings of the 13th International Workshop (Santa Barbara, USA, 1993) 1995 ISBN 0-7923-2851-5
63. J. Peřina, Z. Hradil and B. Jurčo: *Quantum Optics and Fundamentals of Physics.* 1994 ISBN 0-7923-3000-5

Fundamental Theories of Physics

64. M. Evans and J.-P. Vigier: *The Enigmatic Photon.* Volume 1: The Field $B^{(3)}$. 1994
ISBN 0-7923-3049-8
65. C.K. Raju: *Time: Towards a Constistent Theory.* 1994 ISBN 0-7923-3103-6
66. A.K.T. Assis: *Weber's Electrodynamics.* 1994 ISBN 0-7923-3137-0
67. Yu. L. Klimontovich: *Statistical Theory of Open Systems.* Volume 1: A Unified Approach to Kinetic Description of Processes in Active Systems. 1995
ISBN 0-7923-3199-0; Pb: ISBN 0-7923-3242-3
68. M. Evans and J.-P. Vigier: *The Enigmatic Photon.* Volume 2: Non-Abelian Electrodynamics. 1995 ISBN 0-7923-3288-1
69. G. Esposito: *Complex General Relativity.* 1995 ISBN 0-7923-3340-3
70. *Forthcoming*
71. C. Garola and A. Rossi (eds.): *The Foundations of Quantum Mechanics – Historical Analysis and Open Questions.* 1995 ISBN 0-7923-3480-9
72. A. Peres: *Quantum Theory: Concepts and Methods.* 1995 ISBN Pb 0-7923-3632-1
73. M. Ferrero and A. van der Merwe (eds.): *Fundamental Problems in Quantum Physics.* 1995 ISBN 0-7923-3670-4

KLUWER ACADEMIC PUBLISHERS – DORDRECHT / BOSTON / LONDON

MIX
Papier aus verantwortungsvollen Quellen
Paper from responsible sources
FSC® C105338

If you have any concerns about our products,
you can contact us on
ProductSafety@springernature.com

In case Publisher is established outside the EU,
the EU authorized representative is:
Springer Nature Customer Service Center GmbH
Europaplatz 3, 69115 Heidelberg, Germany

Printed by Libri Plureos GmbH
in Hamburg, Germany